T0185597

# Water Science and Technology Library

## Volume 97

The aim of the *Water Science and Technology Library* is to provide a forum for dissemination of the state-of-the-art of topics of current interest in the area of water science and technology. This is accomplished through publication of reference books and monographs, authored or edited. Occasionally also proceedings volumes are accepted for publication in the series. *Water Science and Technology Library* encompasses a wide range of topics dealing with science as well as socio-economic aspects of water, environment, and ecology. Both the water quantity and quality issues are relevant and are embraced by *Water Science and Technology Library*. The emphasis may be on either the scientific content, or techniques of solution, or both. There is increasing emphasis these days on processes and *Water Science and Technology Library* is committed to promoting this emphasis by publishing books emphasizing scientific discussions of physical, chemical, and/or biological aspects of water resources. Likewise, current or emerging solution techniques receive high priority. Interdisciplinary coverage is encouraged. Case studies contributing to our knowledge of water science and technology are also embraced by the series. Innovative ideas and novel techniques are of particular interest.

Comments or suggestions for future volumes are welcomed.

Vijay P. Singh, Department of Biological and Agricultural Engineering & Zachry Department of Civil and Environment Engineering, Texas A&M University, USA Email: vsingh@tamu.edu

More information about this series at http://www.springer.com/series/6689

Ashish Pandey · S. K. Mishra · M. L. Kansal ·
R. D. Singh · V. P. Singh
Editors

# Hydrological Extremes

## River Hydraulics and Irrigation Water Management

Springer

*Editors*
Ashish Pandey
Department of Water Resources
Development and Management
Indian Institute of Technology Roorkee
Roorkee, Uttarakhand, India

S. K. Mishra
Department of Water Resources
Development and Management
Indian Institute of Technology Roorkee
Roorkee, Uttarakhand, India

M. L. Kansal
Department of Water Resources
Development and Management
Indian Institute of Technology Roorkee
Roorkee, Uttarakhand, India

R. D. Singh
Indian Institute of Technology Roorkee
Roorkee, Uttarakhand, India

V. P. Singh
Department of Biological Engineering
Texas A&M University
College Station, TX, USA

ISSN 0921-092X          ISSN 1872-4663   (electronic)
Water Science and Technology Library
ISBN 978-3-030-59150-2       ISBN 978-3-030-59148-9   (eBook)
https://doi.org/10.1007/978-3-030-59148-9

This Springer imprint is published by the registered company Springer Nature Switzerland AG
The registered company address is: Gewerbestrasse 11, 6330 Cham, Switzerland

# Contents

# About the Editors

**Prof. Ashish Pandey** did his B. Tech. (Agricultural Engineering) and M.Tech. (Soil and Water Engineering) from JNKVV, Jabalpur. He received Ph.D. in Soil and Water Conservation Engineering from IIT Kharagpur. Presently, he is working as a Professor at the Department of WRD&M, IIT Roorkee. His research interests include Irrigation Water Management, Soil and Water Conservation Engineering, Hydrological Modeling of Watershed, Remote sensing, and GIS Applications in Water Resources. He has guided 10 Ph.D. and 78 M.Tech. students at IIT Roorkee. He published 155 research papers in peer-reviewed and high impact international/national journals/seminars/conferences/symposia. He has also co-authored a textbook on "Introductory Soil and Water Conservation Engineering" and has edited a book for Water Science and Technology Library (WSTL) Book Series published by Springer Nature Switzerland.

Prof. Pandey also served as Guest Editor for two issues of J. Hydrologic Engineering (ASCE). He is Editor, Indian J. Soil Conservation. His laurels include Eminent Engineers Award-2015 given by the Institute of Engineers (India), Uttarakhand State Centre, DAAD scholarship and ASPEE scholarship. He was offered the prestigious JSPS Postdoctoral Fellowship for Foreign Researchers, Japan and BOYSCAST fellowship of DST, GOI. He is a Fellow member of (1) Institution of Engineers (India), (2) Indian Association of Hydrologists (IAH), and (3) Indian Water Resources Society (IWRS).

**Dr. S. K. Mishra** is a 1984 Civil Engineering graduate of the then Moti Lal Nehru Regional Engineering College (presently MNNIT), Allahabad. He obtained his M.Tech. in Hydraulics and Water Resources from IIT Kanpur in 1986; and doctoral degree from the then University of Roorkee (presently IIT Roorkee) in 1999. He served at the National Institute of Hydrology Roorkee at various scientific positions during 1987–2004. During the period, he also visited Louisiana State University, USA, as a postdoctoral fellow during 2000–2001 and Department of Civil Engineering, IIT Bombay as a Visiting Faculty during 2002–2003.

Dr. Mishra joined IIT Roorkee in 2004 and is presently working as Professor. Besides having been the Head of the Department of Water Resources Development and Management, he is also presently holding Bharat Singh Chair of the Ministry of Jal Shakti, Govt. of India. He is specialized in the fields of hydraulics and water resources, environmental engineering, design of irrigation and drainage works, dam break analysis, surface water hydrology. He has published more 250 technical articles in various international/national journals/seminars/conferences/symposia. His reference book on Soil Conservation Service Curve Number Methodology published by Kluver Academic Publishers, the Netherlands, has received significant attention of the hydrologists/agriculturists/soil water conservationists around the globe. Of late, he has been associated with the Journal of Hydrologic Engineering as a Gust editor of two special issues and the Executive Vice President and Editor of Indian Water Resources Society. He has visited several countries during the period and is on the role of several national/international professional bodies. Among several others, the Eminent Engineers Award and Dr. Rajendra Prasad Award are worth citing.

**Prof. M. L. Kansal** currently working as NEEPCO Chair Professor & Head in the Department of Water Resources Development & Management at Indian Institute of Technology (IIT) Roorkee (India). He is a Civil Engineering graduate with postgraduation in Water Resources Engineering. He obtained his Ph.D. from Delhi University (India) and holds the Post Graduate Diploma in Operations Management. Previously, he worked as Associate Professor in the Department WRD&M, IIT Roorkee and served at Delhi Technical University, Delhi (erstwhile, Delhi College of Engineering, Delhi), NIT Kurukshetra, IIT Delhi, and NIH, Roorkee (India) at various levels. He has published more than 150 research papers and two books. He has got best paper awards from Indian Building Congress, Indian Water Works Association, and received star performer award from IIT Roorkee. He acted as a reviewer for several International Journals and research agencies. He has visited various countries as International Expert and Visiting Professor. Dr. Kansal acted as International expert for RCUWM of UNESCO, IUCN, etc. He is working as Executive Vice President of Indian Water Resources Society and served as expert panel member for All India Council for Technical Education (AICTE), India. He has contributed substantially by providing consultancy services to various national and international agencies of repute. He has worked in various administrative capacities such as Associate Dean of Students Welfare and as Chairman, Co-ordinating Committee of Bhawans at Indian Institute of Technology Roorkee and national expert for several bodies.

**R. D. Singh** did B.E. Civil Engineering and M.E. Civil Engineering with specialization in Hydraulics and Irrigation Engineering from University of Roorkee (Now IIT Roorkee). He did M.Sc. in Hydrology from University College Galway, Ireland. Presently, he is working as Visiting Professor, Department of Water Resources Development & Management, IIT Roorkee. He worked as Director, National Institute of Hydrology (NIH), Roorkee for more than 9 years. Before taking over as Director NIH, he was holding the charge of Nodal Officer Hydrology Project-II, a World Bank Funded Project for Peninsular region of India completed by NIH during the year 2016. During his service at NIH, he had worked on more than 80 sponsored/consultancy projects for solving the real-life problems in water sector. He had also worked as well as guided 11 International Collaborative projects at NIH.

He has research and development experience of more than 40 years in different areas of Hydrology and Water Resources. He has extensive experience in flood estimation, flood management, drought management, hydrological modeling, environmental impact assessment, and climate change and its impact on water resources, etc. He has published more than 316 research papers in the reputed International and National Journals, International and National Seminar/Symposia, Workshops, etc. He has received C.B.I.P. Medal, Institution of Engineers Certificate of merit, Union Ministry of Irrigation award and Best Scientist award form from NIH, Roorkee. He guided 3 Ph.D. and 14 M.E. and M.Tech., and 1 M.Phil dissertations. He has widely travelled abroad for different assignments.

**Prof. V. P. Singh** is a Distinguished Professor, a Regents Professor, and Caroline and William N. Lehrer Distinguished Chair in Water Engineering at Texas A&M University. He received his B.S., M.S., Ph.D. and D.Sc. in engineering. He is a registered professional engineer, a registered professional hydrologist, and an Honorary Diplomate of ASCE-AAWRE. He has published extensively in the area of hydrology and water resources, including 30 textbooks; Handbook of Applied Hydrology; Encyclopedia of Snow, Ice and Glaciers; 71 edited books; 114 book chapters; 1270 refereed journal articles; 330 conference proceedings papers; 13 special edited journal Issues; 50 book reviews; and 72 technical publications and reports. For his seminal contributions, he has received more than 92 national and international awards, as well as 3 honorary doctorates. He is a member of 11 international science/engineering academies. He has served as President of the American Institute of Hydrology (AIH), Chair of Watershed Council of American Society of Civil Engineers, Vice President of Indian Association of Hydrologists, Vice President of Association of Global Groundwater Scientists, and is currently President of American Academy of Water Resources Engineers. He has served/serves as Editor-in-Chief of three journals and two book series and serves on editorial boards of more than 25 journals and 3 book series.

# Part I
# Assessment and Monitoring of Hydrological Extremes

# Chapter 1
# Application of Hydrologic Modelling System (HEC-HMS) for Flood Assessment; Case Study of Kelani River Basin, Sri Lanka

**S. Rajkumar, S. K. Mishra, and R. D. Singh**

## 1.1 Introduction

Hydrology defines that it is a scientific study of water. It is the science that associates with the occurrence, circulation, and distribution of water of the earth and earth's atmosphere. One of the most crucial water sources of the earth is rainfall, extreme of which causes flood disaster. The rainfall characteristics are the temporal and spatial distribution of the rainfall quantity (Jain et al. 2000). The runoff estimation is a crucial aspect of watershed planning (Kumar et al. 2004). Hence the study of transformation from rainfall to runoff also referred to as rainfall-runoff modelling, watershed modelling or hydrological modelling is highly necessary for the academic background of water resources engineering for the mitigation measure against flood disaster and the future development of water resource structure.

There are numerous sources currently available for the application of rainfall-runoff modelling. However, the modern rapid developed technology and software tools assist water resource professionals to model the natural phenomena. The software tools, such as Hydrology Energy Centre-Hydrological Modelling System (HEC-HMS) and Geographical Informatics System (GIS) are simultaneously employed in such modeling tasks nowadays (Nandalal and Ratnayake 2010). The software tool requires various data for its input to run the model systematically.

S. Rajkumar (✉) · S. K. Mishra · R. D. Singh
Department of Water Resources Development and Management, Indian Institute of Technology Roorkee, Roorkee 247667, Uttarakhand, India
e-mail: rajkumar80364@gmail.com

S. K. Mishra
e-mail: skm61fwt01@gmail.com

R. D. Singh
e-mail: rdsingh3@gmail.com

3

A. Pandey et al. (eds.), *Hydrological Extremes*, Water Science and Technology Library 97, https://doi.org/10.1007/978-3-030-59148-9_1

The system of hydrological modeling requires a set of meteorological data (rainfall, etc.), hydrological data (streamflow), and spatial data (topography, land use land cover and soil type) of the relevant basin. Mostly, it is obvious that precise temporal data and high quality of spatial data are not affordable. It is, therefore, a huge challenge in the application of those data with rainfall-runoff modelling. However, the lumped conceptual models which are not much expecting the higher accuracy of data is applied in this assessment. The modelling HEC-HMS is also one of the lumped conceptual model categories (Nandalal and Ratnayake 2010).

According to the available temporal and spatial data of the Kelani river basin, the objectives are (i) Application of HEC-HMS for event base modelling to simulate the Flood Hydrograph of different events in Kelani river basin. (ii) Calibration and validation of various flood simulation models of HEC-HMS. (iii) Comparison of the simulation results based on different objective functions to select a suitable model for flood simulation. (iv) Formulate real-ime flood forecast at the Hanwella gauging site to provide the advance information about the flood for its management.

In order to manage the frequent occurrence of floods, the Govt. of Sri Lanka is planning to take immediate steps to safeguard the capital of the country from this frequent flood menace by adopting suitable structural and non-structural flood mitigation measures, such as (i) Diverting the flood water through a constructed channel at Hanwella gauging station minimizing the floods in the downstream, (ii) Providing embankments and levees along the both riverbanks for flood protection, and (iii) Real-time flood forecasting for the evacuation of the people from the areas likely to be affected during the floods.

For adopting measures, the flood assessments are required analyzing the rainfall-runoff data of flood events occurred in the Kelani river basin. For this purpose, it is required to understand the rainfall-runoff mechanism considering the historical rainfall-runoff events observed in the Kelani river basin. (Silva et al. 2014). Thus, the flood assessment in the Kelani River basin is very much imperative for the water managers and decision-makers since the Kelani river is frequently hit by flood due to south-west monsoon storm.

## 1.2   Study Area

Kalani river is the second largest river of Sri Lanka. Its catchment area up to Hanwella gauging site is 1836 km$^2$ which covers five districts namely Colombo, Gampaha, Kegalle, Ratnapura, and Nuwara-Eliya. In the catchment, the topographical elevations vary between 16 and 2320 m above mean sea level. The contributions of the flow to the river come from the rainfall mostly occur during the two distinct monsoon seasons, i.e. north-east and south-west monsoon. The Administrative Capital 'Sri Jayawardanapura' and the Commercial Capital 'Colombo' are located in the downstream of the Hanwella gauging site. The district Colombo in Western Province with the current population of 300 head per day has an area of 699 km$^2$ with a population of 2.3 million and has population density likely to be 60 times the average population density which is 340 heads per km$^2$.

Figure 1.1 shows the gross basin of Kalani river and the locations of gauging station over the basin. The calibration and validation process is set up with the observed streamflow data of Hanwella gauging station. Therefore, all the data processing is carried out in the upper catchment of Hanwella gauging station (Figs. 1.2 and 1.3). It is referred to as Kalani river basin in this study.

Kelani River in Sri Lanka is being contributed with runoff from both monsoon seasons rainfall, but the major share of its flow contribution is due to the rainfall during south-west monsoon season. The district Colombo frequently experiences

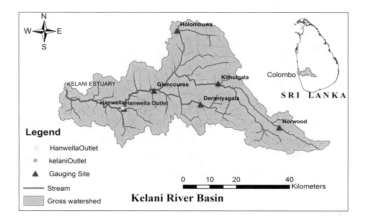

**Fig. 1.1**   Gross basin of Kelani river

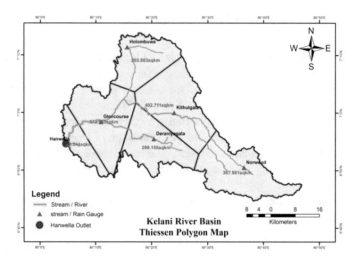

**Fig. 1.2**   Thiessen polygon map

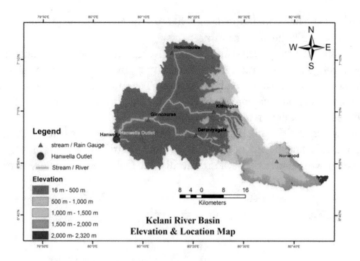

**Fig. 1.3** Gauging station and elevation map

flood menace almost every year. It causes loss to the lives and severe damages to infrastructures, properties and ultimately livelihood of the communities residing in district Colombo. Thus, the economic growth of Sri Lanka dramatically reduces due to the extensive damages caused by frequent flooding.

## 1.3 Data Availability and Data Processing

### 1.3.1 Temporal Data and Processing

Meteorological data (rainfall) on daily basis and hourly basis for selected extreme flood events and Hydrological data (discharge only on daily basis and water level on daily basis and on an hourly basis for selected extreme flood events) at six (06) gauging stations (shown in Fig. 1.1) namely Hanwella, Norwood, Deraniyagala, Kithulgala, Holombuwa, and Glencourse located in the upper basin have been obtained. These data have been obtained from the Dept. of Irrigation, Colombo, Sri Lanka for model calibration and validation. The event-based model calibration and validation is carried out by considering the flood events that occurred in the recent past years (Table 1.1).

The event base modelling calibration requires hourly streamflow data. Because of non-availability of this hourly data, from the observed discharge data and the corresponding stages, stage discharge relationship, which is known as Rating Curve, for the gauging station Hanwella is developed. Then from the computed equation, the hourly discharge is accounted for corresponding stages at the gauging site Hanwella.

**Table 1.1** Flood events considered

| Flood events | Duration in days |
|---|---|
| May 2017 | 14 |
| May 2016 | 30 |
| December 2014 | 8 |
| June 2014 | 16 |
| November 2012 | 9 |

The data of the daily gauge and corresponding observed discharge are available for the Hanwella gauging site. Those data are used to develop the rating curve in the following form using analytical as well as graphical approaches:

$$Q = a \cdot (H - Ho)^b \qquad (1.1)$$

where Q is river discharge in $m^3/s$, H is river stage measured in m at the gauging site, Ho represents the stage reading corresponding to the zero discharge, a and b are the constants which may be computed analyzing the available stage and corresponding discharge data. In the analytical approach, simple linear regression analysis has been carried out transforming the data in the log–log domain whereas in the graphical method the observed stage and corresponding discharges are plotted on arithmetic or log–log scales. The form of the rating curve developed using analytical approach is $Q = 31.48(H - 0.48)^{1.696}$ with ($r^2 = 0.983$) whereas it is in the form of $Q = 21.38(H - 0.48)^{1.63}$ with $r^2 = 0.941$ developed using the graphical method. The rating curve, developed by analytical method, is used to convert the hourly observed gauge values to the hourly discharge values for all the flood events considered for analysis. The hourly average rainfall values for all those flood events are computed from the hourly rainfall values observed at six rain gauge stations using the option of Thiessen polygon method (as shown in Fig. 1.2) available in HEC-HMS programme.

### *1.3.2 Processing of Spatial Data*

The digital elevation model (DEM) and Satellite Image of 30 m resolution, which is available in the United States Geological Survey (USGS) website have been downloaded. The DEM is the fundamental input of the HEC-GeoHMS tool to develop the basin model of this study area and the soil data map. The satellite image has been employed to develop land use land cover map. Figures 1.4 and 1.5 illustrate the spatial distribution of land use/land cover and soil type of the Kelani river basin, respectively.

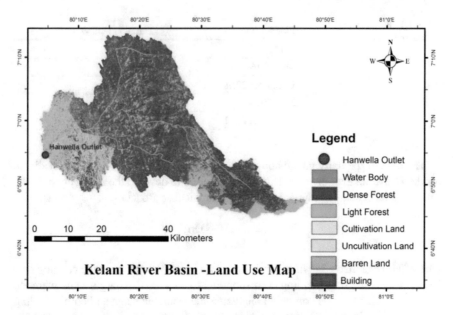

**Fig. 1.4** Land use land cover map

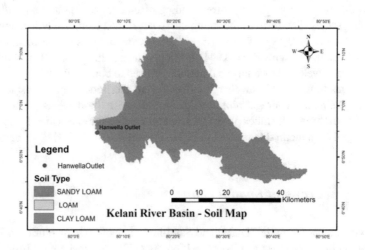

**Fig. 1.5** Soil map

## 1.4 Methodology

The entire modelling exercise has been carried out using software namely HEC-GeoHMS and HEC-HMS. HEC-HMS process along with the steps for its calibration and validation is illustrated in the form of flow chart as shown in Fig. 1.6.

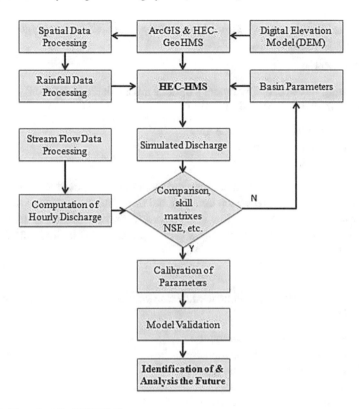

**Fig. 1.6** Flow chart for HEC-HMS process

## 1.4.1 HEC-Geo HMS and HEC-HMS

HEC-Geo HMS is Geospatial Hydrological Modelling System, the extended supplementary application tool of ArcGIS and used to develop basin model and its characteristics from the raw digital elevation model (DEM). Then the developed basin model in HEC-Geo HMS is imported to HEC-HMS for its further application. HEC-HMS is the simulation software developed to simulate all types hydrological processes of river basin (Sampath et al. 2015) It is available as online resource software designed by the US Army Corps of Engineers (USACE) with updated versions and treats user-friendly manner. The model consists of the following components:

**Basin Model and Meteorological Model**

Basin model is developed by HEC-GeoHMS as a single basin and used for the calibration and verification process. The required basin characteristics such as longest flow path, river lengths, upstream and downstream elevations and slopes of each river segments are obtained via this HEC-GeoHMS application process. Meteorological model is created by selecting the gage weight option available in this HEC-HMS model. Gage weights, which have been estimated from Thiessen polygon method, are used for this option.

**Base Flow Model**

The following three base flow model's options are available in HEC-HMS to serve the model running for each event:

- Constant, monthly varying base flow
- Exponential recession model
- Linear reservoir model.

In this application, exponential recession model option is used for the base flow model. Initial discharge, recession constant and threshold discharge are the parameters for this option. The base flow is separated from the ordinates of hourly flow hydrograph in order to obtain the hourly direct surface runoff hydrographs for each flood event.

**Loss Model**

There are numerous runoff volume models also known as loss models are applicable in HEC-HMS modelling.

- Initial and constant rate loss model
- Deficit and constant rate loss model
- SCS curve Number (CN) Loss Model
- Green and Ampt Loss Model
- Soil Moisture Accounting (SMA) Loss Model.

In the application of loss model, the method initial loss and constant rate is used. This method includes two parameters, such as initial loss and constant loss rate. Those parameters depend on the physical properties of the river basin soil, land use and the antecedent moisture condition in the basin. The calibrated loss model is used to estimate the excess rainfall hyetograph subtracting the losses from the average rainfall hyetograph for each flood event.

**Direct Runoff Model (Transform Model)**

The HEC-HMS programme facilitates various transform models to estimate the direct surface runoff hydrograph from the excess rainfall hyetograph. The following three transform models are used in HEC-HMS:

- Clark UH model
- SCS UH model
- Snyder UH model.

**Clark UH Model**

The use of Direct Runoff model—Clark UH (Clark 1945) is requires the following input as initial parameters in addition to the time—area diagram (Singh et al. 2014).

- Time of concentration Tc
- The storage coefficient, R.

The properties of time-area histogram or time-area percent curve of the basin.

The time of concentration Tc and the storage coefficient R are estimated from observed flood hydrograph available to employ as initial parameter values. There is an option available in the software to compute the ordinates of time-area diagram synthetically. Alternatively, the time-area diagram developed for the basin may be supplied as input replacing the synthetic time-area diagram. In this study, the time area diagram has been developed using ArcGIS software tool adopting Kriging Interpolation method.

**SCS UH Model**

For this model, only one parameter, known as basin lag, is available under direct runoff model. The initial value of the parameter is found from the observed flood hydrograph.

**Snyder UH Model**

For this model two parameters, Peaking Coefficient $C_p$ and Standard lag $t_p$ are available under direct runoff model. The initial value of this parameter $C_p$ is taken as 0.3 which is minimum of its possible values described in HEC-HMS technical reference manual. The Standard lag $t_p$ is taken from observed flood hydrograph.

## 1.4.2 Calibration of Model

Calibration of the model is done in two ways such as manual calibration (trial and error) and computer automatic calibration (optimization). However, the initial parameters of the individual model should be fed before simulating the model. Certain parameters have been estimated from observed flood hydrograph and some have been obtained from temporal and spatial data processing. The calibrations of the model have been done based on the various goodness of fit measures derived from the observed and simulated hydrographs in HEC-HMS programme. Based on these measures, the suitable methodologies are selected. The algorithms developed inside this modelling software estimate the optimum model parameters based on the various good of fit measures as an Objective Function. The choice of objective functions depends upon the necessity of the analysis. In the HEC-HMS, there are two options available for optimization which is based on search method. Those are Univariate-Gradient Algorithm and Nelder and Mead Algorithm.

### 1.4.3   Comparison of Three Direct Runoff Models

Although there are plenty of objective functions available in this software, NSE, Percentage Error Peak, Percent Error Time to Peak and Percent Error Discharge volume (Volume Deviation $D_v$) are applied for the quantitative approach of model evaluation.

**Nash–Sutcliffe Model Efficiency Coefficient** is defined as

$$
NSE = 1 - \left[ \frac{\sum_{i=0}^{n}(Q_{obs}(i) - Q_{com}(i))^2}{\sum_{i=0}^{n}(Q_{obs}(i) - \overline{Q})^2} \right] \tag{1.2}
$$

where $Q_{obs}$, $Q_{com}$ and $\overline{Q}$ are the observed, simulated and observed mean discharge over the n hours, respectively. And the optimal value of NSE is 1.

**Volume Deviation**

$$
D_v = \left| \frac{(V_{obs} - V_{com})}{V_{obs}} \right| \times 100\% \tag{1.3}
$$

where $V_{obs}$ and $V_{com}$ are the observed and simulated volume of runoff over the n hours, respectively. And the optimal value of $D_v$ is 0.

**Percent Error in Peak**

$$
Z = \left| \frac{Q_{obs(peak)} - Q_{com(peak)}}{Q_{obs(peak)}} \right| \times 100\% \tag{1.4}
$$

where $Q_{obs(peak)}$ and $Q_{com(peak)}$ are the observed and simulated peak discharge of runoff over the n hours, respectively. And the optimal value of Z is 0.

The NSE was reported as best performance criteria of simulation so far (Cuen et al. 2006). However, in addition to the NSE, percent error in peak, percent error in time to peak, percent error in discharge volume of each direct runoff model was compared in this study individually, of the number of extreme flood events considered.

## 1.5   Analysis and Results

### 1.5.1   Calibration and Validation

For the calibration and validation of HEC-HMS model for Kelani river basin, five number of extreme flood events occurred in the basin, are considered. The details about those flood events are given in Table 1.1. The flood events of May 2017, May 2014, December 2014 and June 2014 are employed for calibration by automatic option also referred to as optimization and the event of November 2012 is employed for validation.

All flood events considered to run the model have been observed during the period of south-west monsoon. The initial loss has been taken as zero as the soil moisture used to be saturated during the south-west monsoon season due to continuous rainfall usually observed in the basin. The constant rate is computed by equating the excess rainfall volume with direct runoff volume from the observed flood hydrograph with separation of base flow. This constant loss rate is found to be approximately 1 mm/hr which is considered as an initial estimate for HEC-HMS for running the optimization option. The imperviousness value in percent is required in the model as input to complete the loss model setup. Land use land cover map (shown in Fig. 1.4), developed in ArcGIS, provides the percent of impervious area as a function of land use. The basin predominantly covers with vegetative of light forest and dense forest with around 85% of the area. Only a small extend of basin covers with urban buildings. This imperviousness percent is obtained as 4% of the land use land cover map.

In addition to that the constant rate may be estimated if the soil type existing over the basin is known. The constant loss rate is the function of soil type of the basin. The soil map is prepared using ArcGIS by downloading the soil base data from the on-line resource of Food and Agriculture Organization (FAO) and shown in Fig. 1.5. This covers predominantly by sandy loam with 93% of the basin area. The range of loss rates for different soil class is described in the Technical Manual of HEC-HMS. However, a suitable value from this applicable range has been adopted to setup in the model as an initial parameter for the constant loss rate.

The direct surface runoff hydrographs are computed separating the base flow from the observed flood hydrograph for each flood events. For this method of base flow separation, the initial value of the recession constant is derived as 0.9 whereas the initial values of the initial discharge and threshold discharge are obtained as 39 m$^3$/s and 214.8 m$^3$/s, respectively from the observed flood hydrograph to employ as base model initial parameters.

For Clark model, the initial parameter values of Tc and R, taken as 18 hrs and 50 h, respectively, extracted from observed direct surface runoff hydrographs of various flood events. The percent curve for the tine-area diagram has been computed using Kriging Interpolation method and used in HEC-HMS software. Figure 1.7 shows the isochrones of Kelani river basin developed using the Kriging interpolation technique option of ArcGIS. For SCS UH model, the initial parameter value of Lag time is obtained as 2350 min from the observed flood hydrograph. For Snyder UH model, initial parameter value of Cp is taken as 0.3 from the technical reference manual of HEC-HMS whereas initial parameter value of standard lag is taken as 22 h from the observed flood hydrograph.

The HEC-HMS programme has been run optimizing the various parameters considering the option of Clark Model as a transform model. Similarly, SCS UH and Snyder UH models are also calibrated using the HEC-HMS programme. The parameter values obtained from the calibration of four events are averaged out and used for the validation for the fifth event for all three transform models. Table 1.2 shows the values of the average parameters obtained from calibration of the three

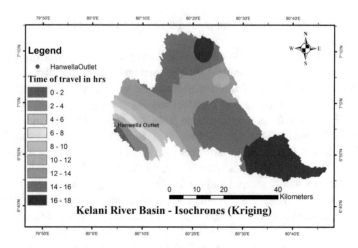

**Fig. 1.7** Isochrones of Kelani river basin (Kriging interpolation method)

**Table 1.2** Average of the parameters obtained from calibration of three transform models

| Parameters | Average of optimized (auto-calibrated) |
|---|---|
| Clark UH—Time of concentration Tc in hrs | 13.0 |
| Clark UH—Storage coefficient R in hrs | 39.0 |
| SCSUH—Lag time in minutes | 1868.2 |
| Snyder UH—Peaking coefficient | 0.3 |
| Snyder UH—Standard lag in hrs | 14.9 |
| Recession—Initial discharge in $m^3/s$ | 45.8 |
| Recession—Recession constant | 0.8 |
| Recession—Threshold discharge in $m^3/s$ | 206.3 |
| Initial and constant—Initial loss in mm | 0.0 |
| Initial and constant—Constant rate mm/hr | 0.9 |

different transform models for four flood events. These average values are used for validating the fifth event.

The performance of the three transform models is judged based on the comparisons of the different goodness-of-fit criteria computed from the simulated and observed flood hydrograph of the flood event at Hanwella gauging site considered for validation. Figure 1.8a, b and c illustrate the rainfall-runoff simulation of Hanwella gauging station after validation for the Clark UH, SCS UH and Snyder UH direct runoff model, respectively.

From the Fig. 1.8a, b and c, it is observed that the Clark UH model performed very well as the simulated flood hydrograph closely matches with the observed flood

**Figure 1.8** **a** Simulation by Clark UH. **b** Simulation by SCS UH. **c** Simulation by Snyder UH

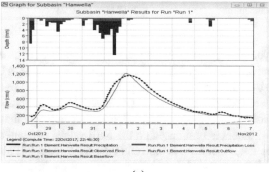

(a)

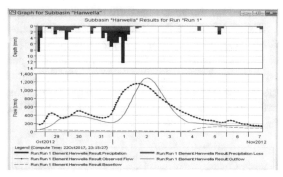

(b)

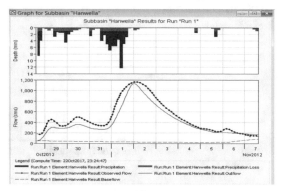

(c)

**Table 1.3** Nash–Sutcliffe model efficiency (NSE) values obtained from calibration and validation

| Computation | Event | Direct runoff model | | |
|---|---|---|---|---|
| | | Clark UH | SCS UH | Snyder UH |
| Calibration (optimization) | 2017 May | 0.97 | 0.468 | 0.97 |
| | 2016 May | 0.88 | 0.62 | 0.84 |
| | 2014 Dec | 0.96 | 0.76 | 0.96 |
| | 2014 Jun | 0.94 | 0.52 | 0.98 |
| Validation | 2012 Nov | 0.93 | 0.61 | 0.88 |

hydrograph during the validation as compared to the other two models, i.e. SCS UH and Snyder UH models. The performance of the SCS UH model is poor as compared to the other two models. To judge the performance of transform models based on the NSE, their computed values obtained during calibration and validation are given in Table 1.3. From the Table 1.3, it is also observed that the performance of the Clark UH model is best based on the NSE values computed as its value is 0.93, which is highest, whereas it is 0.88 and 0.61 for Snyder UH and SCS UH models.

The values of the various Goodness of fit criteria such as NSE, percent error in peak, percent error in time to peak and percent error in discharge volume are compared for the three transform models as shown in Fig. 1.9.

Figure 1.9 shows the value of percent error in discharge volume is minimum for Clark UH model as compared to the other two models. It indicates better simulation of the flood hydrograph. However, the values of percent error in peak and percent error in time to peak for Clark UH model is slightly higher than that of Snyder UH model. Nevertheless, Clark UH model has resulted in the best simulation of the flood hydrograph based on the computed values of NSE for the three models considered for analysis during the validation.

**Fig. 1.9** Comparison of Percent Errors from different transform models for the flood events used for validation

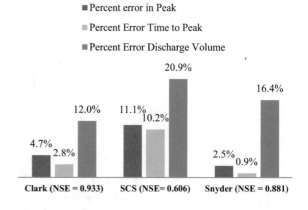

## 1.6 Conclusions and Recommendations

The conclusions are drawn, and recommendations made from the study are as follows:

(i)   ArcGIS software used the USGS satellite's data at 30-m resolution for preparing the land use and land cover map, soil map, and isochrones maps. However, better maps may be generated if the high-resolution data are used.

(ii)  HEC-GeoHMS used for basin modelling is capable of preparing basin maps and providing physiographic and other important geomorphological characteristics of the basin. These maps are the input for HEC-HMS.

(iii) HEC-HMS has the capability of flood simulation using various transform models including Clark model, Snyder model and SCS-CN model. The HEC-HMS has been successfully applied for simulating the five flood events observed in Kelani river basin (up to Hanwella Gauging Site). From the results, Clark model is recommended for simulation of flood events for this basin.

(iv)  The calibrated and validated model may be applied for the estimation of design flood for taking suitable structural measures to protect the important cities and installations in the downstream of Hanwella Gauging site. It may also be used for non-structural measures, such as real-time flood forecasting to issue flood warning to stakeholders to take necessary actions to evacuate the people, likely to be affected due to floods, for saving their lives and properties.

**Acknowledgements**  The authors would like to acknowledge the opportunities provided by Indian Water Resource Society (IWRS) and Department of Water Resources Development and Management, Indian Institute of Technology, Roorkee 247667 to present this paper in the International Conference on" SUSTAINABLE TECHNOLOGIES FOR INTELIGENT WATER MANAGEMENT" during February 16 to 19, 2018. The authors also thank to the Department of Irrigation, Sri Lanka for providing the required data for the study.

## References

Clark CO (1945) Storage and the unit hydrograph. Trans ASCE 110:1419–1446

Cuen RHM, Knight Z, Cutter AG (2006) Evaluation of Nash—Sutcliffe efficiency index. J Hydrol Eng 11(6):597–602

De Silva MMGT Weerakoon SB Srikantha H (2014) Modeling of event and continuous flow hydrographs with HEC–HMS: case study in the Kalani River Basin, Sri Lanka. J Hydrol Eng ASCE 19(4):800–806

Jain SK, Singh RD, Seth SM (2000) Design flood estimation using GIS supported GIUH approach. J Water Res Manage 14:369–376

Kumar RC, Chatterjee C, Singh RD, Lohani LK, Sanjay K (2004) GIUH based Clark and Nash models for runoff estimation for ungauged basin and their uncertainty analysis. Intl J River Basin Manage 2(4):281–290

Nandalal HK, Ratnayake UR (2010) Event based modeling of a watershed using HEC-HMS. J Inst Eng Sri Lanka xxxxiii(2):28–37

Sampath DS, Weerakoon SB, Herath S (2015) HEC-HMS model for runoff simulation in a tropical catchment with intra-basin diversions—case study of the Deduru Oya River Basin, Sri Lanka. J Inst Eng Sri Lanka XLVIII(1):1–9

Singh PK, Mishra SK, Jain MK (2014) A review of the synthetic unit hydrograph: from the empirical UH to advanced geomorphological methods. Hydrol Sci J 59(2):239–261

# Chapter 2
# Developing Strategies for Mitigating Pluvial Flooding in Gurugram

**Abhilash Rawat, M. P. Govind, Jawale Madhuri Vasudev, and Preetam Karmakar**

## 2.1 Introduction

Urban flooding is one of the growing concerns of the governments world wide. In future, it is estimated to increase both in frequency and magnitude. Floods have caused havoc because of inadequate town planning without following the principles of sustainability. Even though the drainage systems are designed for a long design period of 25–30 years, it tends to fail miserably even when the incident rainfall is within the expected limits. Some of the causes of urban flooding are as follows: (1) uneven distribution of rainfall along with rapid urbanization, (2) encroachments on stormwater drains, natural drainage channels, and water bodies, (3) incomplete stormwater drainage network, and (4) improper maintenance of stormwater drainage system. Delhi, Mumbai, and Kolkata are the prime examples where flooding usually takes place. It is observed that many modern builders violate the law and do not follow the norms. The illegal land reclamation of urban water bodies for new construction in these cities is very wide spread. For instance, in Calcutta (Kolkata), the location of Lake Town is very bad w.r.t. floods and it had suffered heavy floods in 1999. The floods were not only observed in 1999 but also with less severity in 1970, 1978, and 1984 (Applo study center 2017). All of these cities grew by leaps and bounds; lack of affordable housing sites has pushed people to any land typically those located on natural drainage channels, water bodies, and flood plains. The settlements thus formed grew and eventually became legalized due to political considerations. Examples are the settlements along Yamuna, Hooghly, and Mithi river flood plains and the drainage channels leading to these. Area under agricultural land use, within the peri-urban areas, which acts as the flood buffer, is also shrinking at a very fast

A. Rawat (✉) · M. P. Govind · J. M. Vasudev · P. Karmakar
Department of Planning, School of Planning and Architecture, Neelbad Road, Near IISER Institute, Bhauri, Bhopal, MP, India
e-mail: rawatabhilash91@gmail.com

A. Pandey et al. (eds.), *Hydrological Extremes*, Water Science and Technology Library 97, https://doi.org/10.1007/978-3-030-59148-9_2

pace everywhere. Farmers sell off their agricultural land for real estate development because of it being more profitable than farming.

## 2.2 Need for the Study

When the country receives normal rainfall of around 94 cm, people in drought hit areas heave a sigh of relief, but for Gurugram, it may sound like a flood warning. In the last years, the city got water-logged and normal life was disturbed even due to 3% less than normal rainfall. Construction of residential, commercial, and industrial projects in low lying areas or where infrastructure is not well developed resulted in stormwater runoff not getting drained out of the city completely. On the other hand, rapid spread of pervious surfaces and built-up areas have caused reduced groundwater recharging. Further the stormwater drains also got choked due to the disposal of solid waste. This paper has been divided into six sections. Introduction is given in first section and aim of the study and objectives are in Sect. 2. Study area characteristics and methodology used are briefed in Sect. 3. Section 4 deals with data collection, fifth section deals with data analysis, and the final section is on recommendations and way forward.

## 2.3 Aim of the Study

The aim of the study is to develop strategies for mitigating pluvial flooding in Gurugram. Objectives of this study are as follows: (1) to study the parameters responsible for urban flood vulnerability (2) to analyze the relative importance of identified parameters vis-à-vis urban flood vulnerability (3) to identify most critical areas vis-à-vis urban flooding in Gurugram, and (4) to propose flood mitigation strategies applicable for the critical areas (Map 2.1).

## 2.4 Study Area

Gurugram District falls in the south-eastern side of the state of Haryana. Its headquarters is located at Gurugram. It lies in between 27° 27′ 20″ and 28° 32′ 25″ latitude, and 76° 39′ 39″ and 77° 20′ 50″ longitude. Being in the vicinity of Delhi, Gurugram falls under National Capital Region (NCR). The total area of Gurugram Planning area is 732 km$^2$ and the population are around 2.3 million (2016).

The boundaries of Gurugram district are shared by following districts of Haryana state: Jhajjar on north, Faridabad on east, Palwal on south-east, Mewat on south, and

Rewari on south-east. Northern boundary is also shared by the national capital territory (NCT) of Delhi (www.gurgaon.nic.in). The study area selected is the Gurugram Planning Area consisting of 80 (eighty) micro-watersheds.

## 2.5   Methodology

This study contains four different stages. In the first stage, existing urban drainage system (stormwater and natural drainage), water-logged areas, and flood situations of recent past in the study area were studied. Also, in this stage, temporal variation of rainfall during monsoon season over 30 years, month-wise rainfall over 30 years in 5-year intervals, average monthly rainfall of 30 years, and annual rainfall of 30 years were studied. In next stage, land use and land cover pattern have been studied over past decades from 2000 to 2016 using Landsat 5, 6, 7, and 8 images. This has helped to find out the spatio-temporal changes in various land uses and land cover. Field visit further helped to find out existing situation along the major drains and the causes of 2016 floods in the city. Runoff depth, weighted curve number, and runoff volume were calculated for year 2016 by using remote sensing (RS) data, Geographic Information System (GIS), and Soil Conservation Service Curve Number (SCS-CN) method in the third stage. An integrated Analytical Hierarchy Process (AHP) and Geographic Information System (GIS) analysis techniques were used to model and predict the extent of flood risk areas. For urban flood vulnerability and risk mapping, the parameters were grouped under following criteria, viz., physical and socio-economic as shown in Fig. 2.1. In the AHP implementation, a pair-wise comparison technique was used to derive the priorities for the criteria in terms of their importance in causing urban flood. In last stage of the study, recommendations were given with respect to sustainable drainage system (SuDS) at high-risk areas which can be applied through source, site, and regional level measures to control and reduce chances of flooding in general and pluvial flooding in particular in Gurugram Planning area.

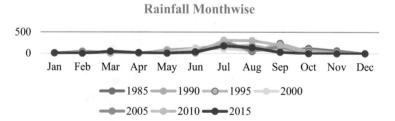

**Fig. 2.1**  Rainfall month wise (*Source* IMD)

## 2.5.1 Data Collection

The data collected for the study were as follows: (1) Remote sensing data: Landsat 5, 6, 7 and 8 images taken from USGS earth explorer. (2) The toposheets: 53D/11, 53D/15, 53H/3, 53H/4, 53D/16, 53D/12 collected from Survey of India, Delhi. (3) The Rainfall data for the period of 30 years (1986–2016) collected from Indian Meteorological Department (4) Haryana soil map collected from National Bureau of Soil Survey and Land Use Planning, Delhi, and (5) Micro-watershed map of Gurugram Planning Area from the Watershed Atlas of India.

## 2.5.2 Data Analysis

### 2.5.2.1 Rainfall

The Gurugram city experiences dry climate except during the summer, monsoon, and winter seasons. The average annual rainfall for five consecutive years: 1995–1999, 1996–2000, 1997–2001 were found out to be 665.2 mm, 628.4 mm, and 560.1 mm, respectively, showing a gradual reduction. In the last week of June, the south-west monsoon sets and draws to an end by the end of September and contributes 85% of the annual rainfall. July and August are the wettest months. Due to western disturbances and thunderstorms, 15% of annual rainfall occurs during the non-monsoon months. In 2010, maximum rainfall was observed in both July and August months as compared to other years as shown in Fig. 2.1. Maximum average monthly rainfall was observed in the month of July, which is around 142 mm. Minimum average monthly rainfall was observed in the month of January, which is around 4 mm as shown in Fig. 2.2.

Maximum rainfall was observed in the year 2010 and then gradual decrease in rainfall was observed until 2014 as shown in Fig. 2.3. In 2010 there was maximum rainfall for continuous 2 months, i.e., in July and August as shown in Fig. 2.4.

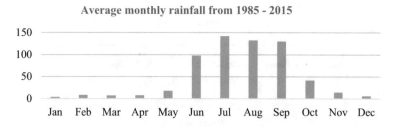

**Fig. 2.2** Average monthly rainfall (1985–2015) (*Source* IMD)

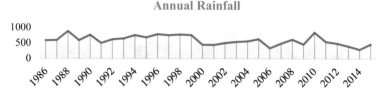

**Fig. 2.3** Annual rainfall (*Source* IMD)

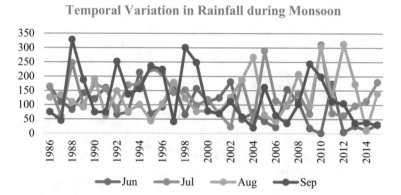

**Fig. 2.4** Temporal variations (*Source* IMD)

#### 2.5.2.2   Topography, Stormwater, and Natural Drainage Network

One of the ends of the Gurugram district is composed of hills and the other is of valleys making it an irregular and diverse nature of topography. In its northern part, Aravalli hills are located and sand dunes exit in the north-western part.

The elevation of the Gurugram planning area ranges from 184 to 348 m above M.S.L. At the time of field survey and data collection, total length of stormwater drainage network within the planning area was 334.09 km, but out of this, only 201.6 km has been completed. Of the rest, total length of drains under stay order, litigation, and encroachments was 97.52 and 24.27 km was the total length of drain under work-in-progress. Remainder of the network were the creeks having a length of 9.67 km (Maps 2.2 and 2.3).

#### 2.5.2.3   Flooding of Gurugram in 2016

The flooding of Gurugram in 2016 had a huge impact on lives and businesses. The worst affected areas included Sectors 14, 15, 17, 31, 38, 44, and 46, Sushant Lok Phase-I, DLF areas, Sohna Road, Golf Course Road, IFFCO Chowk, Sheetla Mata Road, Civil Lines, and Old-Delhi Road. Affluent areas such as Nirvana Country, areas around Huda City Centre, Vatika Chowk, and Medanta Medicity had also faced the

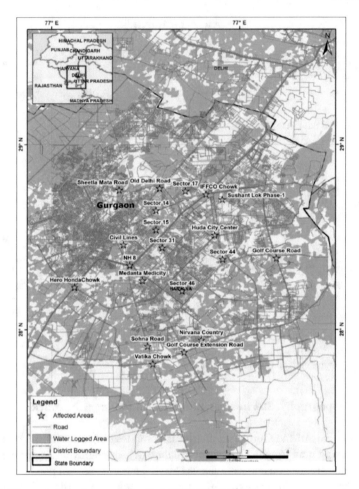

**Fig. 2.5** The water-logged areas of the city in 2016 RSMI (2016)

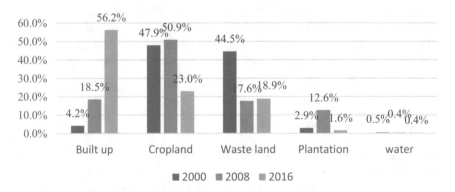

**Fig. 2.6** Percentage distribution of LULC in 2000, 2008, and 2016

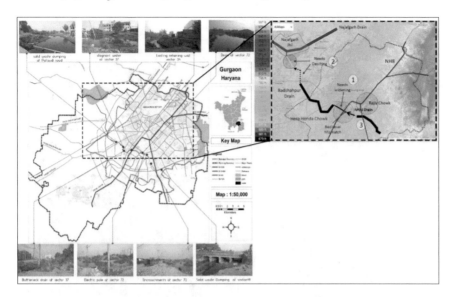

**Fig. 2.7**  Critical issues in Badshapur drain

**Fig. 2.8**  Level of flood hazard

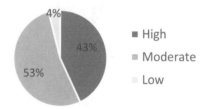

problem of water-logging. Some stretches of NH-8 were submerged under three feet of water.

Over four feet deep water had inundated some places near the Delhi-Gurugram expressway. In addition to these, on NH-8, traffic was completely blocked at Hero Honda Chowk due to the flooding (Fig. 2.5).

#### 2.5.2.4   Land Use Transformation

The population of Gurugram city had increased from 57,000 in 1971 to 1.74 lakh in 2001 and further to 2.3 million in 2016. Rapid population growth means, a decline in cultivable land through its transformation into residential, industrial, and commercial land uses. The pressure on urban land is also increasing, the structure of the city and its surroundings areas are also changing day by day. In this study, changes in land use and land cover pattern have been studied over the past decades from 2000 to 2016. Landsat 8, Landsat 7, and Landsat 5, 6 images were analyzed in ERDAS 2015

| | Permeable paving | Vegetated and Gravel Filter | Flow through and Infiltration Planter | Swales | Rain Gardens | Green gutters |
|---|---|---|---|---|---|---|
| 6mt wide Road | o | | | | | o |
| 9mt wide Road | o | | o | | | o |
| 12mt wide Road | o | o | o | o | | o |
| 18mt wide Road | o | o | o | o | o | o |
| | | | | Only if there is a kerb extension | | |
| 24mt wide Road | o | o | o | o | o | o |
| 30mt wide Road | o | o | o | o | o | o |
| 40mt wide Road | o | o | o | o | o | o |
| 45mt wide Road | o | o | o | o | o | o |
| 60mt wide Road | o | o | o | o | o | o |

**Fig. 2.9** SuDS components feasible for various road types (*Source* Oasis Designs Inc 2012)

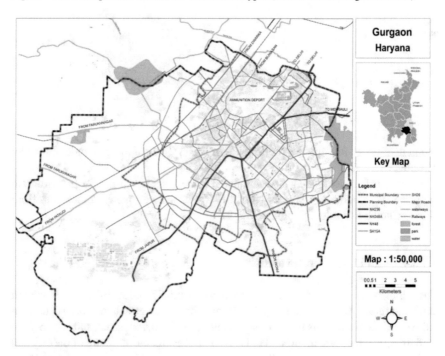

**Map 2.1** Study area: Gurugram planning area

**Map 2.2** Gurugram elevation

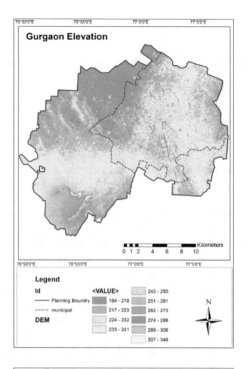

**Map 2.3** Stormwater and natural drainage

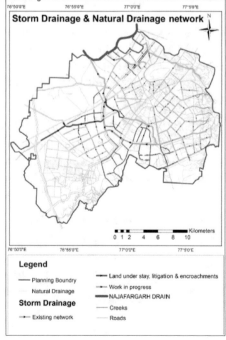

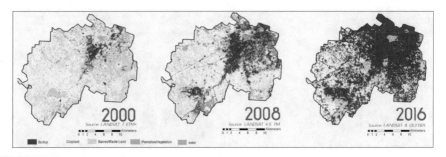

**Map 2.4**  Land use and land cover transformation from 2000 to 2016

software to study the spatial-temporal changes in land use and land cover pattern as shown in Map 2.4.

The analysis revealed that, of the 463.93 km$^2$, which is the total area of Gurugram city, nearly 44.55% was under waste land in 2000. It was reduced to around 17.61% in 2008. The rate of decline has been higher during the last decade. Between 2008 and 2016, 129.45 km$^2$ of agricultural land was lost to development. The built-up area increased from 66.56 km$^2$ in 2008 to 174.48 km$^2$ in 2016, the percentage share having increased from 4.20 in 2000 to 18.55 in 2008, to 56.15 in 2016. The area of water bodies reduced from 2.12 km$^2$ in 2000 to 1.66 km$^2$ in 2008. Figure 2.6 shows the percentage distribution of land use and land cover in 2000, 2008, and 2016.

Due to growth of economic activities and growth of population, agricultural land in peripheral areas of city have been declining rapidly. The development of transportation system and influence of Delhi could be the reasons for rapid land use changes. Growth directions of Gurugram city are toward Delhi and Faridabad. Thus, rise in built-up areas and development of roads in this city prevent water to penetrate into the underground aquifers causing water shortage and drying up of natural water bodies (Table 2.1).

### Causes of Urban Flood in Gurugram in 2016

The width of Badshahpur drain, which is the major drain in Gurugram, reduces after some distance and its capacity becomes one-third of its required capacity due to factors such as encroachments, siltation, and solid waste dumping. The flow transfer

**Table 2.1**  Changes in area under various land use land cover between 2000–2008 and 2008–2016

| Land use class | Changes in area (km$^2$) | | % Changes in area | |
|---|---|---|---|---|
| | 2000–2008 | 2008–2016 | 2000–2008 | 2008–2016 |
| Built up | 66.35 | 174.93 | 340.5 | 203.78 |
| Cropland | 13.92 | −129.46 | 6.3 | −54.81 |
| Waste land | −124.82 | 6.03 | −60.4 | 7.39 |
| Plantation | 45.01 | −51.04 | 334.48 | −87.3 |
| Water | −0.464 | 0 | −20.0 | 0 |

between Badshahpur drain with other drains such as NHAI drain in Gurugram and Najafgarh drain in Delhi have been affected due to heavy siltation and clogging (Fig. 2.7).

### 2.5.2.5  Runoff Generation Model Using SCS-CN Method

There are 80 micro-watersheds in Gurugram planning area as shown in Map  2.5 and runoff generated in all of these watersheds have been estimated. Soil Map has been generated for all micro-watersheds as shown in Map  2.6. Two categories of hydrologic soil groups suitable for the study area were identified as Group C and D. Group "C" has low infiltration rate of 1–4 mm/h whereas group "D" has high runoff potential and very low infiltration rates of 0–1 mm/h. Antecedent moisture condition has been decided as AMC-II on the basis of rainfall in the study area. Finally, weighted curve numbers were determined. Watershed-wise curve numbers have been generated (Map  2.7), where the highest curve number indicates the maximum runoff potential and lowest curve number shows lowest runoff potential. Similarly, runoff depth for all 80 micro-watersheds have been estimated and Map  2.8 has been generated. After estimating the runoff depth, runoff volume of all micro-watersheds has been calculated as shown in Map  2.9 (Table 2.2).

Sub-watershed Number 2C5F2h has got the highest runoff volume due to the presence of highly dense built-up areas. Using GIS, spatial intersection of different land uses and land cover with various hydrological soil groups have been done for all 80 micro-watersheds. The results of spatial intersection were used for calculating

**Map 2.5** Micro-watershed map

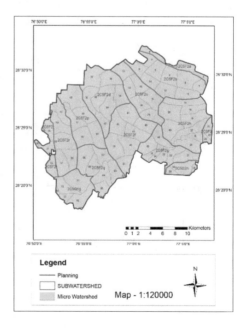

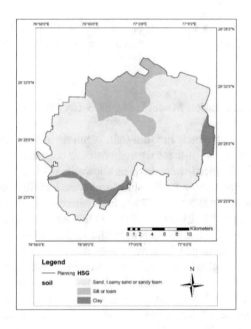

**Map 2.6** Soil map

**Map 2.7** Curve number map

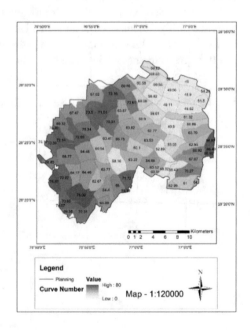

**Map 2.8**  Runoff depth map

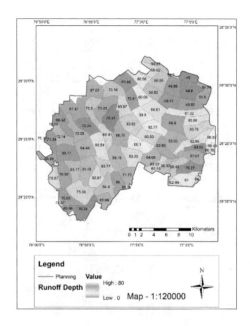

**Map 2.9**  Runoff volume map

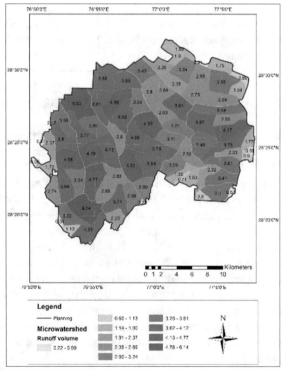

**Table 2.2** Runoff volume at sub-watershed level

| Sub-watershed | Micro-watershed | Storm days | Area (km$^2$) | Runoff volume (Mm$^3$) |
|---|---|---|---|---|
| 2C5F2a | 0, 1, 2, 3 | 60 | 10.02 | 4.50 |
| 2C5F2b | 4, 5, 6, 7, 8, 9, 10, 11, 12 | 60 | 52.39 | 23.30 |
| 2C5F2c | 13, 14, 15, 16, 78, 79 | 60 | 36.86 | 18.61 |
| 2C5F2d | 36, 37, 38, 39, 47, 64, 65 | 60 | 47.92 | 25.57 |
| 2C5F2f | 21, 40, 41, 42, 43 | 60 | 30.32 | 19.63 |
| 2C5F2g | 31, 32, 33, 34, 35, 44, 45 | 60 | 11.89 | 15.50 |
| 2C5F2h | 17, 18, 19, 20, 24, 26, 27, 80 | 60 | 7.84 | 26.98 |
| 2C5F2j | 22, 23, 25 | 60 | 5.37 | 3.26 |
| 2C5D2n | 28, 29, 30 | 60 | 9.53 | 4.72 |
| 2C5F2p | 48, 49, 50, 51, 58 | 60 | 40.62 | 21.16 |
| 2C5F2q | 60, 63, 66, 67, 68, 69 | 60 | 38.03 | 19.46 |
| 2C5F2r | 52, 54, 55, 56, 57 | 60 | 35.50 | 18.95 |
| 2C5F2t | 53, 74, 75, 76, 77 | 60 | 13.61 | 7.46 |
| 2C5G1g | 59, 70, 71, 72, 73 | 60 | 27.20 | 14.93 |

the weighted curve number (CN) in each of the sub-watersheds. On the basis of these, weighted curve numbers (CNw), runoff depth, and volume of runoff have been calculated.

### 2.5.2.6  Urban Flood Vulnerability and Risk Mapping

In order to do prediction and modeling of the magnitude of flood risk areas, integration of following analysis techniques has been done: (1) Analytical Hierarchy Process (AHP) and (2) Geographic Information System (GIS). The flood risk vulnerability mapping follows a multi-parametric approach and integrates some of the causative factors for flooding such as rainfall distribution, elevation and slope, drainage network and density, LULC and soil type (Tateishi 2014). Higher importance is given for DEM cells with lower elevation, which implies that the way in which elevation could be associated with risk is important. Steeper slopes are more susceptible to surface runoff, while flat terrains are susceptible to water-logging. The slope classes having less values were assigned higher rank. Soil infiltration capacity decrease leads to increased surface runoff, which further leads to flooding. Higher is the drainage density, higher is the susceptibility for erosion and sedimentation at lower elevations.

While implementing the AHP, a pair-wise comparison technique was used to derive the priorities for various criteria depending upon their importance in causing urban flooding. Pair-wise comparison matrix of various criteria causing urban flooding is given in Table 2.3. Higher the assigned weightage, higher is the relative importance (Maps 2.10, 2.11, 2.12, 2.13 and 2.14).

**Table 2.3** Pair-wise comparison matrix for flood vulnerability

| Criteria | Drainage density | Rainfall | Elevation | Soil | Land use | Slope | Weightage (%) |
|---|---|---|---|---|---|---|---|
| Drainage density | 1 | 4 | 1/2 | 1/2 | 2 | 4 | 23.75 |
| Rainfall | 1/4 | 1 | 1/2 | 3 | 1/2 | 5 | 10 |
| Elevation | 2 | 2 | 1 | 1/4 | 1/2 | 5 | 21.44 |
| Soil | 2 | 1/3 | 1/3 | 1 | 1/3 | 1/2 | 11.51 |
| Land use | 1/2 | 2 | 4 | 2 | 1 | 2 | 25.73 |
| Slope | 1/4 | 1/3 | 1/5 | 2 | 1/2 | 1 | 7.51 |

**Map 2.10** Digital elevation model

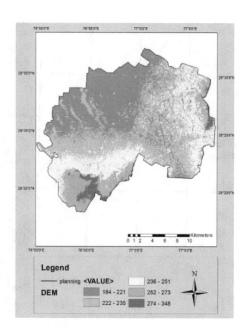

Legend

planning <VALUE>

DEM    184 - 221    236 - 251    252 - 273    222 - 235    274 - 348

N

For assessing ranks for various criteria, weightage values and decision sub-factors have been considered. Then, ranks were assigned based upon influence of various sub-factors in causing floods. Weighted ranks for various criteria are given in Table 2.4.

Finally, flood hazard map for Gurugram Planning area has been prepared. Out of the total, 43% of the areas are highly vulnerable to flooding due to various reasons such as construction activities, built-up areas, location in low lying areas, encroachments, and solid waste dumping on the drainage network and siltation in the drains. All these highly vulnerable areas are within the municipal limits.

Out of the remaining, 53% of the areas located outside the municipal boundary are moderately vulnerable to flooding, due to presence of agricultural land. Micro-watersheds such as 4, 9, 13, 14, 26, 29, 30, 39, 40, 41, 44, 78, 79, and 80 from highly

**Map 2.11**  Slope map

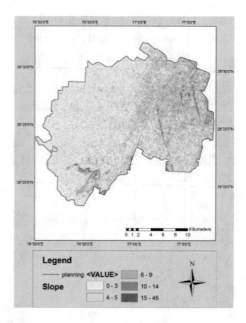

**Map 2.12**  Soil infiltration
capacity

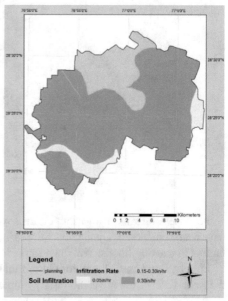

**Map 2.13**   Drainage density

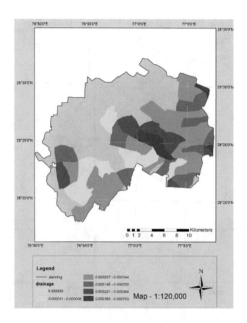

**Map 2.14**   Rainfall distribution

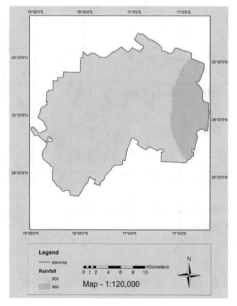

**Table 2.4** Weighted flood vulnerability ranking

| Criteria | Weight age (%) | Decision Sub-factors | Ranking decision |
|---|---|---|---|
| Elevation (meters) | 21.44 | 184–221 | 5 |
| | | 222–235 | 4 |
| | | 236–251 | 3 |
| | | 251–273 | 2 |
| | | 274–348 | 1 |
| Slope (Degree) | 7.51 | 0–3 | 5 |
| | | 4–5 | 4 |
| | | 6–9 | 3 |
| | | 10–14 | 2 |
| | | 15–45 | 1 |
| Soil Infiltration (in/h) | 11.51 | 0.05 | 5 |
| | | 0.15-0.3 | 3 |
| | | 0.30 | 1 |
| Land use | 25.73 | Built-up | 5 |
| | | Barren/waste land | 4 |
| | | Cropland | 3 |
| | | Vegetation | 2 |
| | | Water | 1 |
| Rainfall (mm) | 10 | 600 | 3 |
| | | 800 | 5 |
| Drainage Density (km/km$^2$) | 23.75 | 0.000011–0.000056 | 1 |
| | | 0.000057–0.000144 | 2 |
| | | 0.000145–0.000220 | 3 |
| | | 0.000221–0.000364 | 4 |
| | | 0.000365–0.000703 | 5 |

vulnerable areas have been selected for proposing the interventions (Map 2.15 and Fig. 2.8).

## 2.6 Recommendations and Way Forward

The key strategy proposed here is the series of green infrastructure interventions effecting incremental reduction of water-borne pollution and discharge rates. To accommodate these interventions, the proposal should be through source control, site control, and regional control measures.

**Map 2.15** Urban flood hazard map

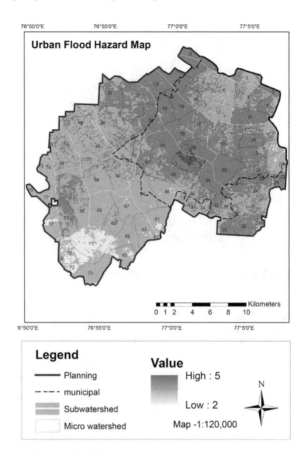

For proposing SuDS management train for tackling flood vulnerability in the selected watershed areas, percentage break-up of various land uses and land cover have been analyzed as shown in Map 2.16.

## 2.6.1   Source Control

Rainwater harvesting (RWH) is the process of collecting stormwater runoff for use. At the source level, the runoff can be collected from roofs and later on get stored in tanks (underground or surface). This water can be treated (where required) and can be used as per requirements. For critical areas, rainwater harvesting structures can be proposed at the building level (Source). Areas suitable for RWH are the residential, commercial, mixed-use, industrial and public semi-public within the selected watersheds (critical) = 2101344.77 km$^2$.

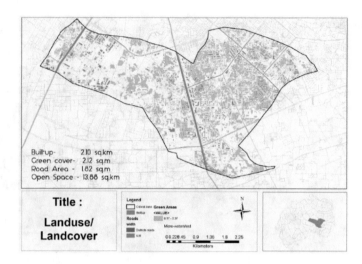

Built-up - 2.10 sq.km
Green cover - 2.12 sq.m
Road Area - 1.82 sq.m
Open Space - 13.88 sq.km

Title :

Landuse/
Landcover

**Map 2.16** Land use/land cover of critical areas

**Table 2.5** Calculation for annual RWH potential (*Source* CPWD 2002)

| Quantity of rainwater that can be harvested | |
|---|---|
| Annual rainfall | 630 mm = 0.63 m |
| Area of roof catchment | 0.105 km$^2$ = 105000 m$^2$ |
| Rainwater endowment of that area | 105000 × 0.63 = 66150 cum |
| Runoff coefficient for roof catchment | 0.85 |
| Coefficient for evaporation, spillage AND FIRST FLUSH | 0.8 |
| Annual RWH potential | = 66.15 × 0.85 × 0.8 = 44.98 ML |

Assuming 5% of area is effective for RWH as roof catchment = 0.105 km$^2$ (Table 2.5).

### 2.6.2 Site Control

Under this, runoff gets conveyed through conventional and sustainable drainage systems to lakes or natural streams, existing or dried up. In the identified critical areas, currently dried up lakes and natural streams can be recharged again through this and this will also help in recharging groundwater. For implementing site control measures, feasible groundwater recharge points were first identified, then sub-catchments and surfacewater flow routes were defined. Further SuDS components of the management

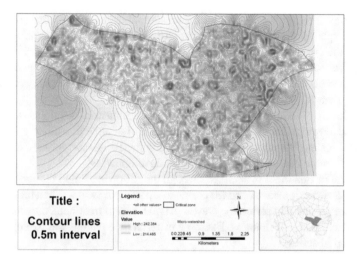

| Title : | Legend | N |
|---|---|---|
| **Contour lines** | | |
| **0.5m interval** | | |

**Map 2.17**  Contour lines

**Map 2.18**  Sub-catchments
in critical areas

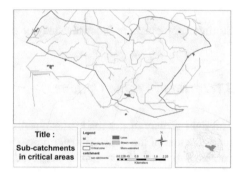

train were selected accordingly. Different SuDS components providing one level of treatment of surface runoff such as permeable paving, bio-swale, rain gutter, detention basins, and filter drains were placed according to site characteristics such as existing land use, topography, and availability of vacant land and its sub-catchment's water holding capacity (Maps  2.17 and  2.18).

## 2.7  Regional Control

Proposed development type is the deciding factor for determining the level of surfacewater treatment required at regional level. As per the SuDS Manual (Ballard 2015), following are the recommendations vis-à-vis green infrastructural (SuDS) interventions.

- One level of treatment is required for residential development consisting of less than or equal to 50 houses and commercial development having a parking capacity of less than or equal to 50 spaces.
- Two levels of treatment are required in cases of residential development having more than 50 houses and commercial development having a parking capacity of greater than 50 spaces.
- Three levels of treatment are required in case of industrial development.
- All road developments would require two levels of treatment, with exception given to small scale developments. All road development schemes typically require two levels of treatment, except for small scale developments.
- One level of treatment is obtained through the following SuDS features: (1) filter drain, (2) detention basin, and (3) permeable paving or swale. When these components are provided in series, they result in two levels of treatment, depending upon the combinations of features. Example is permeable paving and detention basin provided in series.

### 2.7.1  Parking Lot with Swale/Planters

Stormwater planters or bioswales may be provided in the parking lots.

### 2.7.2  Road Profile Options

The right of way of the road determines, how stormwater runoff flows off from a road. Camber of the road also determines the flow direction of surface runoff. SuDS components feasible for various road types based on its right of way are shown in Fig. 2.9.

### 2.7.3  Appropriate Plant Material

Stormwater management can also be done by using appropriate plant material. The plants should be evergreen, locally grown, having aesthetic appeal, and high survival instinct. Plants suitable for Gurugram soil and climatic conditions are Canna, Scripus, Cyperus, Eichhornia crassipes, Phragmites, Solix spp, Potamogeton nodosus, Typha latifolia, Ceratophyllum demersvm, and Sagittaria latifolia.

Sustainable drainage system (SuDS) can be a long-lasting solution to mitigate surface runoff and to diffuse pollution. It can also create an appealing environment with improved biodiversity. Combination of Sustainable drainage system and conventional drainage system is a better practical solution for combating urban flooding in general and pluvial flooding in particular and also for diffusing the pollution load of the surface runoff. Maintenance and management are main issues while implementing the SuDS. So, for its better sustenance, people's awareness and participation are really necessary.

## 2.8  Planning Interventions

In order to implement the SuDS interventions, land use planning system has a great role to play. Land use planning controls the development and helps to make spaces available for public use such as parks, open spaces, green spaces, etc. SuDS interventions at various scales can be incorporated as regulations in the development control regulations (DCRs).

**Acknowledgements**  It gives us immense pleasure and satisfaction in submitting this Paper. Our profound gratitude and deep regard to Suresh Hatkar, Atray Karmahe, Parishit Mehta, and Akkash Chauhaan (M.plan Students, SPA Bhopal).

## References

Applo Study Center TG (2017) 137 marks major issue. PDFappolosupport.com>upload>2014/08
Ballard BW (2015) The SuDS manual. CIRIA, Griffin Court, 15 Long Lane, London, EC1A 9PN, UK
CPWD (2002) Rainwater harvesting and conservation manual. Delhi
Oasis Designs Inc (2012) Retrofitting our urban streets for sustainable drainage. UTTIPEC, Delhi
RSMI (2016) India: flooding in Gurgaon (July 27–18, 2016). RSMI, Nodia
Tateishi YO (2014) Urban flood vulnerability and risk mapping using integrated multi-parametric AHP and GIS: methodological overview and case study assessment. Directory of Open Access Journals

# Chapter 3
# Vulnerability Assessment of Manipur to Floods Using Unequal Weights

H. Sanayanbi, A. Bandyopadhyay, and A. Bhadra

## 3.1 Introduction

Flood is one among the usual and expensive natural disasters which occurs frequently around the world. The degree of flooding is anticipated to increase under the impact of climate change and economic development (Jonkman and Kelman 2005; IPCC 2007). As a result, the whole world will experience an increasing flood risk, particularly in the developing nations. One-fifth of the global death by flood is accounted by India and every year about 30 million people are removed from the affected region. The area vulnerable to flood is 40 M ha and average area affected by flood is eight M ha. The physical and non-physical losses due to floods in India are getting larger because of rapid population growth and larger intrusion into the flood plains for establishment, cultivation, and other blooming projects. In India Ganga and the Brahmaputra are the most persistent flood-prone river basins. The effect of flood relies on susceptibility of the affected socio-economic and ecological systems (Cutter 1996). The vulnerability of Manipur to flood is very important due to its low adaptive capacity. As such, the vulnerabilities of these different environmental systems and socio-economic conditions need to be correctly evaluated. The vulnerability of a place on the earth surface to flood is a function of the region's exposure to the hazard (natural event) and the

H. Sanayanbi (✉)
National Institute of Disaster Management, A-wing, 4th floor, NDCC-II Building, Jai Singh Road, New Delhi 110001, India
e-mail: vivienehod@yahoo.com

A. Bandyopadhyay · A. Bhadra
Department of Agricultural Engineering, North Eastern Regional Institute of Science and Technology, Nirjuli (Itanagar) 791109, Arunachal Pradesh, India
e-mail: arnabbandyo@yahoo.co.in

A. Bhadra
e-mail: aditibhadra@yahoo.co.in

A. Pandey et al. (eds.), *Hydrological Extremes*, Water Science and Technology Library 97, https://doi.org/10.1007/978-3-030-59148-9_3

43

anthropogenic activities carried out within the catchment area, which impedes the free flow of water. Vulnerability is frequently reflected on socio-economic characteristics of the population living in that region. Due to hazard, exposure, and adaptive capacity, water resources systems are quite vulnerable which means that vulnerability of a system is a function of hazard, exposure, and adaptive capacity. Hazard, in terms of flood, is considered as an act of flood posing threat to life, health, property, or environment. Exposure can be considered as the presence of livelihood, infrastructure, physical assets, etc. which are exposed to hazard where floods can occur. Adaptive capacity referred to as the ability of system to cope better with current or possible unfavorable conditions brought about by hazards. High exposure and low coping capabilities would lead to the highest risk from a given flood event and those with low exposure and high coping abilities would lead to the lowest risk (Bhadra et al. 2017).

Vulnerability is dynamic and changes considering social responses as well as new rounds of dangerous events. Vulnerability continually changes because of changes in technology, population behavior, practices, and policies. Thus, vulnerability is dynamic (Wilhite 2000). Quantitative assessment of vulnerability employs perceptive, sensuous-intuitive, and phenomenological methods for studying flood risk from signs, statements, experiences, and evidences (Ciurean et al. 2013). Assessment of the vulnerability to flood is aimed at development of policies that reduce the risks associated with flood. Also, vulnerability assessment will help decision-makers to take on appropriate plans and strategies to lessen the possible damage. The principal aim of vulnerability assessment is to give facts and figures to guide the people living in that region and also increase their adaptive capacity (Ciccarelli et al. 2016). Developing, testing, and implementing indicators for identifying and assessing vulnerability to floods are an important pre-requisite for effective disaster risk reduction (Bhadra et al. 2017). Vulnerability assessment assists to spot people, places, or properties that are vulnerable to suffer harm. Numerous approaches to assess vulnerability can be found in literature (Fedeski and Gwilliam 2007; Taubenbock et al. 2008; Ebert et al. 2009; Fekete 2009; Kienberger et al. 2009; Kubal et al. 2009; Linde et al. 2011).

Number of studies on quantitative vulnerability assessment, viz., Moss et al. (2001), and Luers et al. (2003) exemplified the approach of composite index for vulnerability assessment. For example, an index which is a composite of 16 variables selected from five sensitive sectors (settlement, food security, human health, ecosystem, and water) and three dimensions for coping capacity (economic, human resources, and environmental) to measure vulnerability to climate change for 38 countries was used by Moss et al. (2000) and developed a methodology for the spatial vulnerability assessment of floods in the coastal regions of Bangladesh. Bahinipati (2014) adopted an integrated approach to assess vulnerability across the districts of Odisha, India and provided a better understanding of the adaptive capacity of households toward cyclone and flood. Ousmane et al. (2015) and Liu and Li (2015) also assessed the social vulnerability to flood in Medina Gounas.

In this study, an attempt is made to rank the districts of Manipur in terms of their vulnerability to flood. The vulnerability indices are dependent on number of indicators that contribute to vulnerability of a specific region. The indices are calculated from a set of indicators selected for all the districts and differentiate them with each other or with some reference point on some numerical scale. Based on these indices, different districts are ranked and identified as comparatively less or more vulnerable. Such index-based methods have been used by many researchers, namely, Atkins et al. (1998), Chris (2000), and Ravindranath et al. (2011), and are found to be well acceptable.

## 3.2   Data and Methodology

### 3.2.1   Study Area

Manipur is one of the north-east states of India which is located between latitudes of 23°83'N and 25°68'N, and longitudes of 93°03'E and 94°78'E. The geographical area of this state is 22,327 km² and the population is around 3 million. It lies at an elevation of 790 m (2,590 ft) above sea level. The state is bounded by Nagaland in its north, Mizoram in its south, Assam in its west, and Burma (Myanmar) lies in the east. The maximum temperature goes up to 32 °C in summer months and in winter it goes down below 0 °C (32 °F), bringing frost and snow sometimes in the hilly regions due to the Western Disturbance. The coldest and warmest month is January and July, respectively. Monsoon in the state starts from May and lasts till mid-October with average annual rainfall of 1,467.5 mm. The study area is presented in Fig. 3.1. The points shown in the figure are the centroids of each district labeled

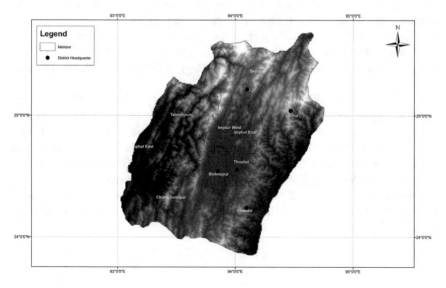

**Fig. 3.1**  The district map of Manipur

**Table 3.1** District headquarters (HQ) of Manipur

| Sl. No. | District | HQ | Latitude, °N | Longitude, °E | Elevation (m) |
|---|---|---|---|---|---|
| 1 | Bishnupur | Bishnupur | 24.32 | 94.00 | 828.00 |
| 2 | Chandel | Chandel | 24.29 | 93.25 | 886.00 |
| 3 | Churachandpur | Churachandpur | 24.81 | 93.96 | 914.40 |
| 4 | Imphal-east | Porompat | 24.82 | 93.91 | 118.00 |
| 5 | Imphal-west | Lamphelpat | 25.32 | 94.15 | 118.00 |
| 6 | Senapati | Senapati | 24.95 | 93.50 | 1050.00 |
| 7 | Tamenglong | Tamenglong | 24.54 | 93.96 | 1260.00 |
| 8 | Thoubal | Thoubal | 23.06 | 87.31 | 785.00 |
| 9 | Ukhrul | Ukhrul | 24.93 | 94.48 | 1662.00 |

with the district names followed by its respective headquarters names. At present, there are 16 districts of Manipur which was just formed recently in the year 2017 but for this study, we have taken only 9 districts since our data period is 2009, and only 9 districts were there during this period. The location and elevation values for all the headquarters of 9 districts are shown in Table 3.1.

### 3.2.2 Data Acquisition

From the Directorate of Economics and Statistics, Government of Manipur, Imphal the values for different indicators of the three components, viz., hazard, exposure, and adaptive capacity were acquired for 9 districts for the year 2009 and some census data were downloaded for year 2001 from https://www.censusindia.gov.in. The rainfall data used in this study were extracted from long period (1901–2010) daily gridded rainfall data set collected from India Meteorological Department (IMD) (Pai et al. 2014).

Trend analysis of these rainfall data was then carried out for the whole 100 years and two sets of 30 years (1971–2000 and 1981–2010). In this study, the Mann–Kendall (MK) test was used for detection of rainfall trend. MK test was a statistical yes/no type hypothesis testing procedure and, therefore, another index, Senslope (Sen 1968) was used to quantify the magnitude of such trend. The toolbar "Arc Trend" of ArcGIS developed by Bandyopadhyay et al. (2011) was used for trend analysis. The Senslope values were directly used as indicators. However, yes/no results of MK test with significance levels 1, 5, and 10% were converted to numeric indicators following Table 3.2.

**Table 3.2** Conversion of MK test result to numeric indicator

| Sl. No. | MK test result | | | Value of indicator |
|---|---|---|---|---|
| | 1% level of significance | 5% level of significance | 10% level of significance | |
| 1 | Y− | Y− | Y− | −3 |
| 2 | N | Y− | Y− | −2 |
| 3 | N | N | Y− | −1 |
| 4 | N | N | N | 0 |
| 5 | N | N | Y+ | 1 |
| 6 | N | Y+ | Y+ | 2 |
| 7 | Y+ | Y+ | Y+ | 3 |

### 3.2.3 Selection of Indicators and Their Functional Relationship with Vulnerability

From the available district-wise data of the state, the indicators for each component of vulnerability were selected based on the definition given by IPCC on its Third Assessment Report (McCarthy et al. 2001). For assessing vulnerability to flood, indicators were selected separately for all the three components in terms of their impact on vulnerability. We selected six indicators for hazard, eight indicators for exposure, and ten indicators for adaptive capacity (Table 3.3). After finalizing the indicators, functional relationships of the indicators with the vulnerability to flood were set up.

The relationships were set up based on how the indicators were affecting the vulnerability to flood. Two types of functional relationship were possible: increased in the value of the indicator increased the vulnerability or decreased the vulnerability. For example, if we take rainfall, it is understandable that larger the value of this indicator, higher will be the vulnerability of that district to flood and we can say there is increasing functional relationship with vulnerability which is shown by ↑ in Table 3.3. On the other hand, more land area means all of it may not get submerged and people can take refuge there, leading to decreasing relationship with vulnerability to flood, which is shown by ↓.

### 3.2.4 Arrangement of Indicators

For the components of hazard, exposure, and adaptive capacity, the collected data were arranged in the form of a rectangular matrix. The districts are represented by rows and indicators are represented by columns. We assumed that M districts and collected K indicators are there. For K number of indicators and M number of

**Table 3.3** Conversion of MK test result to numeric indicator

| Component | Sl. No. | Indicators | Notation | Functional relationship |
|---|---|---|---|---|
| Hazard | 1 | Elevation (m) | (H1) | ↓ |
| | 2 | Rainfall (mm) | (H2) | ↑ |
| | 3 | Rainfall trend for 30 years | (H3) | ↑ |
| | 4 | Slope of rainfall trend for 30 years | (H4) | ↑ |
| | 5 | Rainfall trend of 100 years | (H5) | ↑ |
| | 6 | Slope of rainfall trend for 100 years | (H6) | ↑ |
| Exposure | 7 | Total population | (E1) | ↑ |
| | 8 | % of agricultural land to total land | (E2) | ↑ |
| | 9 | % of rain-fed land | (E3) | ↑ |
| | 10 | % of workforce in agriculture | (E4) | ↑ |
| | 11 | % of rural population | (E5) | ↑ |
| | 12 | Cereal/rice yield (metric tons/ha) | (E6) | ↑ |
| | 13 | Total population of livestock and poultry | (E7) | ↑ |
| | 14 | Consumption of fertilizer | (E8) | ↑ |
| Adaptive capacity | 15 | Land area (km$^2$) | (A1) | ↓ |
| | 16 | % of literacy rate | (A2) | ↓ |
| | 17 | % of literacy rate of people aged 15–24 | (A3) | ↓ |
| | 18 | % or urban population | (A4) | ↓ |
| | 19 | % of household electrified | (A5) | ↓ |
| | 20 | % of female population | (A6) | ↓ |
| | 21 | % of students enrolled in primary education | (A7) | ↓ |
| | 22 | % of students enrolled in secondary education | (A8) | ↓ |
| | 23 | % of students enrolled in tertiary education | (A9) | ↓ |
| | 24 | % of non-worker population | (A10) | ↑ |

districts, will be the value of the indicator j corresponding to district i and the matrix table will have K columns and M rows (i = 1, 2, 3, …, M and j = 1, 2, 3, …, K).

### 3.2.5   Normalization of Indicators

The selected indicators had different scales and units. To get normalized values of indicators without any units, normalization was done so that they all lied between 0 and 1. For normalization, UNDP's Human Development Index (HDI) (UNDP 2006) was followed. Normalization was done based on the functional relationship which had already been set (Table 3.3).

For increasing functional relationship, the following equation was used:

$$x_{ij} = \frac{X_{ij} - \text{Min}\{X_{ij}\}}{\text{Max}\{X_{ij}\} - \text{Min}\{X_{ij}\}} \tag{3.1}$$

And for decreasing functional relationship, the following equation was used:

$$x_{ij} = \frac{\text{Max}\{X_{ij}\} - X_{ij}}{\text{Max}\{X_{ij}\} - \text{Min}\{X_{ij}\}} \tag{3.2}$$

where, $x_{ij}$ is the normalized value of indicator, $X_{ij}$ is the raw value of indicator.

It was lucid that all these values will lie between 0 and 1. The value 1 corresponded to the district with the most vulnerable arising out of a given indicator and 0 corresponded to the district with least vulnerable from an indicator.

### 3.2.6   Methods of Construction of Vulnerability Indices (VIs)

The equal weight method gives equivalent significance to all the indicators which are not inevitably right. Hence many authors (Gbetibouo and Ringler 2009; Swain and Swain 2011; Parekh et al. 2015) prefer to give unequal weights to the indicators.

#### 3.2.6.1   Determination of Unequal Weights

Iyengar and Sudarshan (1982) assumed that the weights vary inversely to the standard deviation of the respective indicators of vulnerability and this method was used to find the weights of the indicators. Hence the weight, $w_j$, is determined by

$$w_j = \frac{c}{\sqrt{V_j}} \tag{3.3}$$

where $V_j = \mathrm{Var}(x_i)$ over all the districts for $j$th indicator and $c$ is the normalizing constant which is obtained as

$$c = \left[ \sum_{j=1}^{K} \frac{1}{\sqrt{V_j}} \right]^{-1} \tag{3.4}$$

The selection of the weights based on the variance of the indicator values ensured that large variation in any indicators would not excessively influence the contribution of the rest of the indicators and thereby misrepresent the inter-district differentiations (Iyengar and Sudarshan 1982).

#### 3.2.6.2 Determination of VIs

The method developed by Iyengar and Sudarshan (1982) to rank the districts in terms of their economic performance was statistically sound and well suited for the development of composite index of vulnerability for this also. The VI of $i$th zone $(\bar{y}_i)$ was considered to be a linear sum of weighted $x_{ij}$ as given below:

$$\bar{y}_i = \sum_{j=1}^{K} w_j x_{ij} \tag{3.5}$$

where $w_j$ $(0 < w_j < 1$ and $\sum_{j=1}^{K} w_j = 1)$ are the weights calculated using Eq. 3.3. The VI $(\bar{y}_i)$ so computed lying between 0 and 1, with 1 indicating maximum vulnerability and 0 indicating minimum vulnerability.

### 3.2.7 Classification of Districts Based on VIs

For meaningful characterization of the different levels of vulnerability, suitable fractal classification from an assumed probability distribution was preferred. Hence, the beta probability distribution, which was generally skewed and takes values in the interval (0, 1) was used. This distribution was also used by Iyengar and Sudarshan (1982) in their study. This distribution has the probability density as given below:

$$f(x) = \frac{1}{\beta(a, b)} x^{a-1} (1 - x)^{b-1}, 0 < x < 1 \text{ and } a, b > 0$$
$$= 0, \text{ otherwise} \tag{3.6}$$

$$\text{with, } \beta(a, b) = \int_0^1 x^{(a-1)}(1-x)^{(b-1)} \tag{3.7}$$

where $x$ is the realization variable of the beta function whose value lies between 0 and 1, $a$ and $b$ are two positive shape parameters.

## 3.3   Results and Discussion

### 3.3.1   Normalization of Indicators

#### 3.3.1.1   Hazard

The maximum and minimum values of each indicator for different districts are presented in Table 3.4. These maximum and minimum values were used for normalizing all the indicator values. For hazard component, taking elevation of the district headquarters indicator as an example, the maximum value is 1662 m (Ukhrul) and minimum value is 118 m (Imphal-east and Imphal-west) above mean sea level. These two values were used in Eq. 3.2 for normalizing the indicators depending on their functional relationship shown in Table 3.4. In case of elevation of the district headquarters indicator, the more is the value, the less would be the vulnerability to flood as most of the district population is normally concentrated near the district headquarters. Hence it has a decreasing functional relationship with vulnerability. Therefore, Eq. 3.2 is used to normalize this indicator. The same procedure is followed for other remaining hazard indicators also. As an example, the map of normalized rainfall indicator is shown in Fig. 3.2. From the map, it can be seen that Tamenglong has the lowest normalized rainfall value of 0.00 and Imphal-east has the highest value of 1.00.

#### 3.3.1.2   Exposure

For exposure component, taking total population as an example, we can see from Table 3.4 that Imphal-west has maximum population with 444,382 and Tamenglong has the lowest with 111,499. By definition of exposure it can be defined that more thickly populated region may be regarded more exposed to flood. This may be because even if the magnitude of the flood is low, a more thickly populated area would mean a larger number of people exposed to the hazard area resulting in more vulnerability. Therefore, for this indicator, there will be an increasing functional relationship with the vulnerability. And hence, Eq. 3.1 was used to normalize this indicator using the maximum and minimum values. Similarly, the other indicators of this component were also normalized taking their respective maximum

**Table 3.4** Maximum and minimum values of the indicators

| Sl. No. | Indicator | Maximum | Minimum |
|---|---|---|---|
| *Hazard* | | | |
| 1 | H1 | 1662 (Ukhrul) | 118 (Imphal-east and Imphal-west) |
| 2 | H2 | 1217.32 (Ukrul) | 1009.36 (Chandel) |
| 3 | H3 | 0 (All districts) | 0 (All districts) |
| 4 | H4 | 0.00486 (Churachandpur) | −0.03608 (Tamenglong) |
| 5 | H5 | 3 (Churachandpur) | 0 (All districts except Churachandpur) |
| 6 | H6 | 0.00709 (Churachandpur) | 0.00024 (Ukrul) |
| *Exposure* | | | |
| 7 | E1 | 444,382 (Imphal-west) | 111,499 (Tamenglong) |
| 8 | E2 | 65.78 (Thoubal) | 0.30 (Churachandpur) |
| 9 | E3 | 38.5452794(Imphal-west) | 2.90974947 (Chandel) |
| 10 | E4 | 30.863953 (Tamenglong) | 6.62403068 (Imphal-west) |
| 11 | E5 | 100 (Ukhrul) | 44.4885256 (Imphal-west) |
| 12 | E6 | 3.02 (Imphal-west) | 1.8 (Churachandpur) |
| 13 | E7 | 699,286 (Senapati) | 260,063 (Churachandpur) |
| 14 | E8 | 10,260 (Thoubal) | 104 (Tamenglong) |
| *Adaptive capacity* | | | |
| 15 | A1 | 4570 (Churachandpur) | 496 (Bishnupur) |
| 16 | A2 | 80.22 (Imphal-west) | 56.23 (Chandel) |
| 17 | A3 | 92.47169629 (Imphal-west) | 68.56370598 (Chandel) |
| 18 | A4 | 55.51147436 (Imphal-west) | 0 (Churachandpur, Senapati, Tamenglong and Ukhrul) |
| 19 | A5 | 77.91926958 (Imphal-west) | 32.98656264 (Tamenglong) |
| 20 | A6 | 50.09226296 (Imphal-west) | 47.8149995 (Ukhrul) |
| 21 | A7 | 18.37953704 (Tamenglong) | 6.099032492 (Churachandpur) |
| 22 | A8 | 7.843927072 (Imphal-west) | 4.330980547 (Tamenglong) |
| 23 | A9 | 14.13851146 (Imphal-west) | 6.407232352 (Tamenglong) |
| 24 | A10 | 60.27056595 (Imphal-east) | 29.83735337 (Senapati) |

and minimum values and considering their functional relationships with vulnerability. As an example, the map of normalized rural population is shown in Fig. 3.3. The map shows that Imphal-west has the lowest normalized rural population and Churachandpur, Senapati, Tamenglong, and Ukhrul have the highest.

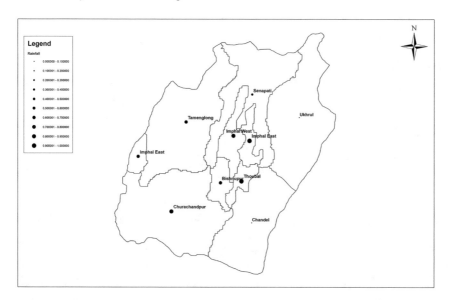

**Fig. 3.2**  Map showing the normalized values of rainfall

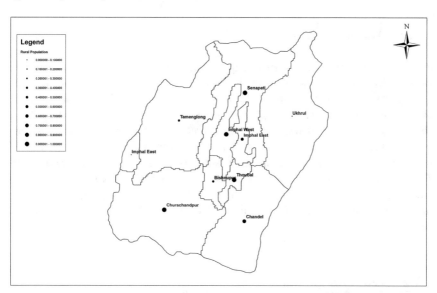

**Fig. 3.3**  Map showing the normalized values of rural population

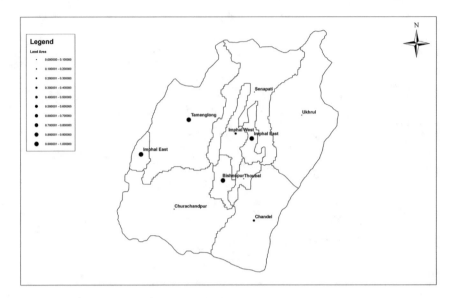

**Fig. 3.4**  Map showing the normalized values of land area

### 3.3.1.3  Adaptive Capacity

For adaptive capacity component, taking literacy rate as an example, from Table 3.4 it can be seen that Imphal-west has the highest literacy rate with 80.22% and Chandel has the least literacy rate with 56.23%. High value of this indicator means more literates in the district which means they are more aware and have more knowledge about how to cope with flood. So the vulnerability will be lower and hence literacy rate has decreasing functional relationship with vulnerability. Therefore, Eq. 3.2 was used for normalizing this indicator using its maximum and minimum values. The other indicators of this component were also normalized following the same procedure. An example map of normalized land area is shown in Fig. 3.4. The map shows that Bishnupur and Thoubal district have the lowest and Churachandpur has the highest land area.

## 3.3.2  Assigning of Weights

Weights for indicators of hazard component were calculated from the normalized hazard indicators. The resulting assigned weights were 0.20, 0.20, 0.00, 0.21, 0.19, and 0.21 for elevation of district headquarters, rainfall, rainfall trend for 30 years, slope of rainfall trend for 30 years, rainfall trend for 100 years, and slope of rainfall trend for 100 years, respectively.

For indicators of exposure component, the weights assigned were 0.12, 0.12, 0.12, 0.12, 0.12, 0.12, 0.16, and 0.13 for total population, percentage of agricultural land, percentage of rain-fed land, percentage of workforce in agriculture, percentage of rural population, cereal yield, total population of livestock and poultry, and consumption of fertilizer, respectively.

For indicators of adaptive capacity component, the assigned weights were 0.08, 0.10, 0.10, 0.09, 0.11, 0.09, 0.12, 0.09, 0.10, and 0.12 for land area, percentage of literacy rate, percentage of literacy rate of people aged 15–24, percentage of urban population, percentage of household electrified, percentage of female population, total students enrolled in primary education, total students enrolled in secondary education, total students enrolled in tertiary education and percentage of non-worker population, respectively.

### 3.3.3   Construction of VIs

The vulnerability indices were estimated using the weights assigned. All the components, viz., hazard, exposure, and adaptive capacity were analyzed individually considering indicators belonging to that component only, and the districts were ranked depending on their relative vulnerability. Also, a composite VI was estimated by considering all the indicators of the three components together. Computing the vulnerability separately for each component helps us comprehend how the districts perform differently in terms of different components of vulnerability (Sharma and Patwardhan 2008). This will ultimately help us identify priority component for remedial measures for districts that rank high in the composite VI.

Using the normalized indicators values and weights assigned in Eq. 3.5, vulnerability indices and ranks for the 9 districts were calculated based on indicators of hazard, exposure, and adaptive capacity components, as well as, composite VI. The VIs and ranks calculated considering all indicators are shown in Table 3.5.

When only hazard component is considered, Churachandpur is the most vulnerable district with Imphal-east on its next, and Tamenglong is the least vulnerable district followed by Chandel. This may be because Churachandpur receives high amount rainfall and has maximum slope in 30 yrs rainfall trend and maximum rainfall trend in 100 yrs and Imphal-east lying in a lower elevation receive highest amount of rainfall; whereas, district headquarters of Tamenglong is located in high elevation with significant negative trend of rainfall for 30 yrs and Chandel has least received rainfall.

It can be seen from Table 3.5 that, when only exposure component is considered, Imphal-west ranks first followed by Thoubal, and Churachandpur is the least vulnerable district followed by Chandel. This can be attributed to the highest population in Imphal-west and Thoubal and Churachandpur having the minimum values of many indicators of exposure component (four out of eight).

When only adaptive capacity is considered, Chandel ranks first followed by Tamenglong while Imphal-west ranks last followed by Imphal-east. This may be

**Table 3.5** Vulnerability indices and ranks for hazard, exposure, adaptive capacity, and composite vulnerability from I&S method

| District | Hazard | | Exposure | | Adaptive capacity | | Composite vulnerability | |
|---|---|---|---|---|---|---|---|---|
| | VI | Rank | VI | Rank | VI | Rank | VI | Rank |
| Bishnupur | 0.28 | 5 | 0.44 | 5 | 0.49 | 7 | 0.42 | 7 |
| Chandel | 0.20 | 8 | 0.26 | 8 | 0.75 | 1 | 0.45 | 5 |
| Churachandpur | 0.86 | 1 | 0.23 | 9 | 0.63 | 4 | 0.55 | 1 |
| Imphal-east | 0.44 | 2 | 0.50 | 3 | 0.43 | 8 | 0.45 | 5 |
| Imphal-west | 0.37 | 3 | 0.60 | 1 | 0.31 | 9 | 0.42 | 7 |
| Senapati | 0.30 | 4 | 0.45 | 4 | 0.67 | 3 | 0.51 | 2 |
| Tamenglong | 0.16 | 9 | 0.32 | 7 | 0.74 | 2 | 0.46 | 4 |
| Thoubal | 0.28 | 5 | 0.59 | 2 | 0.50 | 6 | 0.48 | 3 |
| Ukhrul | 0.23 | 7 | 0.34 | 6 | 0.58 | 5 | 0.41 | 9 |

because Chandel and Tamenglong have very low values for adaptive capacity indicators, e.g., percentage of literacy rate, percentage of literate people ages 15–24, and students enrolled in secondary and tertiary education; whereas, Imphal-west and Imphal-east, a small district in which capital of Manipur is situated, has highest literacy rate, highest percentage of urban population, and highest percentage of household electrified.

In terms of composite VI Churachandpur ranks first followed by Senapati while Bishnupur, Imphal-west, and Ukhrul ranks last.

### 3.3.4 Classifications of Districts

To classify the districts of Manipur based on composite VIs, VIs were further graduated using beta distribution based on the estimated parameters of $a$ and $b$. The values $a$ and $b$ thus calculated from Eqs. 3.6 and 3.7 are 59.92 and 69.93, respectively. With these, the values of $z_1$, $z_2$, $z_3$, $z_4$, and $z_5$ were calculated using the beta distribution online calculator developed by Casio Computer Co. Ltd. (https://keisan.casio.com/exec/system/1,180,573,395) and the resulting 20% cut-off points are 0.42, 0.45, 0.47, and 0.50, respectively. Based on these calculations, the districts of Manipur were finally classified into five clusters (Very high, high, moderate, less, and very less) depending on the levels of composite VIs.

Figure 3.5 presents the district classification map based on indices of adaptive capacity component. In terms of composite vulnerability, Churachandpur and Senapati have the highest VIs and lie in the zone of very high vulnerability followed by Thoubal which lie in the high vulnerability zone. This may be because these districts

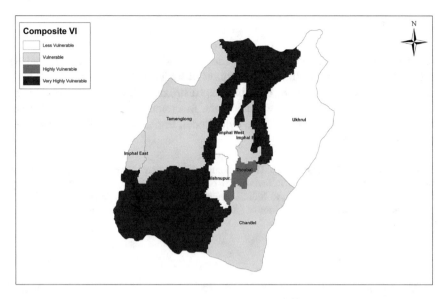

**Fig. 3.5** Classified map of Manipur in terms of composite vulnerability to flood

are low on adaptive capacity indicators with moderate exposure indicators. Bishnupur, Imphal-west, and Ukhrul are least vulnerable to flood followed by Chandel, Imphal-east, and Tamenglong in the vulnerability zone.

### 3.3.5  Validation

While it is important to validate the result obtained from the study, validation for flood vulnerability is a difficult task since vulnerability is dynamic and changes with time incorporating the socio-economic changes. However, validation needs to be done to check the accuracy of the results. Therefore, few available data available with the state government of Manipur, a global flood inventory (Adhikari et al. 2010), and some online compilation of various media reports for different years have been collected and analyzed for validation. Adhikari et al. (2010) has listed Imphal, Bishnupur, Thoubal, and Tamenglong as flood-affected districts. A report given on 25 August 2015 on Manipur Floods (Sphere India 2015) showed that Thoubal, Chandel, and Churachandpur have been worst affected by flood followed by Ukhrul and Imphal. The NDTV (2015) news channel has also reported on 02 August 2015 that Thoubal is severely affected by flood.

## 3.4   Conclusions

The study showed that Churachandpur and Senapati have the highest VIs and lie in the zone of very high vulnerability followed by Thoubal which lie in the high vulnerability zone. This may be because these districts are low on adaptive capacity indicators with moderate exposure indicators. Bishnupur, Imphal-west, and Ukhrul are least vulnerable to flood followed by Chandel, Imphal-east, and Tamenglong in the vulnerability zone.

**Acknowledgements**  The authors would like to thank Indian National Committee on Surface Water, Ministry of Water Resources, Govt. of India (Grant No. 23/73/2012-R&D) for sponsoring this study and also express their appreciation to India Meteorological Department, Directorate of Disaster Management of Manipur, and Directorate of Statistics and Economics of Manipur for providing the needful data.

## References

Adhikari P, Hong Y, Douglas KR, Kirschbaum BD, Gourley J, Adler R, Brankenride GR (2010) A digitized global flood inventory (1998–2008): compilation and preliminary results. Nat Hazards 55:405–422

Atkins J, Mazzi S, Ramlogan C (1998) A study on the vulnerability of developing and island states: a composite index. Technical report, Commonwealth Secretariat, London, UK

Bahinipati CS (2014) Assessment of vulnerability to cyclones and floods in Odisha, India: a district-level analysis. Curr Sci 17(12):1997–2007

Bandyopadhyay A, Pal A, Debanath S (2011) Development of an ArcGIS toolbar for trend analysis of climatic data. World Acad Sci Eng Technol 60:563–570

Bhadra A, Bandyopadhaya A, Hodam A, Yimchungru CY, Debbarma R (2017) Assessment of vulnerability of Arunachal Pradesh (India) to floods. Congress proceedings: papers, posters and presentations. XVI world water congress, 29 May–3 June 2017, Cancun, Mexico

Chris E (2000) The commonwealth vulnerability index. In: Ministerial conference on environment and development in Asia and the Pacific, 31 Aug–05 September 2000, Kitakyushu, Japan

Ciccarelli D, Pinna MS, Alquini F, Cogoni D, Ruocco M, Bacchetta G, Fenu SG (2016) Development of a coastal dune vulnerability index for Mediterranean ecosystems: a useful tool for coastal managers? Estuar Coast Shelf Sci 187. https://doi.org/10.1016/j.ecss.2016.12.008

Ciurean RL, Schroter D, Glade T (2013) Conceptual frameworks of vulnerability assessments for natural disasters reduction. In: Tiefenbacher J (ed) Approaches to disaster management—examining the implications of hazards, emergencies and disasters. https://doi.org/10.5772/55538

Cutter LS (1996) Vulnerability to environmental hazards. Prog Hum Geogr 20(4):529–539

Ebert A, Kerle N, Stein A (2009) Urban social vulnerability assessment with physical proxies and spatial metrics derived from air- and space borne imagery and GIS data. Nat Hazards 48:275–294

Fedeski M, Gwilliam J (2007) Urban sustainability in the presence of flood and geological hazards: the development of a GIS based vulnerability and risk assessment methodology. Landsc Urban Plan 83:50–61

Fekete A (2009) Validation of a social vulnerability index in context to river-floods in Germany. Nat Hazards Earth Syst Sci 9:393–403

Gbetibouo GA, Ringler C (2009) Mapping South African farming sector vulnerability to climate change and variability. IFPRI discussion paper 00885. International Food Policy Research Institute, Washington, DC

IPCC (2007) Climate change (2007): impacts, adaptation and vulnerability. Contribution of working group II to the fourth assessment report of the Intergovernmental Panel on Climate Change (IPCC). In: Parry ML, Canziani OF, Palutikof JP, van der Linden PJ, Hanson CE (eds) Cambridge University Press, Cambridge, UK

Iyengar NS, Sudarshan P (1982) A method of classifying regions multivariate data. Econ Polit Wkly 2047–2052. Special Article

Johnson NL, Kotz S (1970) Continuous univariate distributions. Wiley, New Delhi, India, pp 41–43

Jonkman SN, Kelman I (2005) An analysis of the causes and circumstances of flood disaster deaths. Disasters 29(1):75–97

Kienberger S, Lang S, Zeil P (2009) Spatial vulnerability units—expert-based spatial modelling of socio-economic vulnerability in the Salzach catchment, Austria. Nat Hazards Earth Syst Sci 9:767–778

Kubal C, Haase D, Meyer V, Scheuer S (2009) Integrated urban flood risk assessment—adapting a multicriteria approach to a city. Nat Hazards Earth Syst Sci 9:1881–1895

Linde AH, Bubeck P, Dekkers JEC, de Moel H, Aerts JCJH (2011) Future flood risk estimates along the river Rhine. Nat Hazards Earth Syst Sci 11:459–473

Liu DL, Li Y (2015) Social vulnerability of rural households to flood hazards in western mountainous regions of Henan province, China. Nat Hazards Earth Syst Sci 3:6727–6744

Luers LA, Lobell BD, Sklar LS, Addams CL, Matson AP (2003) A method for quantifying vulnerability, applied to the agricultural system of the Yaqui Valley, Mexico. Glob Environ Chang (13):255–267

McCarthy JJ, Canziani OF, Leary NA, Dokken DJ, White KS (2001) Climate change 2001: impacts, adaptation, and vulnerability. Contribution of working group II to the third assessment report of the intergovernmental panel on climate change. Cambridge University Press, Cambridge, UK

Moss RH, Brenkert AL, Malone EL (2000) Measuring vulnerability: a trial indicator set. Pacific Northwest National Laboratory, Richland, WA, USA

Moss RH, Brenkert AL, Malone EL (2001) Vulnerability to climate change: a quantitative approach. Pacific Northwest National Laboratory, Richland, WA, USA

NDTV (2015) Rain hampers rescue efforts in flood-hit Manipur. New Delhi Television Limited, 03 Aug 2015. https://www.ndtv.com/india-news/rain-hampers-rescue-efforts-in-flood-hit-manipur-1203154. Accessed 28 Sept 2016

Ousmane DS, Gaye AT, Diakhate M, Mawuli A (2015) Social vulnerability assessment to flood in Medina Gounass Dakar. J Geogr Inf Syst 7:415–429. https://doi.org/10.4236/jgis.2015.74033

Pai DS, Sridhar L, Rajeevan M, Sreejith OP, Satbhai NS, Mukhpadyay B (2014) Development of a new high spatial resolution (0.25° × 0.25°) long period (1901–2010) daily gridded rainfall data set over India and its comparison with existing data sets over the region. MAUSAM 65(1):1–18

Parekh H, Yadav K, Yadav S, Shah N (2015) Identification and assigning weight of indicator influencing performance of municipal solid waste management using AHP. J Civ Eng Korean Soc Civ Eng 19(1):36–45

Ravindranath NH, Rao S, Sharma N, Nair M, Gopalakrishnan R, Rao AS, Malaviya S, Tiwari R, Sagadevan A, Munsi M, Krishna N, Bala G (2011) Climate change vulnerability profiles for North East India. Curr Sci 101(3):384–394

Sen PK (1968) Estimates of the regression coefficient based on Kendall's tau. J Am Stat Assoc 63:1379–1389

Sharma U, Patwardhan A (2008) Methodology for identifying vulnerability hotspots to tropical cyclone hazard in India. Mitig Adapt Strat Glob Chang 13:703–717

Sphere India (2015) Manipur floods (2015). Joint needs assessment report: 25th Aug 2015. National Coalition of Humanitarian Agencies in India. https://reliefweb.int/sites/reliefweb.int/files/resources/manipur-jna-flood-2015-report-6-pm-24-08-2015-sphere-india.pdf. Accessed 28 Sept 2016

Swain M, Swain M (2011) Vulnerability to agricultural drought in western Orissa: a case study of representative blocks. Agric Econ Res Rev 24:47–56

Taubenbock H, Post J, Roth A, Zosseder K, Strunz G, Dech S (2008) A conceptual vulnerability and risk framework as outline to identify capabilities of remote sensing. Nat Hazards Earth Syst Sci 8:409–420

UNDP (2006) Human development report. United Nations Development Program (UNDP). https://hdr.undp.org/hdr2006/statistics/

Wilhite DA (2000) Drought as a natural hazard: concepts and definitions (Chapter 1). In: Wilhite DA (ed) Drought: a global assessment. Natural hazards and disasters series, vol 69. Routledge Publishers, UK, pp 3–18

# Chapter 4
# Development of Regional Flood Frequency Relationships for Gauged and Ungauged Catchments of Upper Narmada and Tapi Subzone 3(c)

**Raksha Kapoor, Rakesh Kumar, and Mohit Kumar**

## 4.1 Introduction

Regional flood frequency analysis (RFFA) is seen as a handy tool in settling the issues of short data records and ungauged catchments faced by designers of hydraulic structures, flood protection works, etc. Site-specific quantile estimates of flood can be determined if the flood data for a number of sites constituting a region are available, i.e. they have been gauged earlier, with the help of RFFA. In nutshell, it replaces the time data by space for the computation of floods with different return periods at more than one site, hugely for catchments which are small or at most medium sized, as concluded by Pilgrim and Cordery (1992). Studies on design flood estimation methods have been a priority for hydrologists in India for a long time. Central Water Commission (CWC), in collaboration with India Meteorological Department (IMD), has majorly contributed by carrying out research on the computation of design floods using various methods such as combining synthetic unit hydrograph (SUH) whose parameters are based on the physiographic and meteorological characteristics of the catchment, with design rainfall, for example, CWC (2002) or the RFF studies undertaken by Research Designs and Standards Organization using pooled curve methods and the USGS for few regions in India. RFFA studies have also been carried out by research institutions like National Institute of Hydrology (NIH), Roorkee (e.g. Kumar et al. 1999, 2003; Kumar and Chatterjee 2005).

R. Kapoor (✉) · M. Kumar
Punjab Engineering College (Deemed to be University), Chandigarh 160012, India
e-mail: raksha.kapoor65@gmail.com

M. Kumar
e-mail: mohitkumar255174@gmail.com

R. Kumar
National Institute of Hydrology, Roorkee 247667, India
e-mail: rakeshnih@gmail.com

© The Editor(s) (if applicable) and The Author(s), under exclusive license
to Springer Nature Switzerland AG 2021
A. Pandey et al. (eds.), *Hydrological Extremes*, Water Science
and Technology Library 97, https://doi.org/10.1007/978-3-030-59148-9_4

This study constitutes development of RFF relationships for the catchments which are primarily ungauged and gauged as well, utilizing the L-moments approach for the Upper Narmada and Tapi Subzone 3(c) in India. Firstly, an RFF relationship is generated for gauged catchments with the help of the robustly identified frequency distribution. Various frequency distributions, namely, extreme value (EV1), logistic (LOS), uniform (UNF), generalized logistic (GLO), normal (NOR), generalized normal (GNO), general extreme value (GEV), exponential (EXP), kappa (KAP), generalized Pareto (GPA), Pearson type-III (PE-III), and Wakeby (WAK) with five parameters were employed. Then, the relationship MAPF versus catchment areas derived for gauged catchment is further evolved to the relationship to estimate design flood in the ungauged region of the study area. Conclusively, the study presents two useful RFF relationships which can be utilized to determine various sorts of flood estimates in the gauged and ungauged regions restricted to the study area. Probability weighted moments (PWMs) with their evolved version, i.e. L-moments are briefly described below.

### 4.1.1  Probability Weighted Moments

Probability weighted moments (PWMs) were first established by which can be written as

$$M_{i,j,k} = \int_0^1 x\,(F)^i\,(F)^j\,(1-F)^k\,\mathrm{d}F \tag{4.1}$$

where

$F(x)$  CDF (cumulative density function),
$F(x)$  $\int_{-x}^{x} f\,(x)\,\mathrm{d}x$ and $x(F)$ is the inverse of $F(x)$;

$i, j, k$ are the real numbers. The particularly useful special cases of the PWMs $\alpha_k$ and $\beta_j$ are

$$\alpha_k = M_{1,0,k} = \int_0^1 x\,(F)\,(1-F)^k\,\mathrm{d}F \tag{4.2}$$

$$\beta_j = M_{1,j,0} = \int_0^1 x\,(F)\,(F)^j\,\mathrm{d}F \tag{4.3}$$

One limitation to PWMs is that they can be applicable to the probability distributions which have their possible converses.

## 4.1.2   L-Moments Approach

L-moments as stated by Hosking and Wallis (1997) provide an alternative statistical approach towards delineating shapes of different probability distributions. They evolved by modifying the probability weighted moments (PWMs). The first four L-moments are shown below:

$$\lambda_1 = \alpha_0 = \beta_0 \tag{4.4}$$

$$\lambda_2 = \alpha_0 - 2\alpha_1 = 2\beta_1 - \beta_0 \tag{4.5}$$

$$\lambda_3 = \alpha_0 - 6\alpha_1 + 6\alpha_2 = 6\beta_2 - 6\beta_1 + \beta_0 \tag{4.6}$$

$$\lambda_4 = \alpha_0 - 12\alpha_1 + 30\alpha_2 - 20\alpha_3 = 20\beta_3 - 30\beta_2 + 12\beta_1 + \beta_0 \tag{4.7}$$

where $\lambda_1$ quantifies location, popularly known as mean and $\lambda_2$ quantifies scale or dispersion of random variable, known as standard deviation. Standardization of the higher moments is quite convenient since they become independent of the units of measurement.

$$\tau_r = \frac{\lambda_r}{\lambda_2} \text{ for } r = 3, 4$$

Similar to conventional moment ratios, L-skewness ($\tau_3$) and L-kurtosis ($\tau_4$) reflect the degree of symmetry and measure of peakedness of a sample, respectively. These are defined as

$$\text{L-coefficient of variation (L-CV)}, \ (\tau) = \lambda_2/\lambda_1 \tag{4.8}$$

$$\text{L-coefficient of skewness}, \ \text{L-skewness} \ (\tau_3) = \lambda_3/\lambda_2 \tag{4.9}$$

$$\text{L-coefficient of kurtosis}, \ \text{L-kurtosis} \ (\tau_4) = \lambda_4/\lambda_2 \tag{4.10}$$

Despite the popularity of conventional moments in terms of data description and their more formal statistical procedures, they offer several limitations. Firstly, sample moments show high sensitivity towards extreme observations. Secondly, the asymptotic efficiency of sample moments is quite disappointing particularly for fat-tailed distributions. Also, the asymptotic variances of these sample moments, determined by higher order moments, normally tend to become very large or even unbounded.

## 4.2 Study Area and Data Availability

The location of Upper Narmada and Tapi Subzone 3(c) is specified by the East longitudes of 76° 12′–81° 45′ and North latitudes of 20° 10°–23° 45′. It covers the States of Maharashtra and Madhya Pradesh and is situated near the northern edge of the Deccan plateau. The subzone constitutes about 50% of the entire area of the Narmada and Tapi Basins, combined. The Narmada is a westward flowing river which rises near Amarkantak in the Mahaikala Range in the Shahdol District of the state Madhya Pradesh at an elevation of around 1000 m above sea level. It has a flow length of about 1300 km before falling into the Gulf of Cambay, Arabian Sea (Fig. 4.1). The total catchment area of the Upper Narmada and Tapi Subzone (Fig. 4.2) ranges to 86,353 km². The river Tapi, also a westward flowing river, ascends near Multai under Betwa District in Madhya Pradesh and flows about 725 km before falling into Gulf of Cambay. The net lengths of main Narmada and Tapi Rivers in the upper subzone are 813 km and 219 km, respectively. The important tributaries of Upper Narmada are Burhnar, Banjar, Sher, Shakkar, Dudha, Tawa and Ganjal along left bank and Hiran, Tendori, Barna, Kolar, Jamner and Datuni along right bank, Purna being the main tributary of Tapi. Upper parts of Purna fall in the upper subzone 3(c).

90% of annual rainfall is received from southwest monsoon in the subzone, between June and October, July and August being the wettest of all the months.

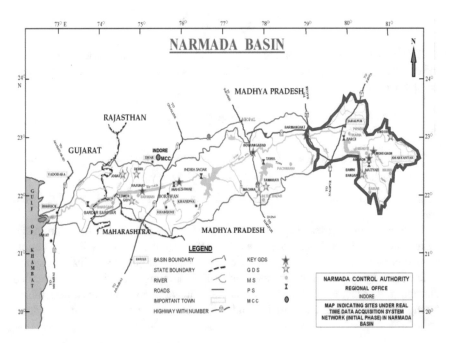

**Fig. 4.1** Image of Upper Narmada Basin (*Sourc* Google images)

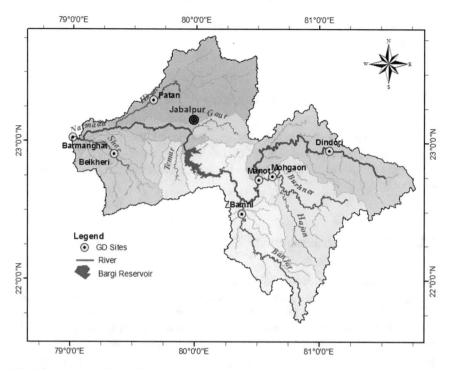

**Fig. 4.2**  Location of Upper Narmada and Tapi Subzone 3(c) (*Source* Google images)

The amount of rainfall ranges from 80 cm in southwestern part of the subzone to greater than 200 cm in the south-central parts of this area. More than 200 cm of highest annual rainfall receiving station in this particular region is Pachmarhi. The rainfall from the south-central part of the subzone decreases sharply and then sores to 160 cm towards western and eastern parts. Further it decreases to less than 80 cm towards the southwest. On the other hand, the far eastern part of the area receives rainfall of the order of 140 cm. Maximum peak flood data recorded annually for the 12 gauging sites of small sized catchments of the study area have been used for the study. The sites with catchment areas ranging from 53.68 to 2,110.85 km$^2$ and MAPF ranging from 209.174 to 1,687.27 m$^3$/s have been used.

## 4.3   Data Analysis and Results

RFFA was executed using the 12 different types of frequency distributions, viz., generalized normal (GNO), uniform (UNF), extreme value (EV1), generalized logistic (GLO), generalized Pareto (GPA), generalized extreme value (GEV), Pearson type-III (PE-III), logistic (LOS), exponential (EXP), Kappa (KAP), normal (NOR), and wakeby (WAK). Sorting the data, checking the homogeneity of the region, establishing the apt distribution for the selected region and development of RFF relationships are indicated below.

### 4.3.1 Discordancy-Based Screening of Data

Discordancy statistic values have been computed using L-moments for each gauging site individually. The highest value of $D_i$, i.e. 1.87 is lesser than 2.757 (critical value for a region composed of 15 or less than 15 sites); accredits all the sites to be suitable for carrying out RFFA.

### 4.3.2 Regional Homogeneity Test

H(1), H(2), and H(3) were calculated using the 12 gauging sites' data by producing 500 regions using Kappa distribution. H(1), H(2), and H(3) were observed to be 1.74, 1.33 and 0.09, respectively. H(1) and H(2) being greater than 1.0 and lesser than 2.0 and H(3) being lesser than 1.0, certify the region as reasonably homogeneous in the study. Hence, all the data taken initially was permitted for RFFA, considering their statistical properties. Table 4.1 contains the catchment characteristics and statistical

**Table 4.1** Catchment areas, statistical parameters, size of the sample and discordance for gauging sites of Upper Narmada and Tapi Subzone 3(c)

| Stream gauging site | Areas of the catchments ($km^2$) | Mean annual peak flood ($m^3/s$) | Size of sample (years) | L-coefficient of variation (t2) | L-skewness (t3) | L-kurtosis (t4) | Discordancy statistic (Di) |
|---|---|---|---|---|---|---|---|
| 731/6 | 115.9 | 252.867 | 30 | 0.2922 | 0.19 | 0.0962 | 1.87 |
| 294 | 518.67 | 919.6 | 30 | 0.347 | 0.1613 | 0.1106 | 0.14 |
| 897/1 | 314.88 | 856.462 | 26 | 0.4119 | 0.3103 | 0.161 | 0.25 |
| 634/2 | 348.92 | 380.103 | 29 | 0.3434 | 0.2221 | 0.2018 | 1.74 |
| 831/1 | 53.68 | 209.174 | 23 | 0.2729 | −0.0222 | 0.0548 | 1.71 |
| 505 | 67.37 | 211.792 | 24 | 0.3104 | 0.1045 | 0.0571 | 0.75 |
| 863/1 | 2110.85 | 1687.273 | 22 | 0.4515 | 0.3783 | 0.1546 | 0.88 |
| 253 | 114.22 | 216.9 | 20 | 0.36 | 0.1247 | 0.0923 | 0.89 |
| 584/1 | 139.08 | 248.783 | 23 | 0.4298 | 0.3415 | 0.1700 | 0.44 |
| 512/3 | 142.97 | 219.955 | 22 | 0.3848 | 0.2383 | 0.0869 | 0.78 |
| 776/1 | 179.9 | 572.778 | 18 | 0.2791 | 0.1842 | 0.1622 | 1.75 |
| 644/1 | 989.89 | 546.250 | 20 | 0.4498 | 0.3764 | 0.1903 | 0.81 |

**Table 4.2** Heterogeneity measures for gauging sites of Narmada and Tapi Subzone 3(c)

| Sl. No. | Heterogeneity measures | Values |
|---|---|---|
| 1. | *H(1)* | |
| | (a) Standard deviation of group L-coefficient of variation (observed) | 0.0603 |
| | (b) Mean of standard deviation of group L-coefficient of variation (simulated) | 0.0436 |
| | (c) Standard deviation of standard deviation of group L-CV (simulated) | 0.0096 |
| | (d) Standard test value of H(1) | 1.74 |
| 2. | *H(2)* | |
| | (a) Average of L-coefficient of variation/L-skewness distance (observed) | 0.1084 |
| | (b) Mean of average L-CV/L-skewness distance (simulated) | 0.0866 |
| | (c) Standard deviation of average L-CV/L-skewness distance (simulated) | 0.0164 |
| | (d) Standard test value of H(2) | 1.33 |
| 3. | *H(3)* | |
| | (a) Average of L-skewness/L-Kurtosis distance (observed) | 0.1063 |
| | (b) Mean of average L-skewness/L-Kurtosis distance (simulated) | 0.1048 |
| | (c) Standard deviation of average L-skewness/L-Kurtosis distance (simulated) | 0.0180 |
| | (d) Standard test value of H(3) | 0.09 |

properties with discordancy measures, for all the 12 gauging sites. However, Table 4.2 contains the values of H(1), H(2) and H(3) derived for the study area.

## 4.3.3 Recognition of Robust Distribution

Validation of the distribution found apt for the region is done by the L-moment ratio diagram criterion accompanied by the $|Z_i^{dist}|$-statistic criterion. L-kurtosis, i.e. $\tau_4 = 0.1282$ and L-skewness, i.e. $\tau_3 = 0.2175$ were found to be the average of all the sites taken and were plotted on the L-moment ratio diagram as shown in Fig. 4.1. The least $|Z_i^{dist}|$-statistic value, i.e. 0.43, corresponding to PE-III, shown in Table 4.3 validates it to be the robust probability distribution for the region (Fig. 4.3).

**Table 4.3** $Z_i^{dist}$ values for various distributions for Narmada and Tapi Subzone 3(c)

| S. No. | Distribution | $Z_i^{dist}$-statistic |
|---|---|---|
| 1 | Pearson Type III (PE-III) | 0.43 |
| 2 | Generalized Pareto (GPA) | −1.85 |
| 3 | Generalized extreme value (GEV) | 1.79 |
| 4 | Generalized normal (GNO) | 1.33 |
| 5 | Generalized logistic (GLO) | 3.35 |

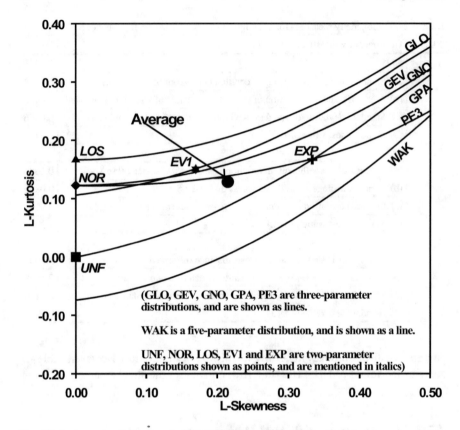

**Fig. 4.3** L-moment ratio diagram for Upper Narmada and Tapi Subzone 3(c)

Therefore, PE-III is recognized as the robustly fit distribution for the region based on the best fit criteria taken. Table 4.4 contains the values of the regional parameters for GNO and PE-III ($Z^{dist}$-statistic $< 1.64$ or accepted at confidence level of 90%) as well as the values of the five parameters of the Wakeby distribution. The Wakeby distribution has been included in the Table 4.4 accounting for the advantage it offers by being a five-parameter distribution, having parameter number higher than maximum distributions which make it able to form a plethora of distributional

**Table 4.4** Regional parameters for GNO, PE-III and WAK distributions for Narmada and Tapi Subzone 3(c)

| Distribution | Statistical parameters | | | | |
|---|---|---|---|---|---|
| GNO | $\xi = 0.863$ | $\alpha = 0.586$ | $K = -0.446$ | | |
| PE-III | $\xi = 1.000$ | $\alpha = 0.671$ | $K = 1.301$ | | |
| WAK | $\xi = 0.025$ | $\alpha = 3.119$ | $\beta = 12.862$ | $\gamma = 0.885$ | $\delta = -0.180$ |

shapes. As a result, the Wakeby distribution becomes helpful in the study of identifying robustness while simulating artificial data. Moreover, Wakeby distribution is also preferred to various distributions in case of heterogeneous regions.

### 4.3.4  RFF Relationship for Gauged Catchments

After identifying PE-III distribution as robust, RFF relationship was generated for determining floods of different return periods in case of catchments which are gauged. The cumulative density function (cdf) of the PE-III distribution is stated as

Let $\alpha = 4/\gamma^2$, $\beta = \sigma\gamma|2|$, $\xi = \mu - 2\sigma/\gamma$ and $\Gamma(.)$ is the gamma function.
Now,

I.   If $\gamma \neq 0$,
(i)   $\gamma > 0$,

Then, $x \in (\xi \leq x < \infty)$ and cdf is

$$F(x) = G\left(\alpha, \tfrac{x-\xi}{\beta}\right)\bigg/ \Gamma(\alpha) \qquad (4.11)$$

and

(ii)   $\gamma < 0$,

Then, $x \in (-\infty < x \leq \xi)$ and cdf is

$$F(x) = 1 - G\left(\alpha, \tfrac{\xi-x}{\beta}\right)\bigg/ \Gamma(\alpha) \qquad (4.12)$$

where incomplete gamma function is expressed as

$$G(\alpha, x) = \int_0^x t^{\alpha-1} e^{-t} dt \qquad (4.13)$$

The study has incorporated the above-expressed quantile functions of the frequency distributions or their inverse forms.

Table 4.5 shows the growth factors of PE-III distribution corresponding to various return periods such as 2, 10, 20, 50, etc. which can be multiplied with MAPF to obtain floods of particular return periods. The growth factor or site-specific scale factor is obtained mathematically on dividing the specific flood quantile ($Q_T$) by the annual mean peak flood ($\overline{Q}$), and is denoted by $Q_T/\overline{Q}$ (Table 4.6).

**Table 4.5** $(Q_T/\overline{Q})$ values for Narmada and Tapi Subzone 3(c)

| Distribution | Return period ($T$) (in years) | | | | | | | | | |
|---|---|---|---|---|---|---|---|---|---|---|
| | 2 | 10 | 20 | 25 | 50 | 100 | 200 | 500 | 1000 | 10000 |
| | Growth factors | | | | | | | | | |
| GNO | 0.863 | 1.875 | 2.285 | 2.417 | 2.831 | 3.256 | 3.692 | 4.289 | 4.759 | 6.444 |
| PE-III | 0.859 | 1.898 | 2.291 | 2.414 | 2.789 | 3.154 | 3.513 | 3.978 | 4.325 | 5.456 |
| WAK | 0.844 | 1.936 | 2.316 | 2.429 | 2.752 | 3.037 | 3.288 | 3.576 | 3.764 | 4.244 |

**Table 4.6** Growth factors of PE-III distribution for Narmada and Tapi Subzone 3(c)

| S. No. | Return period (Years) | Growth factors $\dfrac{Q_T}{\overline{Q}}$ (dimensionless) | Design flood (GF * 526.8281 $(\overline{Q})$ (m³/s) |
|---|---|---|---|
| | | PE-III | PE-III |
| 1 | 2 | 0.859 | 452.5453 |
| 2 | 10 | 1.898 | 999.9197 |
| 3 | 20 | 2.291 | 1206.963 |
| 4 | 25 | 2.414 | 1271.763 |
| 5 | 50 | 2.789 | 1469.324 |
| 6 | 100 | 3.154 | 1661.616 |
| 7 | 200 | 3.513 | 1850.747 |
| 8 | 500 | 3.978 | 2095.722 |
| 9 | 1000 | 4.325 | 2278.531 |

## 4.3.5   RFF Relationship for Ungauged Catchments

A relationship incorporating the physiographic characteristics of the catchment which is still ungauged in terms of observed flow data is needed to determine the MAPF. A plot along with the best fit power equation ($Y = aX^b$) of the MAPF versus catchment area for the 12 gauging sites of Narmada and Tapi Subzone 3(c) is shown in Fig. 4.2.

$$\overline{Q} = 18.615 A^{0.5682} \tag{4.14}$$

where $A$ = area of the catchment (km²) and $\overline{Q}$ = MAPF (m³/s) of the ungauged site to be estimated.

The coefficient of determination for the power equation generated is $r^2 = 0.80$.

Development of RFF relationship for ungauged catchments takes place by

**Table 4.7** $C_T$ values for Upper Narmada and Tapi Subzone 3(c)

| Return period (T) in years | 2 | 10 | 25 | 50 | 100 | 200 | 500 | 1000 |
|---|---|---|---|---|---|---|---|---|
| Regional coefficient $C_T$ | 15.990 | 35.331 | 44.937 | 51.917 | 58.712 | 65.394 | 74.050 | 80.510 |

coupling the RFF relationship developed for gauged catchments with regional relationship between MAPF and catchment area. Following is the regional frequency relationship developed for ungauged catchments:

$$Q^T = c_T A^{0.5682} \qquad (4.15)$$

where $Q_T$ = flood estimate (m$^3$/s) for return period $T$ years and $A$ as before and $C_T$ = regional coefficient. Table 4.7 contains the values of $C_T$ corresponding to commonly used return periods (Fig. 4.4).

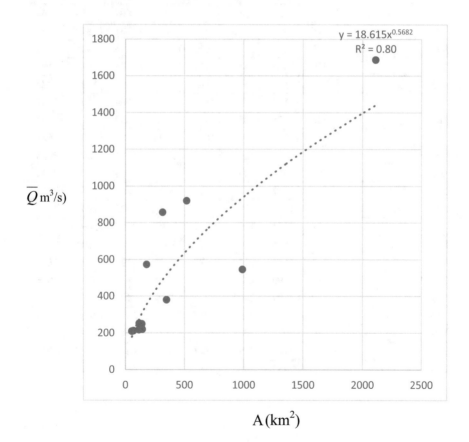

**Fig. 4.4** Variation of MAPF with catchment area

The regional flood formula generated as a result of this study can provide flood estimates for specific return periods, and is tabulated in following table for the convenience of the users (Table 4.8).

**Table 4.8** Variation of floods of various return periods and different catchment areas of Narmada and Tapi Subzone 3(c)

| Catchment area (km$^2$) | Return periods (years) | | | | | | | |
|---|---|---|---|---|---|---|---|---|
| | 2 | 10 | 25 | 50 | 100 | 200 | 500 | 1000 |
| | Floods corresponding to different return periods (m$^3$/s) | | | | | | | |
| 10 | 59 | 131 | 166 | 192 | 217 | 242 | 274 | 298 |
| 20 | 88 | 194 | 247 | 285 | 322 | 359 | 406 | 442 |
| 30 | 110 | 244 | 310 | 359 | 406 | 452 | 511 | 556 |
| 40 | 130 | 287 | 366 | 422 | 478 | 532 | 602 | 655 |
| 50 | 148 | 326 | 415 | 479 | 542 | 604 | 684 | 743 |
| 70 | 179 | 395 | 502 | 580 | 656 | 731 | 828 | 900 |
| 90 | 206 | 456 | 579 | 669 | 757 | 843 | 955 | 1038 |
| 100 | 219 | 484 | 615 | 711 | 804 | 895 | 1014 | 1102 |
| 120 | 243 | 536 | 682 | 788 | 891 | 993 | 1124 | 1222 |
| 140 | 265 | 586 | 745 | 860 | 973 | 1084 | 1227 | 1334 |
| 180 | 306 | 675 | 859 | 993 | 1122 | 1250 | 1416 | 1539 |
| 200 | 325 | 717 | 912 | 1054 | 1192 | 1327 | 1503 | 1634 |
| 300 | 409 | 903 | 1148 | 1327 | 1500 | 1671 | 1892 | 2058 |
| 400 | 481 | 1063 | 1352 | 1562 | 1767 | 1968 | 2229 | 2423 |
| 500 | 546 | 1207 | 1535 | 1774 | 2006 | 2234 | 2530 | 2750 |
| 600 | 606 | 1339 | 1703 | 1967 | 2225 | 2478 | 2806 | 3051 |
| 700 | 661 | 1461 | 1859 | 2147 | 2428 | 2705 | 3063 | 3330 |
| 800 | 713 | 1576 | 2005 | 2317 | 2620 | 2918 | 3304 | 3592 |
| 900 | 763 | 1686 | 2144 | 2477 | 2801 | 3120 | 3533 | 3841 |
| 1000 | 810 | 1790 | 2276 | 2630 | 2974 | 3312 | 3751 | 4078 |
| 1100 | 855 | 1889 | 2403 | 2776 | 3139 | 3497 | 3960 | 4305 |
| 1200 | 898 | 1985 | 2525 | 2917 | 3299 | 3674 | 4160 | 4523 |
| 1300 | 940 | 2077 | 2642 | 3052 | 3452 | 3845 | 4354 | 4734 |
| 1400 | 981 | 2167 | 2756 | 3184 | 3600 | 4010 | 4541 | 4937 |
| 1500 | 1020 | 2253 | 2866 | 3311 | 3744 | 4171 | 4723 | 5135 |

## 4.4 Conclusions

RFF relationships have been developed for gauged and ungauged catchments of Upper Narmada and Tapi Subzone 3(c). Sorting the data utilizing the MAPF of the Narmada and Tapi Subzone 3(c) with the $D_i$ test revealed that each and every site involved was acceptable for carrying out RFFA. Also, all the 12 gauging sites belonged to a homogeneous region accounting to the values of the homogeneity measures 'H(j)' which came out to be lower than their critical values. Probability distributions such as EV1, GNO, GEV, GLO, UNF, WAK EXP, PE-III, NOR, GPA, LOS and KAP have been utilized to find the best fit. Finally, the $|Z_i^{dist}|$-statistic and L-moment ratio diagram criteria are employed to confirm the robust probability distribution as the PE-III type. Then, RFF relationships were developed for the convenience of designers and hydrologists which may be required for the determination of floods corresponding to different return periods for the catchments in the region which are completely ungauged.

Lastly, the analysis carried out is subjected to certain limitations such as the RFFA has been carried out on the catchments with areas ranging from 53.68 to 2,110.85 km$^2$ and hence the developed relationships are restricted to provide flood estimates for catchment areas lying in the same range. The RFF relationship with coefficient of determination value ($r^2 = 0.80$) can explain only 83.4% of initial variance. However, the future scope of these RFF relationships lies in their possible refinement accounting to other available physiographic characteristics such as length of main stream, centroidal length, etc. and climatic characteristics such as temperature, moisture level, etc.

## References

CWC (2002) Flood estimation report for upper Narmada and Tapi Subzone-3c. Central Water Commission, New Delhi

Hosking JRM, Wallis JR (1997) Regional frequency analysis-an approach based on L-moments. Cambridge University Press, New York

Kumar R, Singh RD, Seth SM (1999) Regional flood formulas for seven subzones of zone 3 of India. J Hydrol Eng 4(3):240–244

Kumar R, Chatterjee C, Kumar S (2003) Regional flood formulas using L–moments for small watersheds of Sone subzone of India. Appl Eng Agric 19(1):47

Kumar R, Chatterjee C (2005) Regional flood frequency analysis using L-moments for North Brahmaputra region of India. J Hydrol Eng 10(1):1–7

Pilgrim DH, Cordery I (1992) Flood runoff. Chapter in handbook of hydrology. McGraw-Hill, Inc., New York

# Chapter 5
# Mapping Punjab Flood using Multi-temporal Open-Access Synthetic Aperture Radar Data in Google Earth Engine

**Gagandeep Singh and Ashish Pandey**

## 5.1 Introduction

Floods are one of the most severe catastrophic natural calamities which cause unprecedented destruction all around the world. According to the Global Assessment Report (2019) on Disaster Risk Reduction, between the years 1997 and 2017, floods have affected 76 million people. Floods can be described as the presence of water on dry land. The causes of that flooding can be excessive precipitation, snowmelt that occurs in a short time interval, a dam break, a storm surge, inadequate water management practices, etc.

India is second in absolute terms of people killed by floods, but relatively several other countries have more casualties per million inhabitants by floods than India. India is an agriculture-based economy, and its economic growth has always been under the influence of the weather, especially extreme weather events (Vishnu et al. 2019). Besides heavy agricultural losses, such extreme events also result in huge losses of life, property, and unrest in economic activities.

Punjab is a state in northwestern India. It covers an area of 50,362 km$^2$, i.e., 1.53% of India's total geographical area. Continuous and heavy rainfall in August 2019 caused widespread destruction in several districts of Punjab along the banks of the Sutlej and Beas Rivers. Districts of Amritsar, Fatehgarh Sahib, Ferozepur, Gurdaspur, Jalandhar, Kapurthala, Ludhiana, Moga, Mohali, Patiala, Roopnagar, and Sangrur were among the most severely affected.

G. Singh (✉) · A. Pandey
Department of Water Resources Development and Management, Indian Institute of Technology Roorkee, Roorkee 247667, Uttarakhand, India
e-mail: gsingh@wr.iitr.ac.in

A. Pandey
e-mail: ashish.pandey@wr.iitr.ac.in

A critical element during an ongoing flood event is flood inundation assessment, which forms a crucial component to formulate damage relief plans, damage assessment, estimation and distribution of compensations, and selecting appropriate planning and land use in the flood-affected area (Ran and Nedovic-Budic 2016; Vishnu et al. 2019). Satellite remote sensing data products are exceptional resources in flood mapping as they offer impeccable advantages of synoptic views and reviews. Optical remote sensing data products are not very useful to monitor flood-affected areas in case of an ongoing event due to the presence of cloud cover, which creates a hindrance for data retrieval in visible and near-infrared regions. Microwave remote sensing data products have an advantage of all-weather, day-night coverage with cloud penetration capability (Amitrano et al. 2018; Plank et al. 2017; Martinis 2017; Twele et al. 2016; Uddin et al. 2019). Therefore, active radar sensors operating in the microwave band are the most preferred choice for flood inundation mapping. Hence, sentinel-1 dataset was used in this study to map flood-affected areas in Punjab during and following the August 2019 event.

From a radar perspective, flooding can be defined as an occurrence of temporary or permanent water surface either underneath a tall or short vegetation cover regardless of whether it is forest or agriculture or just open water. Flood maps can help monitor the inundation extent and dynamics for disaster assessment and management. The radar backscatter mechanism that is primarily relevant in terms of flood inundation (Hess et al. 1990) is a key aspect to be discussed at this juncture. The specular scattering occurs in the case of a smooth water surface wherein the signal is scattered away from the satellite sensor, which results in the appearance of open water as very dark in the satellite image (Lillesand et al. 2015). Rough surface scattering occurs when there is some level of roughness in the water surface due to the presence of short floating vegetation, wind, or heavy rainfall resulting in the signal getting scattered in different directions but mostly away from the satellite sensor. Such areas appear dark but not as dark as the completely smooth water surface. The rougher the surface, the larger the signal scattered back to the satellite and brighter that pixel will appear on the image. Double bounce scattering occurs when two smooth surfaces create a right angle and deflect the incoming radiation causing most of the radiation returning to the sensor (Lillesand et al. 2015). These areas appear very bright in the image. This type of scattering is commonly observed in the case of flooded vegetation, which acts as a vertical surface to the horizontal water surface. It is a characteristic of urban areas.

Another important concept related to microwaves is polarization. Polarization is the plane of propagation of the electric field of the signal, which can be either in the horizontal or vertical plane (Lillesand et al. 2015). Irrespective of wavelength, radar signals can be transmitted and/or received in different modes of polarization, and there are four combinations of both transmitted and received polarizations. These are HH-horizontally transmitted, horizontally received, HV-horizontally transmitted, vertically received, VH-vertically transmitted, horizontally received, VV-vertically transmitted, vertically received. The penetration depth of the radar signal is influenced by polarization.

Looking to the aforementioned, the main objective of this study is to demonstrate the potential use of Sentinel-1 SAR images in the cloud-based platform of Google Earth Engine for flood inundation mapping. The operational methodology applied in this study is used to quantify the areal extent of the flooded area.

## 5.2  Material and Methods

### 5.2.1  Study Area

Figure 5.1 shows the study area which comprises 12 districts, namely, Amritsar, Fatehgarh Sahib, Ferozepur, Gurdaspur, Jalandhar, Kapurthala, Ludhiana, Moga, Mohali, Patiala, Roopnagar and Sangrur in the State of Punjab which lies between 29° N to 32°30′ N latitude and the 73° E to 77° E longitude. The total area of the selected districts is 28,386.73 km². The study area is a part of the Indo-Gangetic alluvial plain. The area is drained by two perennial rivers Sutlej and Beas (Chopra and Sharma 1993).

Analysis of SRTM DEM for the flood-affected districts in the state shows that the elevation in the area varies from 143 to 777 m AMSL.

Out of a total area (flooded districts) of 28,386.74 km², 40.03% area lies in the elevation range of 143–230 m. 44.02% of the total area lies in the elevation range of 230–260 m. 14.15% of the total area falls under an elevation range of 260–330 m, and

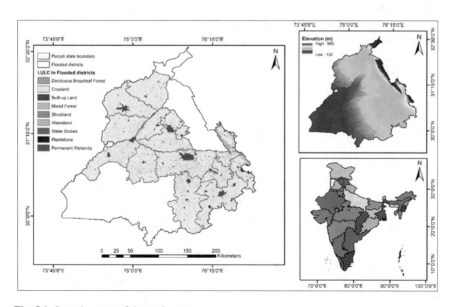

**Fig. 5.1**  Location map of the study area

a mere 1.8% area lies in the high elevation range of 330–777 m. This gives a decent overview of the terrain and signifies that more than 80% of the area has minimal changes in the elevation. Apart from 1.8% area, which can be classified under a high elevation range, the remaining terrain is plain.

### 5.2.2  Data Used

In this study, Sentinel-1 dataset was used to obtain the flood inundation areas in various districts of Punjab. Dual-polarized (VV and VH) Sentinel-1 SAR images acquired from March 13, 2019 to June 13, 2019 were utilized for before the flood event analysis and SAR images acquired from August 21, 2019 to August 31, 2019 were used for after flood event analysis. In this study, dual-polarized (VV and VH) of 5 × 20 m resolution (10-m pixel spacing) Level-1 Ground Range Detected (GRD) Sentinel-1 SAR datasets acquired in Interferometric Wide Swath (IW) mode were used. Additionally, SRTM DEM of 30-m spatial resolution was used for the extraction of elevation zones within the state.

The sentinel-1 data is a C-band synthetic aperture radar data. GEE has the entire Sentinel-1 database. It is being provided by the European Space Agency using two satellites Sentinel-1A and 1B, which individually offer global coverage of 12 days. Besides, global coverage of 6 days over the equator is obtained using data from both satellites. There are four different modes in which the satellite sensors acquire Sentinel-1 data. These are Extra Wide Swath Mode, which is being exclusively used for monitoring oceans and coasts; Strip Mode, which is operated by special order only and is intended for special needs; Wave Mode, which is used for the routine collection for the ocean; and Interferometric Wide Swath Mode, which is used for routine collection for land (this mode is exclusively used for flood mapping applications).

### 5.2.3  Methodology

In this study, the Sentinel-1 SAR datasets were processed using Google Earth Engine, which is a cloud-based geospatial processing platform and is being used widely to analyze the Earth's surface. GEE provides a massive collection of time-series satellite data products and geospatial datasets all for free and hosted on the cloud. GEE also provides a Javascript-based code editor wherein codes were developed to datasets retrieval, processing, and flood inundation mapping.

Methodology flowchart for inundation mapping is presented in Fig. 5.2. The Area of Interest (AOI) was selected in the GEE code editor platform by importing the shapefiles of the flood-affected districts. Once the shapefiles were loaded, the next step was to load the preprocessed sentinel-1 data from the public data archive of the

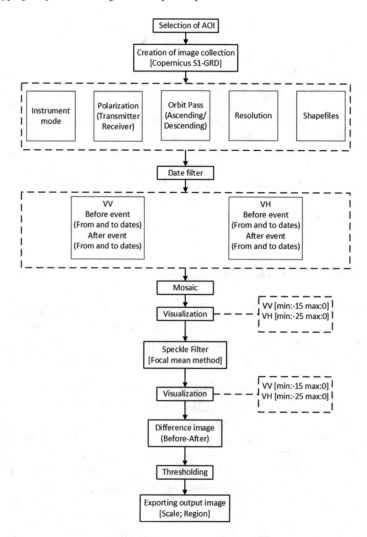

**Fig. 5.2** Flowchart depicting the adopted methodology

GEE. Various filters (instrument mode, transmitter/receiver polarization, orbit pass, resolution, and AOI) were applied to create and load in the image data collection.

The first filtering for Sentinel-1 data collection was defined to load the SAR Ground Range collection with Interferometric Wide (IW) instrument mode, VV and VH polarization, and a descending orbit pass for the previously defined AOI.

After that, a second filter was defined to select the above-filtered data for specific dates. Therefore, the above-obtained dataset collection was further filtered by date for before and after the flood event. To collect the images for "before the event," the "from" and "to" dates were chosen from March 13, 2019 to June 13, 2019. To collect the images for "after the event," the "from" and "to" dates chosen were August 21,

2019 to August 31, 2019. The filtered image collections for before and after flood events were mosaicked, respectively, to create a single image in both VV and VH polarization modes. A total of four images were ready for post-processing as shown in Fig. 5.3a–d. All the four images were then processed to remove speckle (noise reduction). For this, a focal mean smoothing filter was applied with a radius of 50 pixels. The resultant speckle-free images obtained for VH polarization shown in Fig. 5.3e, f were further used to calculate the difference between before and after the flood as shown in Fig. 5.4 to find the inundated areas. The after flood mosaicked image was divided by the before flood mosaicked image to obtain a new image with flood inundated areas. To prominently highlight the flooded areas, a mask was created using a difference threshold value of 1.25. Finally, the VH polarization threshold difference image was exported to compute the areal extent of the flooded area in each district as well as in the entire AOI.

## 5.3   Result and Discussion

The incessant rains in August 2019 caused severe floods in Punjab, wherein 12 majorly affected districts have been mapped for areal assessment of flood inundation. Google Earth Engine (GEE) was used to conduct the entire satellite image processing, as explained in the methodology. A total of 28,386.72 $km^2$ area of Punjab was selected as the area of interest for the study.

The final flood inundation map derived after processing the sentinel-1 SAR images is presented in Fig. 5.5. An area of 205.2 $km^2$ is mapped as flooded in the analysis and is represented in white color. The non-flooded area of 28,181.54 $km^2$ is displayed in black and the river flowing through the study area designated in blue color. Furthermore, a district-wise flooded and non-flooded area assessment was carried out and is presented in Table 5.1. The results obtained are in line and comparable with the inundated area assessment report prepared by the Punjab Agricultural Department. As per this report, 240 $km^2$ were submerged in 12 districts of the state. The report also states that Jalandhar and Kapurthala were the two most affected districts.

Out of the 12 districts Kapurthala, Ferozpur, Jalandhar, and Moga were the most affected ones. Figures 5.6 and 5.7 show the flood inundation in each of the 12 districts.

## 5.4   Conclusion

This study demonstrated the use of Sentinel-1 SAR data for near real-time flood inundation mapping. During monsoon season, the availability of cloud-free optical satellite data products is rare and occasional. SAR data offers a remarkable advantage of capturing data in all weather conditions due to which it serves as the best data source to observe and map flood inundation in near real time.

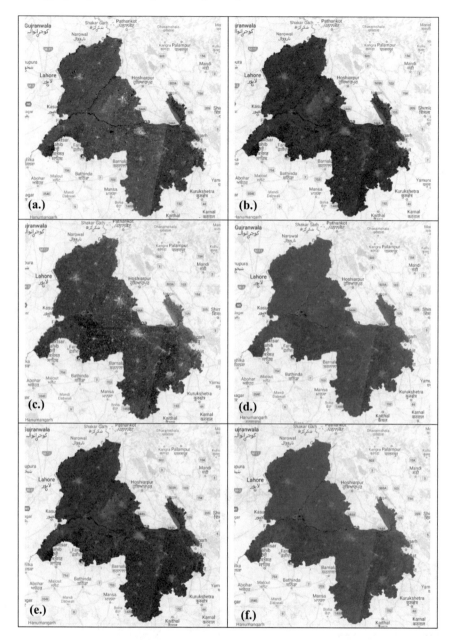

**Fig. 5.3** Outputs of Sentinel-1 data processing in GEE platform **a** before flood VV, **b** before flood VH, **c** after flood VV, **d** after flood VH, **e** before flood speckle removed, **f** after flood VH speckle removed

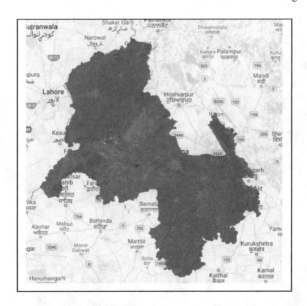

**Fig. 5.4** Difference image VH in GEE platform

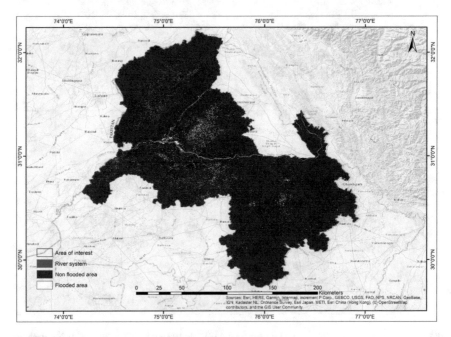

**Fig. 5.5** Flood inundation map of the area of interest

**Table 5.1**  District-wise flooded and non-flooded area assessment

| S. No | District | Flooded area (km$^2$) | Non-flooded area (km$^2$) | Total area (km$^2$) |
|---|---|---|---|---|
| 1 | Amritsar | 3.59 | 2606.42 | 2610.01 |
| 2 | Fatehgarh Sahib | 7.52 | 1165.27 | 1172.79 |
| 3 | Ferozepur | 27.02 | 2471.33 | 2498.34 |
| 4 | Gurdaspur | 1.64 | 2574.46 | 2576.10 |
| 5 | Jalandhar | 45.89 | 2554.90 | 2600.78 |
| 6 | Kapurthala | 42.13 | 1614.42 | 1656.55 |
| 7 | Ludhiana | 35.66 | 3531.71 | 3567.37 |
| 8 | Moga | 25.18 | 2275.36 | 2300.54 |
| 9 | Mohali | 0.91 | 1087.22 | 1088.13 |
| 10 | Patiala | 6.40 | 3328.03 | 3334.43 |
| 11 | Roopnagar | 2.05 | 1378.81 | 1380.86 |
| 12 | Sangrur | 7.21 | 3593.61 | 3600.82 |
| Total | | 205.2 | 28,181.54 | 28,386.72 |

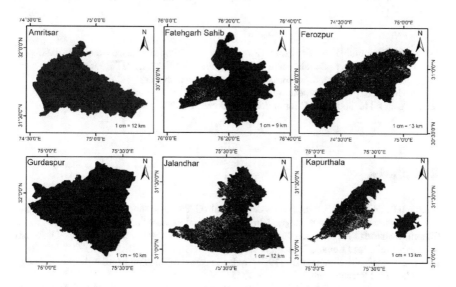

**Fig. 5.6**  Flood inundation maps for Amritsar, Fatehgarh Sahib, Ferozpur, Gurdaspur, Jalandhar, and Kapurthala Districts of Punjab

Based on the results obtained, it can be concluded that the freely available Sentinel-1 SAR data has immense potential for rapid flood mapping and monitoring. GEE can be effectively used for planning disaster risk reduction and, damage assessment in the flood affected areas, and can be used well along with land-use land cover information.

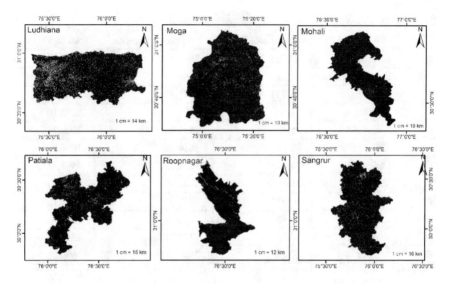

**Fig. 5.7** Flood inundation maps for Ludhiana, Moga, Mohali, Patiala, Roopnagar, and Sangrur Districts of Punjab

**Acknowledgements** We wish to express a deep sense of gratitude and sincere thanks to the Department of Water Resources Development and Management (WRD&M), IIT Roorkee, for providing a conducive environment and resources to conduct the research work.

We acknowledge the European Space Agency (ESA) for providing Sentinel-1 SAR data. We are also grateful to Google LLC for offering the Google Earth Engine platform.

# References

Amitrano D, Di Martino G, Iodice A, Riccio D, Ruello G (2018) Unsupervised rapid flood mapping using Sentinel-1 GRD SAR images. IEEE Trans Geosci Remote Sens 56(6):3290–3299

Chopra R, Sharma PK (1993) Landform analysis and ground water potential in the Bist Doab area, Punjab, India. Int J Remote Sens 14(17):3221–3229

Hess LL, Melack JM, Simonett DS (1990) Radar detection of flooding beneath the forest canopy: a review. Int J Remote Sens 11(7):1313–1325

Lillesand T, Kiefer RW, Chipman J (2015) Remote sensing and image interpretation. Wiley

Martinis S (2017) Improving flood mapping in arid areas using Sentinel-1 time series data. In: 2017 IEEE international geoscience and remote sensing symposium (IGARSS). IEEE, pp 193–196

Plank S, Jüssi M, Martinis S, Twele A (2017) Mapping of flooded vegetation by means of polarimetric Sentinel-1 and ALOS-2/PALSAR-2 imagery. Int J Remote Sens 38(13):3831–3850

Ran J, Nedovic-Budic Z (2016) Integrating spatial planning and flood risk management: a new conceptual framework for the spatially integrated policy infrastructure. Comput Environ Urban Syst 57:68–79

Twele A, Cao W, Plank S, Martinis S (2016) Sentinel-1-based flood mapping: a fully automated processing chain. Int J Remote Sens 37(13):2990–3004

Uddin K, Matin MA, Meyer FJ (2019) Operational flood mapping using multi-temporal sentinel-1 SAR images: a case study from Bangladesh. Remote Sens 11(13):1581

UNDRR (2019) Global assessment report on disaster risk reduction, Geneva, Switzerland. United Nations Office for Disaster Risk Reduction (UNDRR)

Vishnu CL, Sajinkumar KS, Oommen T, Coffman RA, Thrivikramji KP, Rani VR, Keerthy S (2019) Satellite-based assessment of the August 2018 flood in parts of Kerala, India. Geomatics Nat Hazards Risk 10(1):758–767

# Chapter 6
# Study of Drought Characteristics in Ken River Basin in Bundelkhand Region in India

Sudin Moktan, R. P. Pandey, S. K. Mishra, and R. B. Pokharel

## 6.1 Introduction

Drought is a natural recurring phenomenon and its characteristics depend on region's climatic parameters. Usually, droughts influence huge geographical regions for shorter or longer durations and it has significant impact on regional ecosystems, economy, environment, life, livelihoods and all-inclusive human well-being. Droughts diverge from other common hazards like floods, tropical cyclones and earthquakes in a couple of ways (Mishra and Desai 2005). These are defined as a relative deficit in a given region compared to its average or usual water accessibility, either in the form of precipitation, river flow, surface/groundwater storages or due to a combination of these for certain period of time. Thus, a drought is a temporary climatic phenomenon (Pandey et al. 2016). Various researchers/scientists have described drought diversely as per the area of concern. For a meteorologist, it is negative deviation from average precipitation and for a hydrologist a fall in the

S. Moktan (✉)
Ground Water Resources and Irrigation Office, Ministry of Physical Infrastructure and Development, Gandaki Province, Pokhara, Nepal
e-mail: sudina15@gmail.com

R. P. Pandey
National Institute of Hydrology, Roorkee 247667, India
e-mail: rppanndey@gmail.com

S. K. Mishra
Department of Water Resources Development & Management, Indian Institute of Technology Roorkee, Roorkee 247667, India
e-mail: skm61fwt@iitr.ac.in

R. B. Pokharel
Water Resources and Irrigation Department, Ministry of Energy, Water Resources and Irrigation, Kathmandu, Nepal
e-mail: rbpokharel@gmail.com

© The Editor(s) (if applicable) and The Author(s), under exclusive license to Springer Nature Switzerland AG 2021
A. Pandey et al. (eds.), *Hydrological Extremes*, Water Science and Technology Library 97, https://doi.org/10.1007/978-3-030-59148-9_6

stream, lake level or groundwater level. For an agricultural researcher, drought means absence of soil dampness to support crop development, for a financial authority a starvation condition and for an urbanite lack of tap water supply (Dracup et al. 1980).

The droughts are characterized by frequency, severity, duration and spatial extent (Wilhite 2000). How often an area is expected to face a recurrence of drought condition (on an average) is expressed as a frequency. Severity indicates the magnitude of deficit as far as precipitation/accessible water and duration refers to the time span for which a drought condition prevails. Spatial extent refers to the coverage of areas affected by drought. India is one among the most drought vulnerable countries in the world (FAO 2002; World Bank 2003). Failure of monsoon rain is the principal cause of drought in India. Bundelkhand Region is one of the sizeable drought-affected areas in central India. Because of the topographic conditions and poor groundwater availability, this region inherently suffers from water shortages during summer months. During drought years, the condition of water shortages aggravates and the region faces severe water stress even in the domestic water supply.

This research has been carried out to investigate the drought characteristics in Ken Basin of Bundelkhand Region through analysing rainfall departure and critical dry spells using daily rainfall data of period 1980–2016. On an average, drought recurs once in every 5 years in the Bundelkhand Region. Agriculture is the major source of livelihood in this region. About 80% of cultivation in the region depends on rainfall. Normally, the rainfed cropping of less water-needy crops like soybean, maize, jowar and pulse is practiced in the region. Rice and wheat crops are also cultivation but these crops frequently suffer from water stress mainly due to the erratic distribution of seasonal rainfall and lack of irrigation facilities. The agricultural operations for major summer crops (Kharif) start with the onset of southwest monsoon. The major meteorological parameter which influences production in this region is the date of the first occurrence of a rain spell at the onset of the monsoon season that builds a moisture reserve in the soil for the commencement of Kharif crop sowing. The precise prediction of the date of onset of effective monsoon may help the farmers to plan cropping strategy and derive maximum benefits from local soil conditions with the onset of monsoon rains (Sahoo 1993). In order to identify the Kharif sowing rains, the pre-monsoon showers of high intensity cannot be considered as the effective monsoon, particularly because they may be followed by long dry spells. If the sowing is undertaken immediately after these showers, it may affect the germination of seeds and may result in crop failure.

## 6.2   Materials and Methods

### 6.2.1   Description of Study Area (Ken River Basin)

The ken River is an inter-state river system flowing in the middle of Madhya Pradesh and Uttar Pradesh state. The basin is spread between 23° 20′ and 25° 20′ north latitude and 78° 30′ and 80° 32′ east longitude. In total, the Ken River Basin covers 28,058 km² area, out of which 24,472 km² lies in Madhya Pradesh and remaining 3,586 km² lies in Uttar Pradesh. The location map and the rainfall stations used for this research work are presented in Fig. 6.1. The river Ken begins near the Ahirgawan Village in Jabalpur District of M.P. at an altitude of 550.00 m above mean sea level and joins the Yamuna River, near Chilla Village in U.P., at an elevation 95.00 m above m.s.l. It forms the common boundary between Chhattarpur and Panna Districts in M.P. and state boundary between Chattarpur District of M.P. and Banda District of U.P. After forming the boundary between Chhattarpur and Banda Districts, it enters in U.P. at Ganchhar Village of Banda District. The river has a total length of 427 km, out of which 292 km lies in M.P. also, 84 km lies in U.P.; furthermore, 51 km forms the common boundary of M.P. and U.P. The Ken River is one of the vital tributaries of river Yamuna, which joins at the south river bank of Yamuna River. Similarly, the major tributaries of Ken River are Bearma, Sonar, Kopra, Bewas, Katni, Gurne, Patan, Siameri, Banne, etc.

The Ken River Basin lies in semi-arid to dry sub-humid climatic region of India with single monsoon rainy season (June to September) pursued by dry winter (October to January) and after that an extremely dry summer (February to May). The spatial distribution pattern of rainfall is irregular and uneven in the basin. The rainfall pattern within the basin indicates an increasing trend while proceeding from north to south. In Ken Basin, the mean annual rainfall varies from 852 to 1267 mm in a study period of 1980–2016 (Table 6.3). The basin gets the greater part of its precipitation due to southwest monsoon, which begins from the middle of June and keeps going until the finish of September. Up to 90% of the annual precipitation is received throughout the month of June to September. The upper ranges of the basin are spreads with thick forests while the middle and lowest ranges are cultivable lands. The basin experiences the problem of severe lack of water during crop developing period due to the non-uniform nature of average annual rainfall.

## 6.2.2   Data Availability

For carrying out this study, the daily rainfall data (1980–2016) of Ken Basin were collected from India Meteorological Department (IMD), Pune. Figure 6.1 shows the location of meteorological stations in their respective districts.

## 6.2.3   Identification of Drought Years

A drought refers to a deficiency of rainfall over a prolonged duration so as to intervene with some phases of regional economic activities. In this study, daily rainfall records from 1980 to 2016 (37 years) of eight meteorological stations of Ken Basin, viz. Chhatarpur, Damoh, Katni, Panna, Sagar, Banda, Hamirpur and Mahoba have been

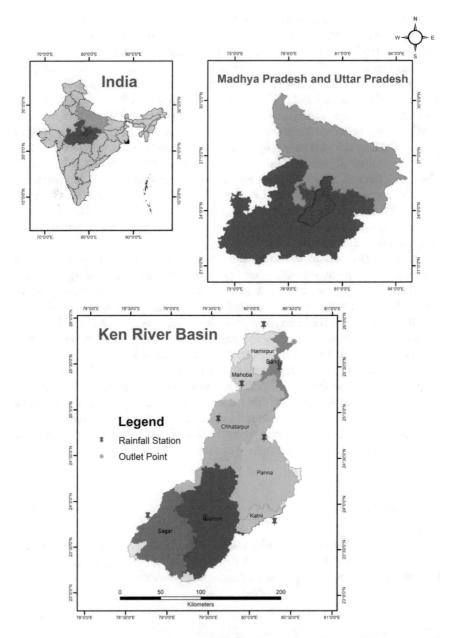

**Fig. 6.1** Location map with rainfall station of Ken River Basin

subjected to various kinds of analyses, where a year and a monsoon season (June–September) are considered as a drought year and drought season, respectively.

### 6.2.4   Computation of Annual and Seasonal Rainfall Departures

Annual rainfall departure analysis has been carried out to identify the meteorological drought years and to estimate the magnitude of a deficit of annual rainfall. Rainfall departure is a good indicator of wet and dry conditions for a given time over the specified area to understand the climatological condition of the basin. As per IMD, 'an area is thought to be drought influenced if it receives total rainfall less than 75% of its normal value (Appa Rao 1986)'. If the entire amount of annual rainfall over an area is deficient by more than 25% then a year/season is considered as a drought year/season. The southwest monsoon contributes almost 90% of the average annual rainfall that occurs during June to September in the region. The annual rainfall departures were computed with the given equation. Similarly, seasonal rainfall departures were also estimated as

$$\text{ADR}_i = \frac{(\overline{\text{AR}}_i - \overline{\text{AR}})}{\overline{\text{AR}}} * 100 \tag{6.1}$$

where $\text{ADR}_i$ = annual rainfall departure; $\text{AR}_i$ = annual rainfall; $\overline{\text{AR}}$ = average annual rainfall; $n$ = number of years of record analysed; $i = 1, 2, \ldots n$. The positive value of rainfall departure represents the excess rainfall and a negative value represents the deficit with respect to the corresponding average. According to annual and seasonal rainfall departures, each of the year or season has been characterized as dry and wet years/season as per defined in Table 6.1.

### 6.2.5   Probability Distribution of Annual and Seasonal Rainfalls

The probability distribution of annual rainfall plays a vital role in predicting the relative frequency of occurrence of a given amount of annual rainfall with reasonable

**Table 6.1** Classification of the year from wet to dry according to IMD criteria

| Received rainfall in any year | >125% of mean | 110–125% of mean | 90–110% of mean | 75–90% of mean | <75% of mean |
|---|---|---|---|---|---|
| Category | Wet | Mild wet | Average | Mild dry | Drought |

accuracy. For calculating the probability distribution of annual rainfall, the annual rainfall values of each station were arranged in descending order and were ranked according to the order of occurrence and their probability distribution was calculated using Weibull's plotting position formula:

$$P_i = \frac{R_i}{(n + 1)}$$ 
(6.2)

where $P_i$ = probability of exceedance of annual rainfall, $R_i$ = rank of particular record, $n$ = total number of observation and $i = 1, 2, ... n$. The percentage probability of occurrence of 75% of mean annual rainfall has been worked out to delineate the drought proneness of various rain gauge stations. An area is considered to be vulnerable to the drought if the probability of occurrence of 75% of normal rainfall is less than 80% (CWC 1982).

### 6.2.6 Estimation of Reference Crop Evapotranspiration

Evapotranspiration refers to the combined loss of moisture to the atmosphere in the form of evaporation from soil, water surfaces and transpiration from the plants. If the supply of water is unlimited, loss of water in the form of evaporation and transpiration will depend on the atmospheric conditions and this loss is termed as potential evapotranspiration. The evapotranspiration from the reference surface (FAO-56) is known as reference crop evapotranspiration (ETo). In this study, ETo has been estimated using CROPWAT 8.0 software, which makes use of the modified Penman method (1963). The computations are based on four meteorological parameters, temperature (max. and min.), relative humidity, wind speed and sunshine hours.

### 6.2.7 Onset and Withdrawal of Effective Monsoon

The date of onset of effective monsoon (OEM) and its termination were computed to determine the average effective length of monsoon period in the basin. The OEM can be determined as the date of commencement of a 7-day dry spell satisfying the following criteria (Ashok Raj 1979; Sahoo 1993):

- The first day's rain in 7-day spell is not less than the average daily evapotranspiration (ET).
- At least 4 out of 7 days are rainy days with not less than 2.5 mm of rain each day.
- The total rain during the 7-day spell is not less than (5ET + 10) mm.

Using the above definition, the rainfall record has been analysed to identify the date of OEM in respective years. A similar criterion was adopted for the termination of monsoon (TOM). 'A 7-day spell which does not satisfy the above one of the three conditions after the onset of monsoon is considered to be the end or termination of monsoon'.

For average dates of OEM and WEM in different districts of the basin, the daily rainfall data were used. To determine the mean date of onset of effective monsoon (OEM), the number of days of OEM in a given year was counted after a reference date (number of days of OEM after 1 June in a given year, $ND_i$). For instance, in a particular year, the monsoon arrived on 05 July, then the $ND_i = 35$ days and so on. The reference date in this study was taken as 1 June. The average of the number of days of OEM was then taken and the corresponding average date was obtained. A similar procedure was adopted to calculate the average date of withdrawal of monsoon from the study area with 30 September as the reference date.

### 6.2.8   Critical Dry Spell (CDS)

After the onset of monsoon, a dry spell is determined as the interval of dry days (none of the days have rainfall more than 2.5 mm) between consecutive wet spells. A wet spell can be defined either as (Ashok Raj 1979; Sahoo 1993; Pandey et al. 2000). A rainy day with rainfall equal to or more than 5ET or a spell of 2 consecutive rainy days with rainfall totaling at least 5ET or a 7-day period having at least 3 or 4 rainy days with a total rainfall not less than 5ET. Based on experience, researchers have reported that 'an effective wet spell of two consecutive rainy days can leave more moisture in the soil profile than that of one effective rainy day having an equal amount of total rainfall'. This is because of more chances of water loss as surface runoff in the latter case. Further, 3 or more rainy days occurring in a week, not necessarily consecutively, having at least a total of rainfall of 5ET are considered to constitute a wet spell.

The intervening periods of dry spells between any two consecutive wet days were identified using the above definition. If the duration of these dry spells exceeds certain limiting period when the moisture stress is experienced by crops (under rainfed conditions), then the dry spell is called as 'critical dry spell'. The occurrences of critical dry spells depend on the rainfall pattern and soil–water–crop composite of the region under consideration (Ashok Raj 1979; Sahoo 1993). Soybean, maize and paddy are the major crops grown in Ken River Basin and the most parts of this region are dominated by mixed-red, medium black and black soils. In case of the soybean crop, the critical stages are an early vegetative stage and flowering stage, while in case of maize crop the critical stages are early vegetative stage plus tasseling and silking stages. Based on the relation between crop and soil, the minimum duration of a dry spell is considered as 10 days that become critical to the paddy crop. However, for the soybean and maize crops, the dry spell may become critical after 14–15 dry days. Thus, in this research, we considered the minimum length of a dry spell as 12 days that become critical to the crop. Meaning that if a dry spell extends for 12 or more days the rainfed Kharif season crops will face water stress and it will affect the crop yield.

### 6.2.9   Calculation of Crop Water Requirement

The total amount of water demand of crops for their growth from sowing stage
to the maturity stage refers to the crop water requirement (CWR). The total water
requirement of a crop depends on the crop type, duration of the crop, sowing time
and the regional climatic conditions. The computation of CWR is largely based on
the estimates of evapotranspiration and crop coefficient at different growth stages of
the crop. The crop coefficient for the selected crops is used as per the effect of crop
properties on $ET_o$ and the crop evapotranspiration (ETcrop) is computed as follows:

$$ETcrop = Kc \times ETo \tag{6.3}$$

where ETcrop = crop evapotranspiration mm/day, ETo = potential evapotranspiration mm/day and Kc = crop coefficient. The factors affecting the value of crop
coefficient (Kc) are mainly crop characteristics, crop planting or sowing period, the
rate of development, length of growing season and climatic condition. The crop
growing season has been divided into four stages, namely, (i) initial stage, (ii) crop
development stage, (iii) mid-season stage and (iv) late-season or ripening stage. The
duration of different growth phases of the selected crops in Ken Basin is presented
in Table 6.2. Based on FAO guidelines (FAO Irrigation and Drainage Paper no. 24),
the crop coefficient curves for selected major crops (rice, soybean and maize) were
developed for Ken Basin.

### 6.2.10   Estimation of Effective Rainfall and Irrigation Requirement

Effective rainfall refers to the amount of moisture available to crops through the rain
during the crop growing periods. Rainfall during the cropping periods fully or partly
fulfil the crop water requirements. Effective rainfall is influenced by characteristics

**Table 6.2**  Probable periods of different growth phases of major crops in Ken Basin

| S. No. | Name of crop | Crop duration (days) | Crop growth phases | | | |
|--------|--------------|----------------------|--------------------|---|---|---|
| | | | Initial stage | Crop development stage | Mid-season stage | Late-season stage |
| 1 | Maize | 110 | Jul 8–Jul 29 | Jul 30–Aug 29 | Aug 30–Oct 4 | Oct 5–Oct 25 |
| 2 | Rice | 90–100 | Jul 1–Jul 29 | Jul 30–Aug 29 | Aug 30–Sep 21 | Sep 22–Oct 8 |
| 3 | Soybean | 120 | Jul 1–Jul 20 | Jul 21–Aug 29 | Aug 30–Oct 5 | Oct 6–Oct 28 |

of rainfall, soil, crop, groundwater, land slope, percolation and runoff. Rainfall for any period varies annually, and therefore, as opposed to utilizing mean rainfall data (saying about 1 year is drier, the following is wetter), a dependable level of rainfall should be preferred. In the present study, the monthly rainfall at 80% probability level has been taken as dependable rainfall during cropping period. As per the guidelines suggested by U.S. Department of Agriculture (1969), monthly effective rainfall has been computed using evapotranspiration/precipitation ratio method (Table 34 in FAO Paper No. 24). Effective rainfall during the probable duration of critical dry spells (CDS) for different stations in Ken Basin has been estimated using interpolation techniques. The exact depth of irrigation that can be stored effectively over root zone is assumed as 75 mm during the time of irrigation.

Crop failure in Kharif season normally happens due to water stress during critical dry spells. Therefore, supplemental irrigation is needed to save crop loss due to dry spells. The irrigation requirement (IR) refers to the additional water needed to be supplied to the crop to save it from injurious water stress either due to deficient rainfall or critical dry spells. A decision on the timing of a supplemental irrigation to Kharif crop is difficult to take due to the unpredictable occurrence of rain. If irrigation is provided at critical stages associated with higher probability of drought, a heavy rain during or at the end of irrigation may drastically reduce the beneficial effects of applied water. Hence, irrigation to Kharif crop at pre-decided times may not lead to the achievement of the full benefit of applied water. It will be judicious to wait for the dry spell to enter into the critical stage of crop growth and start irrigation. The full benefit of applied water can only be obtained if the dry spell continues up to at least 1 week after irrigation.

For irrigation requirement for a critical dry spell, CWR for CDS period is calculated using ET0 and Kc values. Proportionate value of effective rainfall for CDS duration may be calculated using linear interpolation. Afterwards, irrigation requirement of crops is calculated as the difference between CWR (ETcrop) and the effective rainfall. It may be expressed in equation form as follows:

$$IR = ETcrop - ER \qquad (6.4)$$

## 6.3  Results and Discussion

### 6.3.1  Rainfall Departure

Details of rainfall records from 1980 to 2016 of eight meteorological stations of Ken Basin have been given in Table 6.3. From this table, it can be observed that almost 90% of precipitation is experienced at the time of monsoon (June to September) in this basin.

**Table 6.3** Mean monthly, seasonal and annual rainfall at Ken River Basin

| Month | Mean monthly rainfall of districts (mm) | | | | | | | | Ken River Basin as a whole |
|---|---|---|---|---|---|---|---|---|---|
| | Chhatarpur | Damoh | Katni | Panna | Sagar | Banda | Hamirpur | Mahoba | |
| Jan | 12.82 | 13.89 | 17.17 | 16.22 | 19.13 | 13.43 | 29.16 | 16.68 | 17.31 |
| Feb | 14.39 | 15.94 | 18.83 | 17.29 | 17.10 | 12.45 | 36.99 | 10.16 | 17.89 |
| Mar | 7.35 | 8.60 | 10.64 | 10.66 | 11.91 | 10.85 | 31.88 | 4.78 | 12.08 |
| Apr | 4.52 | 4.49 | 5.57 | 6.13 | 6.35 | 3.90 | 18.56 | 2.68 | 6.52 |
| May | 8.47 | 6.32 | 9.59 | 10.01 | 17.27 | 11.70 | 27.86 | 12.53 | 12.97 |
| Jun | 108.77 | 149.10 | 167.22 | 137.17 | 164.96 | 122.90 | 95.53 | 93.06 | 129.84 |
| Jul | 271.67 | 346.97 | 309.18 | 341.84 | 451.43 | 260.65 | 287.08 | 214.97 | 310.48 |
| Aug | 337.36 | 435.28 | 339.01 | 390.95 | 370.87 | 321.75 | 291.76 | 293.05 | 347.50 |
| Sep | 181.66 | 182.35 | 183.26 | 252.98 | 166.94 | 158.37 | 155.75 | 163.35 | 180.58 |
| Oct | 50.58 | 28.14 | 29.77 | 35.32 | 21.28 | 27.30 | 28.27 | 26.78 | 30.93 |
| Nov | 5.39 | 8.65 | 8.22 | 5.27 | 11.40 | 6.76 | 8.60 | 9.37 | 7.96 |
| Dec | 7.82 | 7.19 | 10.42 | 6.03 | 8.77 | 8.66 | 16.18 | 4.72 | 8.72 |
| Mean annual rainfall | 1010.80 | 1206.91 | 1108.88 | 1229.88 | 1267.41 | 958.72 | 1027.63 | 852.13 | 1082.79 |
| Mean seasonal (Jun–Sep) rainfall | 899.46 | 1113.70 | 998.67 | 1122.94 | 1154.19 | 863.67 | 830.11 | 764.43 | 968.40 |
| % at seasonal rainfall | 88.98 | 92.28 | 90.06 | 91.31 | 91.07 | 90.09 | 80.78 | 89.71 | 89.28 |

The rainfall records from 1980 to 2016 of the eight meteorological stations of Ken Basin were analysed using IMD criteria to study the magnitude and frequency of drought in terms of rainfall deficiency. The years during which the Ken Basin suffered from annual and seasonal droughts were identified as shown in Table 6.4 and Table 6.5, respectively.

**Table 6.4**  Identification of drought years in different stations of Ken Basin

| S. No. | Station | Average annual rainfall (mm) | Maximum annual departure (%) | Status for annual drought | | |
|---|---|---|---|---|---|---|
| | | | | Drought year | No. of drought year | Drought return period |
| (1) | (2) | (3) | (4) | (5) | (6) | (7) |
| 1 | Chhatarpur (1980–2016) | 1010.80 | −49.83 | 1995, 1998, 2000, 2006, 2007, 2012, 2014 | 7 | 5 |
| 2 | Damoh (1980–2013) | 1206.91 | −37.43 | 1989, 2002, 2006, 2007 | 4 | 9 |
| 3 | Katni (1982–2013) | 1108.88 | −53.45 | 1982, 1987, 1991, 1992, 1993, 2006, 2007, 2009, 2010, 2011 | 10 | 4 |
| 4 | Panna (1980–2013) | 1229.88 | −56.06 | 1981, 2000, 2006, 2007, 2009, 2010 | 6 | 6 |
| 5 | Sagar (1980–2016) | 1267.41 | −42.98 | 1981, 1984, 1986, 1988, 1989, 1992, 2002, 2007, 2010, 2012, 2014, 2015 | 12 | 3 |
| 6 | Banda (1980–2016) | 958.72 | −52.71 | 1988, 1989, 2006, 2007, 2009, 2014 | 6 | 6 |
| 7 | Hamirpur (1980–2016) | 1027.63 | −55.02 | 1993, 1994, 1995, 2001, 2002, 2011, 2015 | 7 | 5 |
| 8 | Mahoba (1980–2012) | 852.13 | −60.49 | 1987, 1995, 2000, 2004, 2006, 2007, 2009, 2010, 2012 | 9 | 4 |
| | Ken Basin as a whole | 1082.79 | −51.00 | | 8.00 | 5 |

**Table 6.5** Identification of drought seasons in different stations of Ken Basin

| S. No. | Station/basin | Status for seasonal drought | | | | |
|---|---|---|---|---|---|---|
| | | Average rainfall (mm) | Maximum departure (%) | Drought year | No. of drought year | Drought return period |
| 1 | Chhatarpur (1980–2016) | 899.46 | −58.5 | 1986, 1995, 1998, 2000, 2006, 2007, 2009, 2014, 2015 | 9 | 4 |
| 2 | Damoh (1980–2013) | 1113.70 | −43.5 | 1986, 1989, 1996, 2002, 2006, 2007, 2009 | 7 | 5 |
| 3 | Katni (1982–2013) | 998.67 | −52.82 | 1987, 1991, 1992, 1993, 2006, 2007, 2009, 2010 | 8 | 5 |
| 4 | Panna (1980–2013) | 1122.94 | −60.1 | 1981, 2000, 2002, 2006, 2007, 2009, 2010 | 7 | 5 |
| 5 | Sagar (1980–2016) | 1154.19 | −44.52 | 1981, 1984, 1986, 1988, 1989, 1992, 1995, 1997, 2002, 2007, 2010, 2012, 2014, 2015 | 14 | 3 |
| 6 | Banda (1980–2016) | 863.67 | −62.71 | 1986, 1988, 1989, 2006, 2007, 2009, 2014, 2015 | 8 | 5 |
| 7 | Hamirpur (1980–2016) | 830.11 | −45.33 | 1987, 1994, 2001, 2002, 2011, 2015 | 6 | 6 |
| 8 | Mahoba (1980–2012) | 764.43 | −63.99 | 1987, 1995, 2000, 2004, 2006, 2007, 2009, 2010, 2012 | 9 | 4 |
| Ken Basin as a whole | | 968.40 | −53.93 | | 9 | 4 |

It is evident from Tables 6.4 and 6.5 that Mahoba Station has the maximum value in both annual and seasonal rainfall departures as compared to other stations of the basin but Sagar Station has a higher frequency of drought. In the year 2004, the value of annual rainfall departure in the Mahoba Station was observed as −60.49% while the corresponding maximum seasonal rainfall departure was recorded as −63.99%. However, all seasonal droughts did not continue to become annual droughts, revealing that some may get terminated due to rain in the non-monsoon period. From the rainfall departure analysis, the maximum annual and seasonal rainfall deficiencies of 51.00% and 53.93%, respectively, for the Ken Basin have been observed as a whole. It also exhibits that the inadequacy of monsoon season rainfall is principally responsible for the incident of drought and ensuing water stress in the Ken River Basin.

The annual and seasonal rainfall departures from their respective means are given in Fig. 6.2a and b, respectively, for Damoh and Mohoba Districts. In these plots, the month in which the rainfall departure is more than 25% and 50% of its corresponding mean value was taken as moderate drought and severe drought, respectively. From the figure, it is also noticeable that the significance of seasonal rainfall departure is more prominent than that of the annual value in most of the drought years.

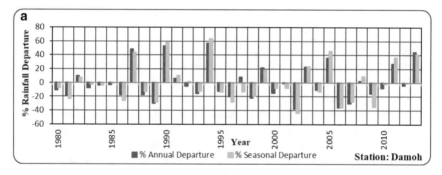

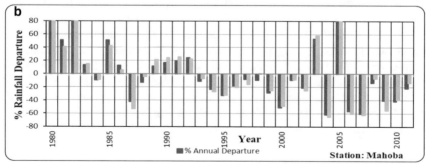

**Fig. 6.2** Plots of annual and seasonal rainfall departures for Damoh and Mahoba

**Table 6.6** Probability distribution of annual and seasonal rainfall at different stations

| S. No. | Station/basin | Annual | | | Seasonal | | |
|---|---|---|---|---|---|---|---|
| | | Mean rainfall (mm) | 75% of mean rainfall (mm) | Probability of occurrence of 75% of mean rainfall (%) | Mean rainfall (mm) | 75% of mean seasonal rainfall (mm) | Probability of occurrence of 75% of mean rainfall (%) |
| 1 | Chhatarpur | 1010.73 | 758.04 | 80 | 899.44 | 674.58 | 74 |
| 2 | Damoh | 1206.86 | 905.14 | 86 | 1113.69 | 835.26 | 79 |
| 3 | Katni | 1124.93 | 843.69 | 70 | 1016.81 | 762.6 | 73 |
| 4 | Panna | 1229.83 | 922.37 | 81 | 1122.93 | 842.19 | 78 |
| 5 | Sagar | 1261.11 | 945.83 | 66 | 1147.96 | 860.97 | 61 |
| 6 | Banda | 958.66 | 718.99 | 82 | 863.65 | 647.73 | 82 |
| 7 | Hamirpur | 1027.57 | 770.67 | 81 | 830.09 | 622.56 | 82 |
| 8 | Mahoba | 852.06 | 639.04 | 66 | 764.40 | 573.3 | 68 |
| Ken Basin as a whole | | 1083.97 | 812.97 | 76.50 | 969.87 | 727.40 | 74.625 |

## 6.3.2 Probability Distribution of Rainfall

The analytical comparison of occurrence probabilities of normal and 75% of normal rainfall for annual and seasonal rainfall series for different districts for the period of 1980–2016 is presented in Table 6.6.

It can be observed from Table 6.6 that the probability of receiving 75% of mean annual rainfall varies from 66% in Mahoba and Sagar Districts to 86% in Damoh District, which indicates that Damoh has very good overall chances of receiving a good quantity of rainfall. In other words, the Mahoba and Sagar Districts are more susceptible to the drought than Damoh. Similarly, normal seasonal rainfall varies from 61% in Sagar District to 82% in Banda District. The districts Katni, Mahoba and Sagar are drought prone to both annual and seasonal rainfall occurrences to the probability value. Finally, it also revealed that on as average the probability of occurrence of 75% of mean annual and seasonal rainfalls show that the Ken Basin is in drought condition. Also, from the plots of probability of occurrence of annual and seasonal rainfalls, it can be observed that all 75% of mean annual and seasonal rainfalls correspond to less than 80%, which reveals the occurrence of drought in that region. The sample plots are presented in Fig. 6.3a, b.

## 6.3.3 Estimation of ETo

The daily ETo was estimated using CROPWAT 8.0 software for preparing monthly as well as weekly evapotranspiration. The average daily ETo for 52 standard weeks

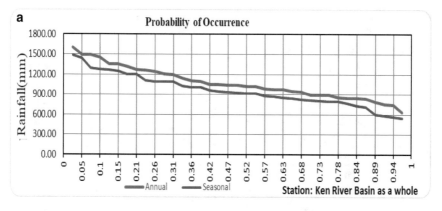

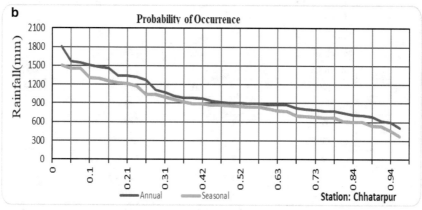

**Fig. 6.3**  Plots of probability of exceedance of annual and seasonal rainfall

is useful for finding CWR during entire crop duration and average daily ETo for months are useful for determining the onset and withdrawal of effective monsoon. The daily ETo of Sagar District for 52 standard weeks and 12 months are presented in Tables 6.7 and 6.8, respectively.

### 6.3.4   Onset and Withdrawal of Effective Monsoon

Onset and withdrawal of effective monsoon have been analysed using the definitions and methodology described in Sect. 6.7. The results are shown in Table 6.9.

From Table 6.9, it is observed that the mean dates of OEM at various stations vary from 17 June to 27 June and mean date of WEM varies from 16 September to 23 September. It can thus be concluded that the mean dates of OEM and WEM for Ken Basin as a whole were 22 June to 19 September having 89 days in a wet season. The longest length of average monsoon period is 97 days in Sagar District while shortest

**Table 6.7** Average daily ETo for 52 standard weeks in mm/day (averaged over 35 years, 1983–2016 for Sagar Station)

| Week no. | Time period | Eto | Week no. | Time period | Eto |
|---|---|---|---|---|---|
| 1 | Jan 1–7 | 2.75 | 27 | July 2–8 | 4.12 |
| 2 | Jan 8–14 | 2.77 | 28 | July 9–15 | 4.03 |
| 3 | Jan 15–21 | 2.97 | 29 | July 16–22 | 4.18 |
| 4 | Jan 22–28 | 3.32 | 30 | July 23–29 | 3.67 |
| 5 | Jan 29–Feb 4 | 3.40 | 31 | Jul 30–Aug 5 | 3.91 |
| 6 | Feb 4–11 | 3.75 | 32 | Aug 6–12 | 3.58 |
| 7 | Feb 12–18 | 3.87 | 33 | Aug 13–19 | 3.43 |
| 8 | Feb 19–25 | 3.97 | 34 | Aug 20–26 | 3.46 |
| 9 | Feb 26–March 4 | 4.23 | 35 | Aug 2–Sep 2 | 3.60 |
| 10 | March 5–11 | 4.74 | 36 | Sep 3–9 | 3.94 |
| 11 | March 12–18 | 5.00 | 37 | Sep 10–16 | 3.90 |
| 12 | March 19–25 | 5.17 | 38 | Sep 17–23 | 4.05 |
| 13 | March 26–Apr 1 | 5.58 | 39 | Sep 24–30 | 4.04 |
| 14 | Apr 2–8 | 5.79 | 40 | Oct 1–7 | 3.68 |
| 15 | Apr 9–15 | 6.16 | 41 | Oct 8–14 | 4.09 |
| 16 | Apr 16–22 | 6.08 | 42 | Oct 15–21 | 3.78 |
| 17 | Apr 23–29 | 6.65 | 43 | Oct 22–28 | 3.64 |
| 18 | Apr 30–May6 | 6.77 | 44 | Oct 29–Nov 4 | 3.49 |
| 19 | May 7–13 | 6.86 | 45 | Nov 5–11 | 3.12 |
| 20 | May 14–20 | 6.65 | 46 | Nov 12–18 | 2.93 |
| 21 | May 21–27 | 7.01 | 47 | Nov 19–25 | 2.97 |
| 22 | May 28–Jun 3 | 6.88 | 48 | Nov 26–Dec 2 | 2.94 |
| 23 | Jun 4–10 | 6.61 | 49 | Dec 3–9 | 2.64 |
| 24 | Jun 11–17 | 5.95 | 50 | Dec 10–16 | 2.65 |
| 25 | Jun 18–24 | 5.75 | 51 | Dec 17–23 | 2.57 |
| 26 | Jun 25–July 1 | 4.79 | 52 | Dec 24–30 | 2.56 |

**Table 6.8** Average daily ETo for various months in mm/day

| Station: | Sagar | RL: | 551 m |
|---|---|---|---|
| Latitude | 23.85 N | Longitude | 78.75 E |
| Month | Eto (mm/day) | Month | Eto (mm/day) |
| Jan | 2.72 | Jul | 3.93 |
| Feb | 3.67 | Aug | 3.47 |
| Mar | 4.9 | Sep | 3.87 |
| Apr | 6 | Oct | 3.63 |
| May | 6.86 | Nov | 3.02 |
| Jun | 5.56 | Dec | 2.52 |

**Table 6.9** Onset and withdrawal of effective monsoon in Ken River Basin of Bundelkhand Region

| S. No. | Name of district | Data used | | Average date of | | Duration of the wet season (Days) |
|---|---|---|---|---|---|---|
| | | From | To | Onset of monsoon | Withdrawal of monsoon | |
| 1 | Banda | 1980 | 2015 | 24-Jun | 18-Sep | 86 |
| 2 | Chhatarpur | 1980 | 2015 | 23-Jun | 18-Sep | 87 |
| 3 | Damoh | 1980 | 2013 | 21-Jun | 21-Sep | 92 |
| 4 | Hamirpur | 1980 | 2015 | 25-Jun | 16-Sep | 83 |
| 5 | Katni | 1982 | 2013 | 21-Jun | 22-Sep | 93 |
| 6 | Mahoba | 1980 | 2012 | 27-Jun | 16-Sep | 81 |
| 7 | Panna | 1980 | 2013 | 22-Jun | 23-Sep | 93 |
| 8 | Sagar | 1980 | 2016 | 17-Jun | 22-Sep | 97 |
| Mean date of OEM and WEM for the Ken River Basin as a whole | | | | 22-Jun | 19-Sep | 89 |

length is 81 days in Mahoba District. For getting maximum benefit from monsoon rain and to finish the sowing in time, it is crucial to know the date of OEM.

### 6.3.5 Identification of Critical Dry Spell (CDS)

Based on the definition of CDS, the available rainfall records were analysed for identifying the dates of start and end of dry spells and their duration, as shown in Table 6.10. During the analysis of intervening dry spells, it was observed that in some years the critical spells were either nil or there was only one critical spell. However, in some years, more than one number of critical spells had occurred. The average number of times CDS occurred during monsoon season was one. The numbers of CDS in different stations were not the same during a year. These predictions about CDSs are important in selection of suitable crops and their varieties to achieve the required level of drought tolerance or to plan for supplemental irrigation at appropriate times of crop need. On an average, CDSs in September month had been of shorter duration in Ken Basin as a whole as than that in July and August months.

### 6.3.6 Estimation of Crop Water Requirement

The crop water requirement (ETcrop) for different crops grown in Ken Basin was computed by multiplication of crop coefficient (Kc) with ETo. Thus, the computed consumptive use (ETcrop) for different crops in different growing stages during the growing period of the selected Kharif crops for Ken Basin is presented in Table 6.11.

**Table 6.10** Occurrence of CDS during monsoon season in Ken River Basin

| S. No. | Name of district | Critical dry spells (CDS) | | | | | | No. of CDS |
|--------|------------------|------|---|------|---|------|---|-----|
| | | July | | August | | September | | |
| | | Onset Date | Duration (Days) | Onset date | Duration (Days) | Onset Date | Duration (Days) | |
| 1 | Banda | 14-Jul | 20 | 19-Aug | 16 | | | 1 |
| 2 | Chhatarpur | 12-Jul | 18 | 19-Aug | 19 | 7-Sep | 12 | 1 |
| 3 | Damoh | 18-Jul | 14 | 22-Aug | 17 | 6-Sep | 13 | 1 |
| 4 | Hamirpur | 8-Jul | 18 | 21-Aug | 17 | 1-Sep | 18 | 1 |
| 5 | Katni | 6-Jul | 17 | 22-Aug | 15 | | | 1 |
| 6 | Mahoba | 16-Jul | 18 | 15-Aug | 16 | 2-Sep | 17 | 1 |
| 7 | Panna | 15-Jul | 15 | 22-Aug | 14 | 2-Sep | 15 | 1 |
| 8 | Sagar | 14-Jul | 14 | 20-Aug | 21 | 5-Sep | 15 | 1 |
| Mean date of CDS for the Ken Basin as a whole | | 13-Jul | 17 | 20-Aug | 17 | 3-Sep | 15 | 1 |

## 6.3.7   Effective Rainfall and Irrigation Requirement

Rainfall for any period varies annually, and therefore, rather than utilizing mean rainfall data, the monthly rainfall at 80% probability level has been taken as dependable rainfall during cropping period, i.e. July, August and September. As per the guidelines suggested by U.S.D.A. (1969), monthly effective rainfall has been computed using evapotranspiration/precipitation ratio method (Table 34 in FAO Paper No. 24). Effective rainfall and irrigation requirement during the duration of critical dry spells (CDS) for different stations in Ken basin are presented in Table 6.12.

Table 6.12 for irrigation requirement revealed that the rice required relatively more amount of irrigation during every CDS period in all districts of the basin. In order to ensure the benefits of at least one-season crops in this region, the irrigation requirement shown in this table can be referred as guidelines for the planning of supplemental irrigation.

## 6.4   Summary and Conclusions

The occurrence of drought, its duration, severity and frequency during the period 1980–2016 in the Ken River Basin have been analysed using rainfall departure from the corresponding mean of annual and seasonal rainfalls, probability distribution of annual and seasonal rainfalls, and onset and withdrawal of effective monsoon and examination of critical dry spells with its duration. All types of analysis which is conducted for examining the drought characteristics of Ken River Basin revealed that

**Table 6.11** Weekly crop water requirement (ETcrop) in mm/day during the growing period

| S. No. | Standard week No. | Time Period | Eto mm/day | Crop coefficient (Kc) values | | | Crop water requirement (mm) | | |
|---|---|---|---|---|---|---|---|---|---|
| | | | | Maize | Rice | Soybean | Maize | Rice | Soybean |
| 1 | 27 | July 2–8 | 4.12 | | 1.10 | 0.28 | | 31.72 | 8.08 |
| 2 | 28 | July 9–15 | 4.03 | 0.42 | 1.10 | 0.30 | 11.84 | 31.01 | 8.46 |
| 3 | 29 | July 16–22 | 4.18 | 0.42 | 1.10 | 0.47 | 12.29 | 32.19 | 13.75 |
| 4 | 30 | July 23–29 | 3.67 | 0.42 | 1.10 | 0.63 | 10.79 | 28.26 | 16.18 |
| 5 | 31 | July 30–Aug 5 | 3.91 | 0.42 | 1.10 | 0.79 | 11.51 | 30.14 | 21.65 |
| 6 | 32 | Aug 6–12 | 3.58 | 0.55 | 1.05 | 0.93 | 13.78 | 26.30 | 23.30 |
| 7 | 33 | Aug 13–19 | 3.43 | 0.63 | 1.05 | 1.10 | 15.14 | 25.23 | 26.43 |
| 8 | 34 | Aug 20–26 | 3.46 | 0.89 | 1.05 | 1.10 | 21.56 | 25.43 | 26.64 |
| 9 | 35 | Aug 27–Sep 2 | 3.60 | 1.05 | 1.05 | 1.10 | 26.45 | 26.45 | 27.71 |
| 10 | 36 | Sep 3–9 | 3.94 | 1.14 | 1.05 | 1.10 | 31.45 | 28.97 | 30.35 |
| 11 | 37 | Sep 10–16 | 3.90 | 1.14 | 0.95 | 1.05 | 31.09 | 25.91 | 28.63 |
| 12 | 38 | Sep 17–23 | 4.05 | 1.14 | 0.95 | 1.05 | 32.33 | 26.94 | 29.78 |
| 13 | 39 | Sep 24–30 | 4.04 | 1.14 | 0.95 | 0.90 | 32.27 | 26.89 | 25.48 |
| 14 | 40 | Oct 1–7 | 3.68 | 0.90 | 0.95 | 0.86 | 23.18 | 24.46 | 22.15 |
| 15 | 41 | Oct 8–14 | 4.09 | 0.75 | 0.95 | 0.80 | 21.46 | 27.18 | 22.89 |
| 16 | 42 | Oct 15–21 | 3.78 | 0.63 | | 0.68 | 16.66 | | 17.99 |
| 17 | 43 | Oct 22–28 | 3.64 | 0.43 | | 0.30 | 10.94 | | 7.64 |
| Total water requirement | | | | | | | 322.73 | 417.09 | 357.09 |

the meteorological droughts frequently occurred in the basin and few drought events persisted for 2 consecutive years. A scrutiny of rainfall data revealed that the droughts in the basin occurred due to either delay in the onset of monsoon or prolonged dry days during the monsoon period or deficit amount of rainfall and sometime due to early withdrawal of monsoon in different years. The drought condition of the basin

Table 6.12 Weekly crop water requirement (ETcrop) in mm/day during the growing period

| S. No. | Name of the station | Effective (CDS) | | | Crop water requirement (ETcrop) during CDS (mm) | | | Effective rainfall (ER) during CDS (mm) | Irrigation requirement (IR) during CDS (mm) | | |
|---|---|---|---|---|---|---|---|---|---|---|---|
| | | No. of CDS | Time period | Duration (Days) | Maize | Rice | Soybean | | Maize | Rice | Soybean |
| 1 | Chhatarpur | Ist CDS | 12 Jul–30 Jul | 18 | 37.20 | 97.42 | 48.49 | 64.39 | | 33.02 | |
| | | IInd CDS | 19 Aug–7 Sep | 19 | 75.89 | 85.91 | 90.00 | 60.14 | 15.74 | 25.77 | 29.86 |
| | | IIIrd CDS | 7 Sep–19 Sep | 12 | 60.47 | 52.16 | 56.58 | 24.12 | 36.35 | 28.04 | 32.46 |
| Total | | | | | 173.55 | 235.48 | 195.06 | | 52.09 | 86.83 | 62.32 |
| 2 | Damoh | Ist CDS | 18 Jul–1 Aug | 14 | 27.07 | 70.91 | 35.29 | 63.45 | | 7.45 | |
| | | IInd CDS | 22 Aug–8 Sep | 17 | 65.16 | 73.76 | 77.28 | 49.84 | 15.32 | 23.92 | 27.44 |
| | | IIIrd CDS | 6 Sep–19 Sep | 13 | 63.28 | 54.58 | 59.21 | 22.34 | 40.94 | 32.24 | 36.87 |
| Total | | | | | 155.51 | 199.25 | 171.78 | | 56.26 | 63.62 | 64.30 |
| 3 | Katni | Ist CDS | 6 Jul–23 Jul | 17 | 33.13 | 86.77 | 43.19 | 63.87 | | 22.90 | |
| | | IInd CDS | 22 Aug–6 Sep | 15 | 57.35 | 64.92 | 68.01 | 46.30 | 11.05 | 18.62 | 21.72 |
| Total | | | | | 90.48 | 151.69 | 111.20 | | 11.05 | 41.53 | 21.72 |
| 4 | Panna | Ist CDS | 15 Jul–30 Jul | 15 | 30.30 | 79.37 | 39.50 | 72.15 | | 7.21 | |
| | | IInd CDS | 22 Aug–5 Sep | 14 | 54.77 | 62.00 | 64.96 | 46.61 | 8.16 | 15.40 | 18.35 |
| | | IIIrd CDS | 2 Sep–17 Sep | 15 | 74.21 | 64.02 | 69.44 | 27.77 | 46.44 | 36.25 | 41.67 |
| Total | | | | | 159.29 | 205.38 | 173.90 | | 54.61 | 58.86 | 60.02 |
| 5 | Sagar | Ist CDS | 14 Jul–28 Jul | 14 | 23.11 | 60.52 | 30.12 | 55.02 | | 5.50 | |
| | | IInd CDS | 20 Aug–10 Sep | 21 | 71.30 | 80.71 | 84.56 | 51.99 | 19.30 | 28.72 | 32.56 |
| | | IIIrd CDS | 5 Sep–20 Sep | 15 | 66.18 | 57.08 | 61.92 | 20.74 | 45.44 | 36.35 | 41.19 |
| Total | | | | | 160.58 | 198.32 | 176.60 | | 64.75 | 70.57 | 73.75 |
| 6 | Banda | Ist CDS | 14 Jul–3 Aug | 20 | 41.74 | 109.33 | 54.42 | 73.07 | | 36.25 | |

(continued)

**Table 6.12** (continued)

| S. No. | Name of the station | Effective (CDS) | | | Crop water requirement (ETcrop) during CDS (mm) | | | Effective rainfall (ER) during CDS (mm) | Irrigation requirement (IR) during CDS (mm) | | |
|---|---|---|---|---|---|---|---|---|---|---|---|
| | | No. of CDS | Time period | Duration (Days) | Maize | Rice | Soybean | | Maize | Rice | Soybean |
| | | IInd CDS | 19 Aug–4 Sep | 16 | 65.67 | 74.34 | 77.88 | 53.85 | 11.81 | 20.49 | 24.03 |
| Total | | | | | 107.41 | 183.67 | 132.30 | | 11.81 | 56.74 | 24.03 |
| 7 | Hamirpur | Ist CDS | 8 Jul–26 Jul | 18 | 39.46 | 103.36 | 51.44 | 79.89 | | 23.47 | |
| | | IInd CDS | 21 Aug–7 Sep | 17 | 72.05 | 81.56 | 85.45 | 54.86 | 17.19 | 26.70 | 30.59 |
| | | IIIrd CDS | 1 Sep–19 Sep | 18 | 95.21 | 82.13 | 89.09 | 35.23 | 59.98 | 46.90 | 53.86 |
| Total | | | | | 206.72 | 267.05 | 225.98 | | 77.17 | 97.07 | 84.44 |
| 8 | Mahoba | Ist CDS | 16 Jul–3 Aug | 18 | 37.45 | 98.08 | 48.82 | 43.91 | | 54.17 | 4.91 |
| | | IInd CDS | 15 Aug–31 Aug | 16 | 64.85 | 73.42 | 76.91 | 45.35 | 19.50 | 28.07 | 31.56 |
| | | IIIrd CDS | 2 Sep–19 Sep | 17 | 87.99 | 75.89 | 82.33 | 20.97 | 67.02 | 54.93 | 61.36 |
| Total | | | | | 190.28 | 247.39 | 208.05 | | 86.52 | 137.17 | 97.83 |
| Average value for Ken Basin | | | | | 155.48 | 211.03 | 174.36 | | 51.78 | 76.55 | 61.05 |

calls for supplemental irrigation during critical dry spells and genuine water resources planning and management to utilize optimally the accessible water resources in the Ken Basin. This can be accomplished through the development of conservation of monsoon runoff in reservoir/ponds and tanks, in situ rainwater harvesting and moisture conservation practices and watershed intercession measures by the construction of medium-sized dams along with check dam and artificial recharge structure for the essential utilization and conservation of the available water resources. In brief, the following conclusions can be drawn from the study:

- The magnitude of annual rainfall deficiency varied from −37 to −60% with erratic distribution in time and space and had caused severe water stress and crop loss in the Ken Basin. The probability of occurrence of 75% of normal annual rainfall varied from 66 to 86%. Katni, Mahoba and Sagar Districts experienced more frequent droughts than did the other districts of the basin. Thus, the Ken Basin being mainly rainfed is subjected to occurrence of frequent droughts.
- The southwest monsoon period starts around June 17–27 and terminates in September between 16 and 23, with an average length of the wet season of about 89 days. Therefore, crop varieties with the duration of 85–95 days are preferred for rainfed cultivation.
- The average length of critical dry spells in the month of July, August and September is expected in the order of 17, 17 and 15 days, respectively. Therefore, the provisions for supplemental irrigation during critical dry spells may ensure the success of Kharif season crops.
- The water requirement of crops, viz. maize, rice and soybean were 323 mm, 417 mm and 357 mm, respectively. The irrigation requirement during the critical dry spells is lowest for maize crop as compared to rice and soybean crops. The estimates of irrigation requirement may be useful in planning of supplemental irrigation for Kharif cropping season in Ken Basin.

# References

Appa Rao G (1986). Drought climatology. Jal Vigyan Samiksha, Publication of High-Level Technical Committee on Horology, National Institute of Hydrology, Roorkee

Ashok Raj PC (1979) Onset of effective monsoon and critical dry spells, a computer-based forecasting technique. Water Technology Centre, IARI, New Delhi

Central Water Commission (CWC) (1982) Report on the identification of drought-prone areas for 99-districts, New Delhi

Dracup JA, Lee KS, Paulson EG Jr (1980) On the definition of droughts. Water Resour Res 16(2):297–302

FAO (2002) Report of FAO-CRIDA expert group consultation on farming system and best practices for drought-prone areas of Asia and the Pacific region. Central Research Institute for Dryland Agriculture, Hyderabad, India

Mishra AK, Desai VR (2005) Spatial and temporal drought analysis in the Kansabati River Basin, India. Int J River Basin Manag 3(1):31–41

Pandey RP, Soni B, Ramasastri KS (2000) Drought investigations and crop water need in Kalahandi district in Orissa. In: International conference on integrated water resources management for sustainable development, ICWRM-2000, Dec 19–21, at Delhi, National Institute of Hydrology, Roorkee, India, vol II, pp 1139–1152

Pandey RP, Daradur Mihail, Jain Vinit K, Jain Manoj K (2016) Assessment of vulnerability to drought towards effective mitigation planning: the case of Ken River Basin in India. J Indian Water Resour Soc IWRS 36(4):36–48

Sahoo MK (1993) Analysis of drought phenomenon of pre-divided Kalahandi district of Orissa. M.Tech. thesis, Dept. of Soil and Water Conservation Engineering, Orissa University of Agriculture and Technology, Bhubaneswar

Wilhite DA (2000) Drought as a natural hazard: concept and definition. In: Wilhite DA (eds) Drought: a global assessment. Natural hazards and disaster series, vol 1, Chap 1. Routledge Publisher, UK

World Bank (2003) Report on financing rapid onset natural disaster losses in India: a risk management approach, Report No. 26844

# Chapter 7
# Meteorological Drought Characteristics in Eastern Region of India

Kumar Amrit, R. P. Pandey, and S. K. Mishra

## 7.1  Introduction

Drought is a natural disaster occurring due to less than average precipitation over a prolonged duration at a particular place, and subsequently, leading to water deficiency and socioeconomic loss. Droughts have most devastating effect, occur in every regions and have larger areal extent than any other natural hazards (i.e. earthquakes and floods) (Wilhite 2000; Ponce et al. 2000; Samadi et al. 2004; Amrit et al. 2018a). The various researches on hydrological extremes indicated that many regions around the globe are vulnerable to severe impacts of droughts (Dracup et al. 1980; Gregory 1989; Tallaksen and Van Lanen 2004; Pandey et al. 2008; Mishra et al. 2019). During 2011–2012, severity of drought placed 50% area of the USA, Eastern Africa and Korean Peninsula into the category of disaster zone (Mosley 2012; Dutra et al. 2012; USDA 2012). Droughts are very complex than any other natural hazard, as their onset and termination are quite difficult to identify (Wilhite 1993; Amrit et al. 2018a, b). Mishra and Liu (2014) investigated the variation in pattern of precipitation and risk of drought events and observed that the risk of drought events over Indian regions gets increased due to increase in the length of dry spells and total number of dry days, as a consequence of global warming. Water quality is also affected by

K. Amrit (✉)
Research & Innovation Center, CSIR-National Environmental Engineering Research Institute, Mumbai 400018, Maharashtra, India
e-mail: amrit27upadhyay@yahoo.com

K. Amrit · S. K. Mishra
Department of Water Resources and Management, Indian Institute of Technology Roorkee, Roorkee 247667, Uttarakhand, India

R. P. Pandey
National Institute of Hydrology, Roorkee 247667, Uttarakhand, India

© The Editor(s) (if applicable) and The Author(s), under exclusive license
to Springer Nature Switzerland AG 2021
A. Pandey et al. (eds.), *Hydrological Extremes*, Water Science
and Technology Library 97, https://doi.org/10.1007/978-3-030-59148-9_7

111

the drought because the moderate climatic fluctuations lead to changes in hydrologic regimes, which has significant effects on the lake chemistry (Webster et al. 1996). Depletion in agricultural production, reduction in employment opportunities in agricultural sector, scarcity of food, drinking water and fodder, high inflation rate and death due to starvation/malnutrition/diseases are the major impacts of drought. Increase population and industrial developments have increased water demand by many fold compared to past decades. Drought has no single definition, which could be accepted everywhere (Wilhite 1993; Zhang et al. 2012).

Hisdal and Tallaksen (2003) done a case study on Denmark for regional meteorological and hydrological drought characteristics. In their study brings out new approach to find the possibility of a specific region to get affected by a drought of given magnitude of deficit and its efficacy was checked to find both the meteorological and hydrological droughts. The comparison of drought characteristics indicates that hydrological droughts are less homogeneous over the area, less frequent and its duration is longer than meteorological droughts. Moradiet al. (2011) investigated the characteristics of meteorological drought for Fars province in Iran. In this study, effort has been made to predict the frequency, magnitude of deficit, duration and its extent in the Fars province. Jain et al. (2015) have done the comparative study of Rainfall Departure, SPI, EDI, China Z-index, statistical Z-score and Rainfall Decile Based Drought Index for the assessment of different characteristics of drought in the Ken river basin of central India. The study had been done on the five-time scale viz. 1, 3, 6, 9 and 12 months using the various indices for the computation of the severity and compared with each other. The analysis revealed that the drought indices estimated for the 9-month time scale were found in good agreement with each other. Khadr (2017) investigated the spatio-temporal characteristics of meteorological droughts using the rainfall data of 48 years (1960–2008) from 22 stations in Upper Blue Nile river basin to plan the strategy for sustainable management of water resources. The analysis indicated that the extent and impact of drought were very severe during the year 1965 and 1980s. The analysis revealed that droughts with shorter return period were limited to only small part of the basin while major part of the basin was affected by drought with higher return period.

The assessment of severity and frequency of droughts in addition to their persistency in various districts of eastern India is the major goal of this study. The analysis done in this study will be very useful for making plans for effective mitigation for droughts in eastern India.

## 7.2  Study Area and Data Used

The study area comprising of the states of Bihar, Jharkhand, West Bengal and a part of Uttar Pradesh (Fig. 7.1). Bihar located in east India covering the area of about 94,163 km$^2$ and recognized as the 13th largest state of the country according to area. It extends between latitude of 24° 20′ 10″ and 27° 31′ 15″ N and longitude of 83° 19′ 50″ and 88° 17′ 40″ E. The state has sub-tropical climate and receives 1200 mm

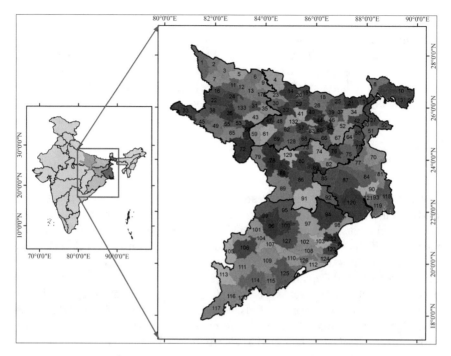

**Fig. 7.1**  Index map of study area

of rainfall annually. The area has hot summers and cold winters with the average temperature of 27 °C. The region is large stretch of fertile plains and drained by the River Ganges along with its tributaries originating from Himalayan region of Nepal viz. Gandak and Koshi. Jharkhand is the state of India that came into existence in 2000 when it was divided from Bihar covering the area of 79,710 km². Most of the portion of state covered with forest and received 1300 mm of rainfall annually. Odisha is the state lies in eastern region of India covering 155,707 km² area with the coastline of 450 km. The coastal plain lies in the eastern part of the Odisha. It extends from the Subarnarekha River in the north to the River Rushikulya in the south. The region becomes fertile due to the deposition of silt by the Subarnrekha, Brahmani, Rushikulya, Baitarani, Budhabalanga and Mahanadi rivers flowing into Bay of Bengal. About three-fourth part of the state covered with the forest and hills. Most of the part of the region includes the hills and mountains of Eastern Ghats. Major part of state comes under sub humid region and receives 1450 mm of rainfall annually. The state of West Bengal is located in eastern region of India extending from Himalayas to Bay of Bengal covering the area of about 88,752 km². There are four main seasons viz. summer, rainy, autumn and winter in the state. The average annual rainfall in West Bengal is 1500 mm except the northern part which includes Darjeeling, Jalpaigudi and Cooch Behar districts, which receives heavy rainfall of more than 3000 mm. The eastern Uttar Pradesh located in eastern part of India includes the districts of the proposed Purvanchal state. The region receives 1000 mm of rainfall annually. The region lies in the Gangetic plain.

For assessing the characteristics of droughts across eastern India, monthly rainfall records of 113 years (1901–2013) of Bihar, Jharkhand, Orissa, West Bengal and Uttar Pradesh states comprising of 132 districts have been used. The monthly rainfall data of these states were procured from India Meteorological Department, Pune.

## 7.3  Methodology

The characteristics of drought (return period, severity and persistence) for every district of the eastern part of the country have quantified by estimating the percentage deviation in annual and seasonal rainfall from average of corresponding rainfall records. In this study, the percentage deviation in seasonal rainfall has been calculated and drought years were identified for the period of 1901–2013 using the definition of India Meteorological Department (IMD) for every district falling in the study area. As per IMD, the season/year is said to be drought season/year if it receives rainfall lower than 75% of corresponding mean. The classification of annual and seasonal droughts as moderate, severe and extreme was done on the basis of deviation in rainfall from corresponding (annual or seasonal) long-term mean. The three different categories (moderate, severe and extreme) of drought based on the percentage departure of rainfall from long-term average are presented in Table 7.1.

The drought return period can also be defined as the ratio of the total number of years of rainfall records analyzed to the number of years with 25% or more deficit in rainfall.

$$\text{Return period} = \frac{\text{Number of years of rainfall analyzed}}{\text{Number of deficit years}}$$

Similarly, return period of severe and extreme droughts has been estimated across the districts of eastern India.

The drought event is said to be persistent if it continues for two or more years in a row (Amrit et al. 2018b). Further, the study propagates with identification of droughts of 2, 3 and 4 consecutive years for every district during the period of 113 years.

| Table 7.1 Categorization of drought based on rainfall departure (Amrit et al. 2018a) | Rainfall deviation from mean (%) | Drought category |
|---|---|---|
| | <−25 to −45 | Moderate |
| | <−45 to −60 | Severe |
| | <−60 or less | Extreme |

## 7.4  Result and Discussion

### 7.4.1  Return Period

The states of Orissa, West Bengal, Jharkhand, Bihar and eastern part of Uttar Pradesh together constitute the eastern region of India. Figure 7.2 shows the spatial representation of drought frequency across the districts of eastern region of the country. Figure 7.2 clearly shows that the average frequency of drought in eastern region of the country is than once in 5 years or more. Figure 7.2 clearly indicates that the larger part of study area has return period in the range of 7–9 years. The north-west that covers the eastern UP and some districts in north which is the part of central Bihar experienced drought one in 5–6 years. Sonbhadra is only district in eastern UP, which experienced drought one in 9 years. Return period of drought in north,

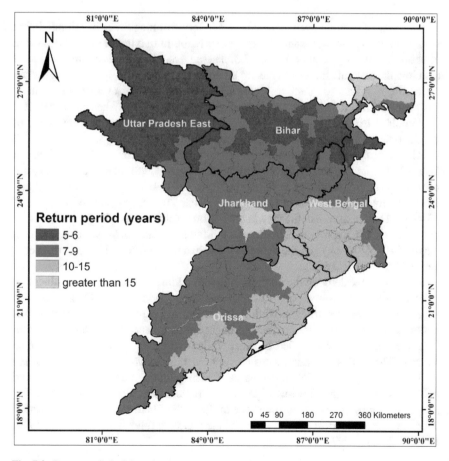

**Fig. 7.2** Return period of drought across eastern India

central and south-west part of the region has been estimated to be 7–9 years. The southern part of the region that includes the eastern Orissa and southern West Bengal has drought return period in the range of 10–14 years. The two districts Ranchi in Jharkhand and Darjeeling in West Bengal have the average return period of 16 years and 19 years, respectively. The district Sonbhadra in UP east, Kishanganj in Bihar, Ranchi in Jharkhand, Mayurbhanj in Orissa and Darjeeling in West Bengal differs in drought characteristics than other districts of the corresponding states due to their topography, geographical location, forest cover and average annual rainfall. The study revealed that the topography, morphology and land use of a particular place also affects the drought characteristics of that place. The return period of drought across eastern region of the country is presented in Fig. 7.2.

## 7.4.2 *Drought Severity*

The severity of drought events is expressed in terms of magnitude of rainfall deficit from the corresponding mean. The spatial variation in maximum rainfall departure in eastern India is presented in Fig. 7.3. The analysis indicated that magnitude of deficit in the state of Bihar estimated to be the order of 40–60%, however, it reached up to 71% in Jehanabad district in year 1966. The maximum severity in Jharkhand has been estimated to be the order of 45–55% however; it reached up to 63% for the Gumla district in year 1979. The analysis revealed that, in year 1966, most of the districts of the state have faced the maximum severity. The maximum rainfall deficiency in Orissa is in the order of 40–50% but once it reached up to 61% for Debgarh district in year 1979. In addition of the above, the major part of the state has rainfall deficit condition in following years 1901, 1918, 1923, 1974, 1979 and 1987. The magnitude of deficit in the eastern Uttar Pradesh region is of the order of 50–65%. The maximum deficiency in the region reached up to 75% for Mau district in year 1998. The maximum deficiency of rainfall in the state of West Bengal is in the range of 40–55%, however, the deficiency reached up to 62% for Hooghly district in year 1982.

The study revealed that the districts of eastern Uttar Pradesh are more vulnerable to severe droughts. Some parts of Bihar and the western part of Orissa had also experienced the severe droughts while in Jharkhand and West Bengal, severe droughts rarely occur. The incidence of extreme drought events is generally confined to eastern Uttar Pradesh and some districts of Bihar while in rest of the region, extreme events are rare. The analysis also indicates that approaching to east from west side of the region the magnitude of deficit decreases. The magnitude of deficit had been found to be maximum in the north-west part of the region, which covers the eastern UP and some portion of Bihar in the order of 50–65%. There had been huge rainfall deficiency found in most of the districts of the region in following years 1908, 1966, 1979, 1982 and 1997.

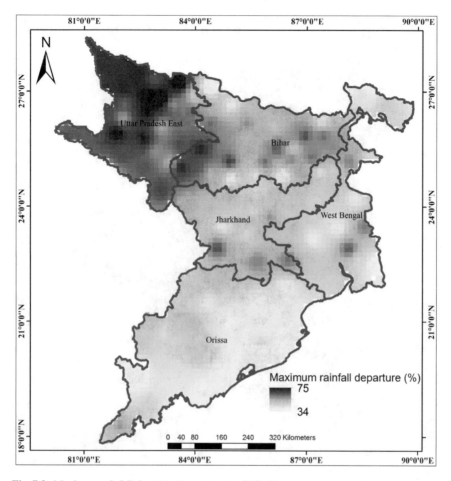

**Fig. 7.3** Maximum rainfall departure across eastern India

## 7.4.3 *Maximum Drought Persistence*

Persistence of drought event means the occurrence of an event consecutively for 2 or more years in a row. The total number of various persistent droughts (2, 3 and 4 consecutive years) has been estimated for each district of eastern part of the country. Figure 7.4 represents the spatial distribution of maximum persisted drought events in the eastern part of India. Figure 7.4 clearly shows that the larger part of the study area has experienced the persistent drought event for maximum of 2 and 3 years. The drought events of 4 consecutive years rarely persisted in the region except in east Uttar Pradesh. The northern part of the West Bengal never experienced any persistent event ever in 113 years. The major portion of Orissa and West Bengal has faced the maximum persistence of 2 years.

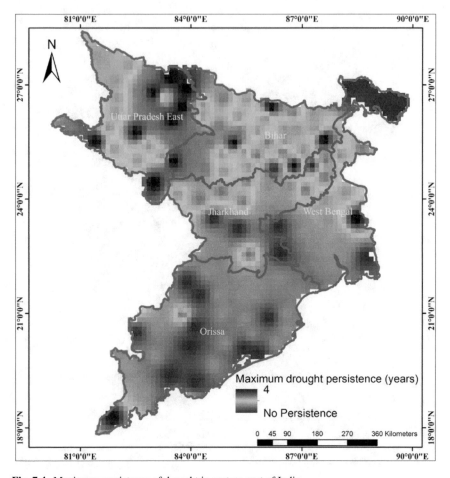

**Fig. 7.4** Maximum persistence of drought in eastern part of India

## 7.5 Conclusion

In this study, using the huge rainfall data (1901–2013), the frequency, severity and persistence of droughts have been assessed for 132 districts of eastern India located in states of Bihar, West Bengal, Jharkhand, Orissa and a part of Uttar Pradesh. The analysis revealed that north and north-west part has return period of drought in between 5 and 6 years. The central, eastern, southern and south-west portion of study area have return period of 7 years or more. The eastern Orissa and southern West Bengal has drought return period varies from 10 to 14 years. The two districts Ranchi in Jharkhand and Darjeeling in West Bengal rarely experienced droughts once in 16 years and 19 years, respectively. The estimates indicated that the severe droughts occur in once in 19 years or more. West Bengal and Jharkhand are not susceptible to severe droughts. The various districts in the region experienced the maximum drought persistency of 2, 3 and 4 consecutive years. The major part of eastern India experienced the maximum drought persistence of 2 years. The occurrence of droughts of extreme severity is much rare in the region. The presented study can be useful for making strategies and plans for effective mitigation of droughts in eastern India.

**Acknowledgements** The authors are very thankful to IIT Roorkee, for providing all resources to carry out this study. The authors would like to acknowledge the support from India Meteorological Department, Pune for providing data for study. Financial support to the first author was given by Ministry of Human Resources and Development, Government of India, New Delhi.

# References

Amrit K, Pandey RP, Mishra SK (2018a) Assessment of meteorological drought characteristics over Central India. Sustain Water Resour Manag 4(4):999–1010

Amrit K, Pandey RP, Mishra SK (2018b) Characteristics of meteorological droughts in northwestern India. Nat Hazards 94(2):561–582

Dracup JA, Lee KS, Paulson EG (1980) On the definition of droughts. Water Resour Res 16(2):297–302

Dutra E, Magnusson L, Wetterhall F, Cloke HL, Balsamo G, Boussetta S, Pappenberger F (2012) The 2010–2011 drought in the horn of Africa in ECMWF reanalysis and seasonal forecast products. Int J Climatol

Gregory S (1989) The changing frequency of drought in India, 1871–1985. Geogr J 322–334.

Hisdal H, Tallaksen LM (2003) Estimation of regional meteorological and hydrological drought characteristics: a case study for Denmark. J Hydrol 281(3):230–247

Jain VK, Pandey RP, Jain MK, Byun HR (2015) Comparison of drought indices for appraisal of drought characteristics in the Ken River Basin. Weather Clim Extrem 8:1–11

Khadr M (2017) Temporal and spatial analysis of meteorological drought characteristics in the upper Blue Nile river region. Hydrol Res 48(1):265–276

Moradi HR, Rajabi M, Faragzadeh M (2011) Investigation of meteorological drought characteristics in Fars province, Iran. Catena 84(1–2):35–46

Mosley J (2012) Translating famine early warning into early action: an east Africa case study.

Mishra SK, Amrit K, Pandey RP (2019) Correlation between Tennant method and standardized precipitation index for predicting environmental flow condition using rainfall in Godavari Basin. Paddy Water Environ 17(3):515–521

Mishra A, Liu SC (2014) Changes in precipitation pattern and risk of drought over India in the context of global warming. J Geophys Res Atmos 119(13):7833–7841

Pandey RP, Dash BB, Mishra SK, Singh R (2008) Study of indices for drought characterization in KBK districts in Orissa (India). Hydrol Process 22(12):1895–1907

Ponce VM, Pandey RP, Ercan S (2000) Characterization of drought across climatic spectrum. J Hydrol Eng 5(2):222–224

Samadi S, Jamali JB, Javanmard S (2004) Drought early warning system in IR of Iran

Tallaksen LM, Van Lanen HA (2004) Hydrological drought: processes and estimation methods for streamflow and groundwater, vol 48. Elsevier

USDA (2012) World agricultural supply and demand estimates. Technical report, United States Department of Agriculture

Webster KE, Kratz TK, Bowser CJ, Magnuson JJ, Rose WJ (1996) The influence of landscape position on lake chemical responses to drought in northern Wisconsin. Limnol Oceanogr 41(5):977–984

Wilhite DA (1993) Planning for drought: a methodology. In: Drought assessment, management, and planning: theory and case studies. Springer, Boston, MA, pp 87–108

Wilhite DA (2000) Drought: a global assessment, vol 1. Routledge, New York

Zhang Q, Xiao M, Singh VP, Li J (2012) Regionalization and spatial changing properties of droughts across the Pearl River basin, China. J Hydrol 472:355–366

# Chapter 8
# Meteorological Drought Assessment in Tripura of Humid Northeast India Using EDI

Annu Taggu and Salil K. Shrivastava

## 8.1 Introduction

According to India Meteorological Department (IMD), drought happens in any area when the mean annual rainfall is less than 75% of the normal rainfall. It is very difficult to detect a drought before it becomes a serious problem, because of its slow developmental nature. Drought is a natural hazard that affects the agricultural production and due to which it affects the economy of the country. Numerous methods have been implemented by various researchers in an attempt to identify the droughts such as Standardized Precipitation Index (SPI), Effective Drought Index (EDI), China Z Index (CZI), Rainfall Anomaly Index (RAI), Percent of Normal Index (PNI), etc. It is necessary to monitor the drought and implement appropriate measures to control its negative impacts. Limited studies have been carried out to study the occurrences of drought in northeast region of India. Jhajharia et al. (2007) studied the drought proneness at Guwahati and reported that there were six drought events observed between 1995 and 2003 and the worst drought was observed in the year 1952. Meteorological droughts at North Lakhimpur district of Assam using rainfall data for 18 years (1981–1998) were analyzed by Shrivastava et al. (2008) considering various timescale viz., weekly, monthly, seasonal and yearly time scales and it was observed that the maximum frequency of droughts occurred in 23rd and 39th week, November month and postmonsoon season. The observed frequency of drought was minimum in 24th week, July and August months; and monsoon; and summer seasons. The years 1981, 1986, 1992 were found to be deficit years with annual rainfall below 81, 68 and 79% of its mean value. The intensity of drought was moderate/mild and no severe drought was occurred during the period of analysis.

A. Taggu (✉) · S. K. Shrivastava
Department of Agricultural Engineering, North Eastern Regional Institute of Science and Technology, Nirjuli, Itanagar 791109, Arunachal Pradesh, India
e-mail: annutaggu1525@gmail.com

© The Editor(s) (if applicable) and The Author(s), under exclusive license
to Springer Nature Switzerland AG 2021
A. Pandey et al. (eds.), *Hydrological Extremes*, Water Science
and Technology Library 97, https://doi.org/10.1007/978-3-030-59148-9_8

By using the reconnaissance drought index (RDI) for the period from 1985 to 2012, the droughts in the East Districts of Sikkim were characterized during the winter, and from the analysis, it was found that the drought was experienced for 66.7 and 59.3% during October–December and January–March, respectively (Kusre and Lalringliana 2014). In North East (NE) India, Assam and Meghalaya experienced drought consecutively during 2005 and 2006 and about 10–14 districts were affected in 3 consecutive years during 2009–2011, which suggests that frequent droughts were observed in the recent decade (Parida and Oinam 2015). These findings suggest that a detailed study is warranted over the northeastern region of India so as to provide a better monitoring of the drought.

For this study, the districts of Tripura were selected and Effective Drought Index (EDI) was used, which measure the daily water accumulation based on weighted current and antecedent rainfall. Various studies have used EDI to investigate and quantify the meteorological droughts and flood as well such as (Deo et al. 2015; Kalamaras et al. 2010; Kim et al. 2010; Oh et al. 2013; Padhee et al. 2014; Swain et al. 2017). EDI can overcome the limitation of Standardized Precipitation Index (SPI) as antecedent effect of rainfall is incorporated in EDI. EDI helps in detecting long term, extremely long-term and short-term drought, short-term rainfall and also dealing with the problem of overestimation and underestimation (Byun and Wilhite 1999; Byun and Kim 2010). Several studies agree that the assessment of droughts by EDI is more accurate and its predictions carry less error than that of other meteorological indices (Padhee et al. 2014; Byun and Kim 2010; Morid et al. 2006, 2007).

## 8.2   Study Area and Data

The State of Tripura, with an area of 10,492 km$^2$ is situated between the latitudes of 220 56′ N and 240 32′ N and the longitudes of 900 09′ E and 920 10′ E with elevation ranging from 600 to 900 m above mean sea level. The average precipitation of the area is about 2100 mm and the climate is humid tropical. The study was carried out district wise, which includes four districts viz. North Tripura, South Tripura, West Tripura and Dhalai. The details of the districts are given in Table 8.1 and the location of districts is shown in Fig. 8.1.

The gridded rainfall data having spatial resolution of 0.25° × 0.25° for the period 1901–2013 were acquired from India Meteorological Department (IMD). The

**Table 8.1** Description of the districts under investigation

| Districts | Latitude °N | Longitude °E | Elevation (m) | Period |
|---|---|---|---|---|
| North Tripura | 24.13 | 92.16 | 56 | 1901–2013 |
| South Tripura | 23.45 | 91.59 | 97 | 1901–2013 |
| West Tripura | 23.88 | 91.49 | 74 | 1901–2013 |
| Dhalai | 23.91 | 91.92 | 97 | 1901–2013 |

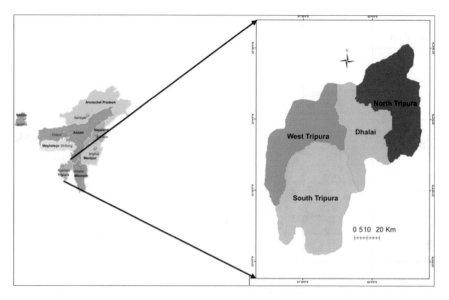

**Fig. 8.1** Map showing districts of Tripura

required rainfall data for the four districts of Tripura were extracted using Kriging interpolation and ArcGIS® 10. The long-term mean, maximum and minimum rainfall of each of the four districts during premonsoon, monsoon, first half and second half of the year are given in Table 8.2.

## 8.3 Methodology

The gridded rainfall data were interpolated using Kriging technique and then extracted for the four districts of Tripura. Since, most of the rainfall occurs during monsoon season, meteorological droughts were inspected using EDI at medium-range time scales (premonsoon, monsoon and 6-monthly, i.e., two halves of the year). The months of March to May were considered premonsoon and the months from June to September were taken as monsoon season which is generally being accepted in this region. Effective Drought index (EDI) (Oh et al. 2010) measures the daily water accumulation based on weighted current and antecedent rainfall. In this method, effective precipitation (EP) is calculated by summing precipitation over time, considering the loss of rainfall due to runoff or evaporation with the passage of time and is presented in Eq. (8.1):

$$EPi = \sum_{n=1}^{i}\left[\left(\sum_{m=1}^{n} P_m\right)/n\right] \tag{8.1}$$

**Table 8.2** Mean, maximum and minimum rainfall (mm) of the four districts of Tripura during 1901–2013

| | Premonsoon | Percentage of the total annual average rainfall | Monsoon | Percentage of the total annual average rainfall | First half | Percentage of the total annual average rainfall | Second half | Percentage of the total annual average rainfall |
|---|---|---|---|---|---|---|---|---|
| *Tripura* | | | | | | | | |
| Mean | 658 | 27 | 1461 | 61 | 1150 | 48 | 1245 | 52 |
| Max | 2012 | 39 | 3522 | 69 | 2404 | 47 | 2698 | 53 |
| Min | 58 | 6 | 661 | 72 | 410 | 44 | 513 | 56 |
| *Dhalai* | | | | | | | | |
| Mean | 802 | 31 | 1552 | 59 | 1332 | 51 | 1290 | 49 |
| Max | 1633 | 38 | 2346 | 54 | 2526 | 58 | 1809 | 42 |
| Min | 230 | 17 | 937 | 71 | 680 | 51 | 649 | 49 |
| *North Tripura* | | | | | | | | |
| Mean | 793 | 29 | 1627 | 60 | 1335 | 50 | 1355 | 50 |
| Max | 1569 | 36 | 2247 | 51 | 2479 | 56 | 1911 | 44 |
| Min | 233 | 16 | 941 | 64 | 605 | 41 | 859 | 59 |
| *South Tripura* | | | | | | | | |
| Mean | 595 | 26 | 1434 | 63 | 1069 | 47 | 1206 | 53 |
| Max | 1199 | 30 | 2260 | 56 | 2092 | 52 | 1910 | 48 |
| Min | 53 | 6 | 815 | 96 | 152 | 18 | 701 | 82 |
| *West Tripura* | | | | | | | | |
| Mean | 658 | 29 | 1389 | 61 | 1137 | 50 | 1158 | 50 |
| Max | 1347 | 36 | 2172 | 58 | 2041 | 55 | 1680 | 45 |
| Min | 144 | 13 | 870 | 78 | 357 | 32 | 760 | 68 |

where, $EP$ = Effective precipitation, mm; $P_m$ = Precipitation m days ago and $i = 1$ to 365. The EDI is calculated as follows:

(1) Calculation of the daily $EP$.
(2) Calculation of the 30-year mean EP (MEP) for each calendar day.
(3) Determination of DEP, which is the difference between the EP and MEP:

$$DEP = EP - MEP \tag{8.2}$$

When the DEP is represented by a negative number, this signifies that it is drier than the average, and while this dry period continues, add the days of prolonged dryness to the existing period ($i = 1$ to 365) and the EP will be recalculated for that specific period. Once again, MEP and DEP will be calculated. Then DEP will be divided for each calendar day by the standard deviation (SD) of the DEP over the past 30 years as given below

**Table 8.3** Classification of drought based on the EDI values (Oh et al. 2010)

| Drought class | EDI |
|---|---|
| Normal | EDI > −1.0 |
| Moderately dry | −1.5 < EDI ≤ −1.0 |
| Severely dry | −2 < EDI ≤ −1.5 |
| Extremely dry | EDI ≤ −2 |

$$EDI = \frac{DEP}{\sigma_{DEP}} \qquad (8.3)$$

where, $\sigma_{DEP}$ is the standard deviation of each day DEP. Table 8.3 shows the classification of effective drought index.

The EDI values so obtained using Eq. (8.3) were classified into four different classes such as Normal, Moderately dry, severely dry and extremely dry. The EDI ranges of these classes are shown in Table 8.3 (Oh et al. 2010). The areal average rainfall for each day of the Tripura state was calculated using rainfall of the four districts. The area of each four districts is known and can be used Theisen polygon method to calculate the areal rainfall of the state. The whole calculation was done in Microsoft excel. Using Kriging technique in ArcGIS model builder, spatial maps of Tripura for monsoon and premonsoon seasons for each year were developed.

## 8.4 Results and Discussion

As illustrated in Table 8.2, on an average, Tripura receives rainfall of about 1461 mm and 658 mm during monsoon and premonsoon seasons, respectively, as against total annual rainfall of 2395 mm. Among the four districts, North Tripura receives the highest monsoon rainfall while West Tripura receives the least. EDI was calculated for the various time scales and has been classified according to Table 8.3. The premonsoon and monsoon rainfall accounted for more than 85% of the total average annual rainfall in Tripura.

### 8.4.1 Analysis of Drought

The frequency of drought occurred in the study area is presented in Table 8.4. During the period of analysis in the pre-monsoon season, west Tripura was mostly affected, which has witnessed total 18 drought comprising of 2 extreme, 2 severe and 14; north Tripura experienced 15 droughts with more number of extreme droughts. From the analysis of monsoon EDI (Table 8.4) it was revealed that the north Tripura has faced maximum number of droughts (22). During first half of the year, Dhalai district was the most affected district by drought (18 times). During the analysis, we have found

**Table 8.4** Frequency of drought in the study area at different time scales during 1901–2013

Premonsoon

|              | Extreme | Severe | Moderate | Number of droughts |
|--------------|---------|--------|----------|--------------------|
| Tripura      | 0       | 4      | 7        | 11                 |
| North Tripura| 5       | 3      | 7        | 15                 |
| South Tripura| 1       | 4      | 9        | 14                 |
| West Tripura | 2       | 2      | 14       | 18                 |
| Dhalai       | 0       | 8      | 8        | 16                 |

Monsoon

|              | Extreme | Severe | Moderate | Number of droughts |
|--------------|---------|--------|----------|--------------------|
| Tripura      | 0       | 4      | 12       | 16                 |
| North Tripura| 1       | 3      | 18       | 22                 |
| South Tripura| 1       | 5      | 9        | 15                 |
| West Tripura | 1       | 4      | 11       | 16                 |
| Dhalai       | 1       | 3      | 16       | 20                 |

January–June

|              | Extreme | Severe | Moderate | Number of droughts |
|--------------|---------|--------|----------|--------------------|
| Tripura      | 0       | 4      | 6        | 10                 |
| North Tripura| 7       | 2      | 5        | 14                 |
| South Tripura| 0       | 7      | 5        | 12                 |
| West Tripura | 1       | 7      | 5        | 13                 |
| Dhalai       | 2       | 4      | 12       | 18                 |

July–December

|              | Extreme | Severe | Moderate | Number of droughts |
|--------------|---------|--------|----------|--------------------|
| Tripura      | 0       | 2      | 13       | 15                 |
| North Tripura| 2       | 5      | 11       | 18                 |
| South Tripura| 1       | 3      | 10       | 14                 |
| West Tripura | 2       | 3      | 10       | 15                 |
| Dhalai       | 1       | 5      | 11       | 17                 |

seven cases of extreme drought in north Tripura, which is highest among all the other districts. Similarly, in the second half of the year, more number of droughts was found in north Tripura. When all the districts of Tripura are visualized under one window, the frequency of drought was found to be reduced at all the time scales with no case of extreme drought, however, four severe drought each in premonsoon, monsoon and first half and two severe droughts in second half of the year were found.

The annual variation in the computed values of EDI is shown in Fig. 8.2 through six for all the studied districts and Tripura as a whole. The EDI values of Dhalai district at different time scales along with the upper limit of drought classification are shown in Fig. 8.2. It may be seen that after 1961 the drought is occurring almost after every ten years with the highest being in 1962. During the monsoon season, the extreme drought occurred in the year 1962 and the cycle of its occurrence was 20 years. The first half of the year follows the same pattern of occurrence of drought as that of premonsoon has been revealed. The second half of the year that also contains

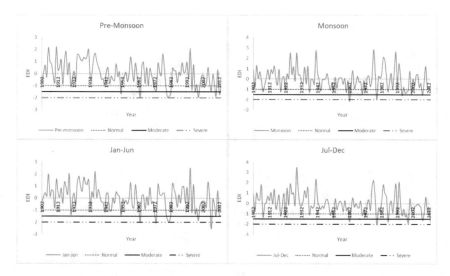

**Fig. 8.2** Annual variation of EDI in Dhalai district during 1901–2013

majority of monsoon months also experienced drought in the year 1979. Figure 8.3 exhibited the annual variation of EDI values for North Tripura district and it was found that in the year 1962, 1972 and 2011, extreme droughts have been observed during premonsoon. During monsoon and first half of the year, extreme drought was noticed in the year 1962 and 2009, respectively. For South Tripura district, the annual variation in EDI values is shown in Fig. 8.4. Extreme droughts were observed in the

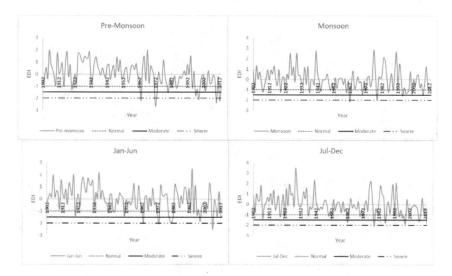

**Fig. 8.3** Annual variation of EDI in North Tripura district during 1901–2013

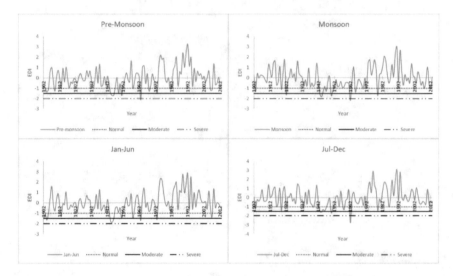

**Fig. 8.4** Annual variation of EDI in South Tripura district during 1901–2013

years 1962 for premonsoon season, monsoon season and second half of the year, respectively.

Similarly, the variation of EDI for West Tripura district is shown in Fig. 8.5. Two extreme droughts were observed in the years 1951 in premonsoon season and first half of the year and in 1962 during monsoon and second half of the year. The analysis was also extended over the whole Tripura state. Figure 8.6 shows the EDI fluctuation

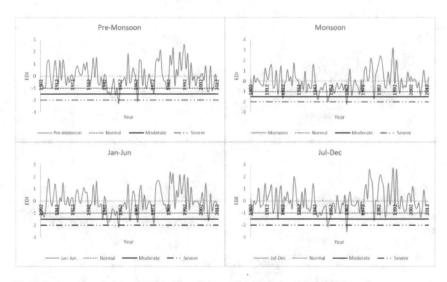

**Fig. 8.5** Annual variation of EDI in West Tripura district during 1901–2013

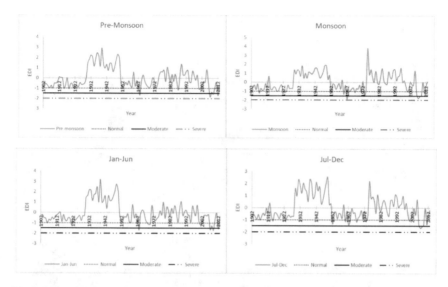

**Fig. 8.6** Annual variation of EDI in Tripura during 1901–2013

over all the considered time scale. It was significant to not from Fig. 8.6 that the years of 1962, 1973, 2007 and 2008 were severely dried in the state. Within this period, January–June, 1963 experienced severe drought. Again, Tripura as a whole experienced a prolonged moderate meteorological drought during, 2006 (May) to 2009 (July).

## 8.4.2  Spatial Analysis

Spatial maps of EDI for Tripura were developed using Kriging technique in ArcGIS model builder. Out of these maps, we selected maps for 1922, 1944, 1950, 1962, 1979 and 2008 during which monsoon period faced drought condition and are depicted in Fig. 8.7a through Fig. 8.7f. In 1922, the whole Tripura experienced moderate to severe drought during monsoon. As shown in Fig. 8.7a, more than 50% of West Tripura and South Tripura were affected with severe drought while rest of the area was under moderate drought. The condition of a rainfall was found to be better in 1944 during monsoon (Fig. 8.7b) as North Tripura and part of Dhalai showed no drought. However, West Tripura and South Tripura continued with severe drought condition as it was in 1922. Monsoon season witnessed severe drought in more than 50% of South and West Tripura districts as shown in Fig. 8.7c. Monsoon season was affected by extreme drought in 1962, which was spread over all the districts as shown in Fig. 8.7d. In 1979, the situation improves slightly and due to increase in the rainfall the drought condition has changed from extreme to severe throughout the Tripura (Fig. 8.7e). During the monsoon of 2008, the North Tripura and part of

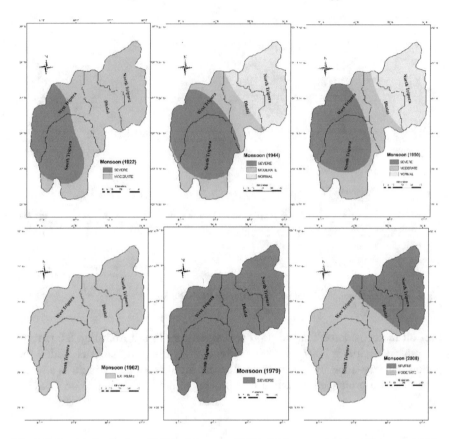

**Fig. 8.7** Spatial distribution of EDI: **a** Monsoon (1922), **b** Monsoon (1944), **c** Monsoon (1950), **d** Monsoon (1962), **e** Monsoon (1979), **f** Monsoon (2008)

Dhalai remained under the severe drought while the intensity of drought over South Tripura and West Tripura changed from severe to moderate in comparison to the monsoon of 1979 (Fig. 8.7d).

## 8.5   Conclusions

In this study, EDI was used to quantify the severity of meteorological droughts in districts of Tripura as well as over entire Tripura. The following inferences can be drawn from the results of the study.

1. Moderate to extreme droughts occurred over different time scales.

2. The maximum number of droughts was observed in the Dhalai district; however, North Tripura has faced maximum numbers of extreme droughts during premonsoon.
3. The monsoon rainfall that is very important from the agricultural point of view is worst affected by drought.
4. The occurrence of droughts in monsoon season is more as compared with premonsoon season.
5. Tripura as a whole experienced a prolonged meteorological drought during May 2006 to July 2009.
6. All the droughts that have been occurred in the region, the drought of monsoon in 1962 was the most severe till date.
7. The results of this study may be useful for proper drought planning of the Tripura.

# References

Byun HR, Kim DW (2010) Comparing the effective drought index and the standardized precipitation index. Opt Méditerranéennes 95:85–89
Byun HR, Wilhite DA (1999) Objective quantification of drought severity and duration. J Clim 12:2747–2756
Deo RC, Byun HR, Adamowski JF, Kim DW (2015) A real-time flood monitoring index based on daily effective precipitation and its application to Brisbane and Lockyer Valley flood events. Water Res Manage 29:1–19
Jhajharia D, Shrivastava SK, Tulla PS, Sen R (2007) Rainfall analysis for drought proneness at Guwahati. Indian J Soil Conser 35(2):163–165
Kalamaras N, Michalopoulou H, Byun HR (2010) Detection of drought events in Greece using daily precipitation. Hydrol Res 126–133
Kim DW, Choi KS Lee J, Byun HR (2010) Impact of climate change on the East Asia droughts. Adv Geosci 165–177
Kusre BC, Lalringliana J (2014) Drought characterization and management in the East District of Sikkim, India. Irrigat Drain 63:698–708
Morid S, Smakhtin V, Moghaddasi M (2006) Comparison of seven meteorological indices for drought monitoring in Iran. Int J Climatol 26:971–985
Morid S, Smakhtin V, Bagherzadeh K (2007) Drought forecasting using artificial neural networks and time series of drought indices. Int J Climatol 27:2103–2111
Oh SB, Byun HR, Kim DW (2013) Spatiotemporal characteristics of regional drought occurrence in East Asia. Theo Appl Climatol 1–13
Oh SB, Kim DW, Choi KS, Byun HR (2010) Introduction of East Asian drought monitoring system. SOLA 6A:9–12
Padhee SK, Nikam BR, Aggarwal SP, Garg V (2014) Integrating effective drought index (EDI) and remote sensing derived parameters for agricultural drought assessment and prediction in Bundelkhand region of India. In: The International Archives of the Photogrammetry, Remote Sensing and Spatial Information Sciences, vol. XL-8, ISPRS Technical Commission VIII Symposium, Hyderabad, India, pp 8–89
Parida BR, Oinam B (2015) Unprecedented drought in North East India compared to Western India. Curr Sci 109(11):2121–2126
Shrivastava SK, Rai RK, Pandey A (2008) Assessment of meteorological droughts in North Lakhimpur district of Assam. J Indian Water Res Soc 28(2):26–30

Swain S, Patel P, Nandi S (2017) Application of SPI, EDI and PNPI using MSWEP precipitation data over Marathwada, India. In: 2017 IEEE international geoscience and remote sensing symposium (IGARSS). IEEE, pp 5505–5507

# Chapter 9
# Multifractal Description of Droughts in Western India Using Detrended Fluctuation Analysis

S. Adarsh and K. L. Priya

## 9.1  Introduction

The knowledge of long memory persistence in hydrological variables helps for improving the accuracy of hydrologic simulations. The attempts for detecting the long-range correlations in time-series data have started in the 1950s (Hurst 1951). Hydrological time series often exhibit self-similarity and fractal behaviors within specific time scale ranges and capturing such information may help for commenting on their long memory characteristics. Mandelbrot (1974) presented the conceptual outline of scaling analysis. Different researchers have used diverse approaches to investigate the multifractality of hydrological time series. The rescaled range (R/S) method (Hurst 1951), double-trace moments (DTM) and Fourier spectrum (Hurst 1965; Tessier et al. 1996; Pandey et al. 1998), multifractal detrended fluctuation analysis (MFDFA) (Kantelhardt et al. 2002) and wavelet transform modulus maxima (Kantelhardt et al. 2003) are some among them.

Droughts have a significant impact on the water resources and agricultural productivity of a region. For the monitoring, forecasting and estimation of drought severity, many quantitative indices were proposed in the past (Zargar et al. 2011). These indices mainly differ in the key drivers, which can lead to drought, which includes the precipitation, evapotranspiration, extent of vegetation, etc. The Rainfall Anomaly Index (RAI) (Van-Rooy 1965), Palmer Drought Severity Index (PDSI) (Palmer1965), Standardized Precipitation Index (SPI) (McKee et al. 1993), Standardized Precipitation Evapotranspiration Index (SPEI) (Vicente-Serrano et al. 2010), etc., are some of them. SPI is the most widely accepted index because of the features to account for different time scales, the requirement of very less number of input information and the

---

S. Adarsh (✉) · K. L. Priya
Department of Civil Engineering, TKM College of Engineering Kollam, Kollam 691005, Kerala, India
e-mail: adarsh_lce@yahoo.co.in

A. Pandey et al. (eds.), *Hydrological Extremes*, Water Science and Technology Library 97, https://doi.org/10.1007/978-3-030-59148-9_9

easiness in computation (Bazrafshan et al. 2013). In the Indian context, few studies considered the changeability of droughts considering SPI as the index (Pai et al. 2011; Ganguli and Reddy 2014; Thomas et al. 2015; Thomas and Prasannakumar 2016; Joshi et al. 2016).

As the drought indices (DIs) are quantified based on variables like precipitation or evapotranspiration, streamflow, the natural processes of droughts may be considered as complex systems, the behavior of which can also be considered to be fractal (Wei et al. 2018). The detrended fluctuation analysis (DFA) propounded by Peng et al. (1994) was one popular approach for detecting the fractal behavior of the time series. Many studies in the past reported that a single scaling exponent may not be sufficient to investigate the fractal behavior (Tessier et al. 1996; Pandey et al. 1998). Stemming from this hypothesis, Kantelhardt et al. (2002) extended DFA into MFDFA, which enables to detect the MF behavior of time series. The MFDFA technique attained wide popularity owing to its easiness in implementation and robust character in the estimation of fractality of precipitation and streamflow records from different river basins across the globe (Kantelhardt et al. 2003; Koscielny-Bunde et al. 2003; Zhang et al. 2008, 2009; Yu et al. 2014). However, the multifractal analysis of extremes low flows like droughts is really scarce in literature while few studies applied such analysis upon the SPI series of different provinces of China recently (Zhang et al. 2016; Wei et al. 2015, 2018). In this context, this paper aims to investigate the multifractal properties of SPI-3, 6 and 12 time series of four meteorological subdivisions in India using MFDFA analysis. The next section presents the algorithm of MFDFA analysis and section thereafter briefly presents the study area and data details. The results of MFDFA of 12 SPI time series along with discussions are presented in Sect. 9.4. The final section concludes the findings of the study.

## 9.2    MFDFA Algorithm

MFDFA is a practical method for detecting the multifractality of nonstationary time-series signals. The steps of MFDFA computational procedure are (Kantelhardt et al. 2006; Zhang et al. 2008):

1.  Let a signal $X$ $(x_1, x_2... x_N)$, is having length $N$ and mean $\overline{x}$. The accumulated deviation of the signal is calculated as

$$D(i) = \sum_{k=1}^{i} [x_k - \overline{x}] \qquad (9.1)$$

where $i = 1, 2, ..., N, k = 1, 2, ..., N$

2.  Divide the profile $D(i)$ into $NS = \text{int}(N/S)$ nonoverlapping segments of length, here $S$ is the length of segments (called as scale) and int() is the integer part. To

avoid any possible exclusion of data points from the tail end (as $N$ may or may not be a multiple of $S$), the calculation is repeated from tail end also.

3. Estimate the local trend for each segment by the method of least squares as:

$$F^2(S, \upsilon) = \frac{1}{S} \sum_{i=1}^{S} \{D[(\upsilon - 1)S + j] - d_\upsilon(j)\}^2 \text{ for } \upsilon = 1, 2, \ldots, NS \quad (9.2)$$

and

$$F^2(S, \upsilon) = \frac{1}{S} \sum_{j=1}^{S} \{D[N - (\upsilon - NS)S + j] - d_\upsilon(j)\}^2 \quad (9.3)$$

for $\upsilon = N_S + 1, \ldots, 2N_S$

Here $d_\upsilon(j)$ is the polynomial used in segment $\upsilon$.

4. Determine the $q$th order fluctuation function (FF) as given below:

$$F_q(S) = \left\{ \frac{1}{2NS} \sum_{\upsilon=1}^{2NS} [F^2(S, \upsilon)]^{q/2} \right\}^{1/q} \quad (9.4)$$

Here the moment order $q$ can take any real value other than zero. For the index $q = 0$, FF can be estimated by a log-transformation cum averaging (Kantelhardt et al. 2002)

$$F_0(S) = \exp \left\{ \frac{1}{4NS} \sum_{\upsilon=1}^{2NS} \ln[F^2(s, \upsilon)] \right\} \quad (9.5)$$

5. Develop logarithmic plots of $F_q(S)$ versus $S$ for different $q$-orders. For power-law correlated series, $F_q(S)$ bears a relationship $F_q(S) \sim S^{h(q)}$ and $h(q)$ is the generalized Hurst exponent (GHE).

The GHE value for $q = 2$ is called as Hurst exponent (Hurst 1951) for stationary signals. For signals with nonstationarity, the GHE value for $q = 2$, may exceed unity and for such cases, H is computed by subtracting 1 from GHE ($q = 2$) (Zhang et al. 2009). For a pure noise series, $H$ will be 0.5. If H is between 0.5 and 1, the time series possesses long-term persistence (LTP) and if it lies between 0 and 0.5 indicate a short-term persistence or STP. LTP implies that the time series is predictable, as the value of data point at a time instant may influence the subsequent values. Moreover, such effect prevails for a longer length within the series. Appropriate fixing of maximum and minimum scale ($S_{max}$ and $S_{min}$), polynomial order, range of $q$, etc., are crucial in the implementation of MFDFA. The value of $S_{min}$ should be much larger than the order of polynomial order and should be much smaller than $N/2$ (Ihlen 2012). Also, the polynomial order can be chosen between 1 and 3 to eliminate any plausible over-fitting problems (Ihlen 2012). In this study, FF for $q = 2$ (in log scale) is developed by fixing $S_{max}$ as $N/2$ (Hajian and Movahed 2010).

From the GHE, the exponents like Reyni exponent (mass exponent, $\tau(q)$) and singularity exponent ($\alpha$) can be deduced, which helps to confirm mono or multifractal behavior of the signal. The mathematical computation involves

$$\tau(q) = qh(q) - 1 \tag{9.6}$$

$$\alpha = \frac{d\tau(q)}{dq} \tag{9.7}$$

and

$$f(\alpha) = q\alpha - \tau(q) \tag{9.8}$$

The $f(\alpha)$ versus $\alpha$ plot is the multfractal spectrum, which will be a parabola if the signal is multifractal. More the base width of the spectrum, higher will be its strength of multifractality. The $\tau(q)$ versus $q$ plot is called as Reyni exponent plot, which is characterized with a breakpoint at $q = 0$. The slope difference before and after the breakpoint helps to state about the multifractality of the series.

## 9.3   Study Area and Database

As per the definition of Indian Institute of Tropical Meteorology (IITM) Pune, there are 36 meteorological subdivisions in India. First, the monthly precipitation (rainfall) data of these subdivisions of 1871–2016 period are collected from http://www.tropmet.res.in/. This data was used for computing SPI at different accumulation time scale. In this work, multifractal properties of SPI of four subdivisions Western Rajasthan (WR), Saurashtra-Kutch (SK), Marathwada (MW) and Kerala (KER) are examined using MFDFA.

## 9.4   Results and Discussion

SPI is an important index for management of meteorological drought. It is computed in the following steps:

(i) prepare rainfall series for specific aggregation time scale (say 3, 6 or 12 months); (ii) find Cumulative Distribution Function (CDF) by Gamma distribution for this series based on the mixed distribution formula $F_X(x) = q + (1 - q)G_X(x)$, accounting for the zero values, in which $q$ is the probability of zero rainfall in the series; $G_X(x)$ is the CDF of nonzero rainfall records and $F_X(x)$ is the CDF of original time series; (iii) perform an equiprobability transformation between $F_x(x)$ and $Z = \psi^{-1}(F_x(x))$ where Z is the standard normal distribution, $\psi^{-1}(\cdot)$ is the inverse of the CDF.

The precipitation at monthly scale of the four subdivisions was used first for obtaining the SPI of different time scales (3, 6 and 12 months), which represents the drought conditions. The short-term drought (SPI-3) has direct implications on the soil moisture conditions, SPI-6 has direct implications on streamflow and SPI-12 (long-term drought) has direct implications on the groundwater flow. The multifractality of the three SPI series is examined by plotting $h(q)$ versus $(q)$ (Generalized Hurst Exponent plot or GHE plot), mass exponent $(\tau(q))$ versus $q$ and the multifractal spectra. The same procedure is used for the data of all the four subdivisions. A typical plot between $q$-order fluctuation function and scale for SPI-6 series (an intermediate case) of the four subdivisions along with GHE plots are presented in Fig. 9.1. Figure 9.1 shows that different time series are multifractal in character, which is inferred from the unequal slopes corresponding to different $q$-values. The nonlinear dependency between $h(q)$ versus $q$ also decipher the multifracatlity of SPI-6 time series and identical comments hold good for in the relationship between $h(q)$ versus $q$ of SPI-3 and SPI-12 series.

The Hurst exponent (H) for the $q = 2$ is noted down for the four time series. The H-values of different SPI series were also determined by the Dispersional Analysis (DA) method (Bassingthwaighte and Raymond 1995). The results are presented in Table 9.1. In all cases, it is found that the values range (0.5, 1), indicating the long-term persistence (long memory dependency structure).

The Reyni exponent was also considered to examine the multifractal behavior of different time series. The results of SPI3, SPI6 and SPI12 series are presented in

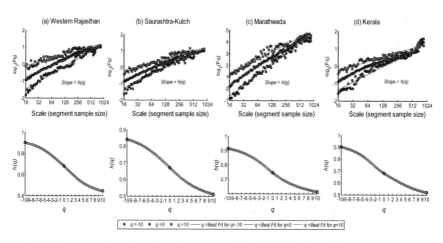

**Fig. 9.1** Fluctuation function plots (upper panels) and GHE plots (lower panels) depicting multifractality of SPI-6 series of four subdivisions

**Table 9.1** Values of Hurst exponents of SPI series of different subdivisions, estimated by MFDFA and DA methods

| Subdivision | SPI-3 | | SPI-6 | | SPI-12 | |
|---|---|---|---|---|---|---|
| | MFDFA | DA | MFDFA | DA | MFDFA | DA |
| WR | 0.5442 | 0.6236 | 0.5528 | 0.6466 | 0.5574 | 0.6756 |
| SK | 0.5397 | 0.4669 | 0.5722 | 0.5591 | 0.608 | 0.6109 |
| MW | 0.5959 | 0.6236 | 0.6598 | 0.6701 | 0.7082 | 0.7167 |
| KER | 0.5907 | 0.6873 | 0.5918 | 0.7116 | 0.6508 | 0.7512 |

Figs. 9.2, 9.3, 9.4. Different types of behavior before and after $q = 0$ can be noted from the plots, which also infers multifractality of the time series.

The slopes indicated in Figs. 9.2, 9.3, 9.4 show that the slope difference between the two segments of mass exponent plots of SPI series of WR region is 0.332,0.489 and 0.767, respectively, for SPI-3, SPI-6 and SPI-12 series. The slope difference between the segments is found to be 0.225, 0.351 and 0.666 for SPI series of SK; 0.248, 0.3203 and 0.591 for SPI series of MW. For Kerala region, the values are 0.236, 0.4089 and 0.5608, respectively, for SPI-3, 6 and 12. The difference in slopes is an indicative of strength of multifractality of the series and the numerical figures indicate highest multifractal strength is possessed by the SPI-12 of WR subdivision. Interestingly, it is noted that multifractal strength of drought indices is more for larger accumulation time scale.

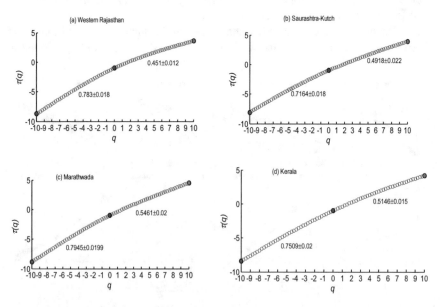

**Fig. 9.2** Reyni exponent plots of SPI-3 of different subdivisions **a** Western Rajasthan; **b** Saurashtra-Kutch; **c** Marathwada; **d** Kerala. Slopes of the segments are provided in different panels

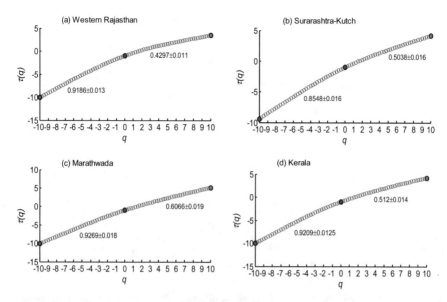

**Fig. 9.3**  Reyni exponent plots of SPI-6 of different subdivisions **a** Western Rajasthan; **b** Saurashtra-Kutch; **c** Marathwada; **d** Kerala. Slopes of the segments are provided in different panels

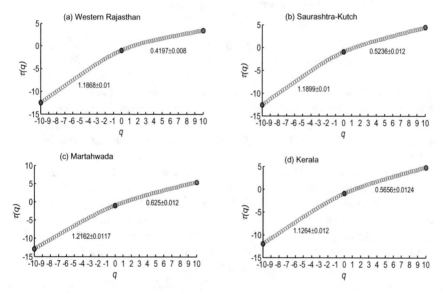

**Fig. 9.4**  Reyni exponent plots of SPI-12 of different subdivisions. Slopes of segments are provided in different panels

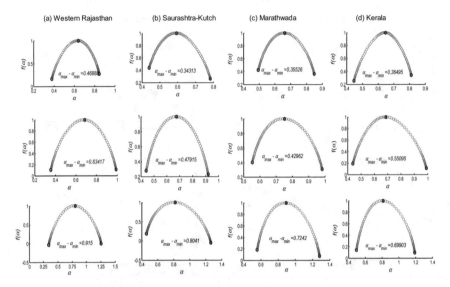

**Fig. 9.5** Multifractal spectra of SPI series of different subdivisions **a** Western Rajasthan; **b** Saurashtra-Kutch; **c** Marathwada; **d** Kerala. The upper panel shows the plots for SPI-3 series, middle panels show the plots for SPI-6 series and lower panels show the plots for SPI-12 series. The base width of the multifractal spectra is given in the respective figures

The multifractal spectra of different SPI series of WR are presented in Fig. 9.5. The parabolic shaped curves in Fig. 9.5 indicate the multifractality of SPI time series. On examining the plots, it is noted that the spectrum of SPI-6 and SPI-12 is near symmetrical. This infers that the long-term droughts are more stable or persistent at WR subdivision. The long right limb of the spectrum indicates that time series is not sensitive to local high amplitude fluctuations. On the other hand, the long left limb indicates that the series is not susceptible to local small amplitude fluctuations (Ihlen 2012).

The ranges of base width of the spectrum ($\alpha_{max}$–$\alpha_{min}$) of SPI-3 series are 0.34–0.469 (available in different panels). The range is 0.7–0.92 for the SPI-12 and 0.42–0.63 for SPI-6 time series. On considering a particular subdivision, the multifractal strength is found to be more for SPI with larger accumulation time scale. For WR subdivision, the ranges are 0.469, 0.634 and 0.915, respectively, for SPI series with 3, 6 and 12 month scales. On considering SPI series of a specific time scale, the dataset of WR subdivision possess the highest multifractal degree. The average annual rainfall based on dataset of 1871–2016 period is ~29.9, 47.8, 82.8 and 282 cm for WR, SK, MW and Kerala. This also infers that the climatic conditions of these four subdivisions are distinctly different. In short, despite the difference in climatology, the degree of multifractality increases with accumulation scale.

This paper performed the multifractal analysis of drought properties of meteo-rological subdivisions of western India. The understanding of long- or short-range correlations (persistence) of SPI provides better insights into the prediction selection in time series modeling of droughts in India. The signatures of multifractality and the derived parameters may help in developing multifractal models for drought predictions and subsequent risk assessment studies.

## 9.5   Conclusions

MFDFA of SPI series with different aggregated time scales of four different subdivisions Western Rajasthan, Saurashtra and Kutch (SK), Marathwada and Kerala at different temporal scales is performed in this study. The nonlinear $q$-dependency of generalized Hurst exponent ($h(q)$) showed multifractal behavior in all of the time series and the value of Hurst exponent (H) indicated the long-term persistence of different SPI time series. For all the subdivisions, the multifractal degree of SPI-12 series is found to be the highest followed by that for SPI-6 and SPI-3. This concludes that multifractal degree increases with aggregation time scale. Among the SPI time series of different subdivisions, the series of WR displayed the highest strength of multifractality. The present study gives a foundation for modeling and risk assessment of droughts in India in multifractal framework.

## References

Bazrafshan J, Hejabi S, Rahimi J (2013) Drought monitoring using the Multivariate Standardized Precipitation Index (MSPI). Water Resour Manag 28(4):1045–1060

Bassingthwaighte JB, Raymond GM (1995) Evaluation of the dispesional analysis method for fractal time series. Ann Biomed Eng 23:491–505

Ganguli P, Reddy MJ (2014) Evaluation of trends and multivariate frequency analysis of droughts in three meteorological subdivisions of Western India. Int J Climatol 34(3):911–928

Hurst HE (1965) Long-term storage: an experimental study. Constable, London

Hurst HE (1951) Long-term storage capacity of reservoirs. Trans Am Soc Civil Eng 116:770–808

Hajian S, Movahed MS (2010) Multifractal detrended cross-correlation analysis of sunspot numbers and river flow fluctuations. Phys A 389(21):4942–4957

Ihlen EAF (2012) Introduction to multifractal detrended fluctuation analysis in MATLAB. Front Physiol 3:141. https://doi.org/10.3389/fphys.2012.00141

Joshi N, Gupta D, Suryavanshi S, Adamowski J, Madramootoo CA (2016) Analysis of trends and dominant periodicities in drought variables in India: a wavelet transform based approach. Atmos Res 182:200–220

Kantelhardt JW, Zschiegner SA, Koscielny-Bunde E, Halvin H, Bunde A, Stanley HE (2002) Multifractal detrended fluctuation analysis of non-stationary time series. Phys A 316:87–114

Kantelhardt JW, Rybski D, Zschiegner SA, Braun P, Koscielny-Bunde E, Livina V, Havlin S, Bunde A (2003) Multifractality of river runoff and precipitation: comparison of fluctuation analysis and wavelet methods. Phys A 330:240–245

Kantelhardt JW, Koscielny-Bunde E, Rybski D, Braun P, Bunde A, Havlin S (2006) Long-term persistence and multifractality of precipitation and river runoff records. J Geophys Res: Atmos 111(D1). https://doi.org/10.1029/2005JD005881

Koscielny-Bunde E, Kantelhardt JW, Braun P, Bunde A, Havlin S (2003) Long-term persistence and multifractality of river runoff records: detrended fluctuation studies. J Hydrol 322:120–137

Mandelbrot BB (1974) Intermittent turbulence in self-similar cascades: divergence of high moments and dimension of the carrier. J Fluid Mech 62:331–358

McKee TB, Doesken NJ, Kleist J (1993) The relationship of drought frequency and duration to time scales. In: Preprints 8th conference on applied climatology, vol 17–22, pp 179–184

Pai DS, Sridhar L, Guhathakurta P, Hatwar HR (2011) District wise drought climatology of the Southwest monsoon season over India based on standardized precipitation index. Nat Hazards 59:1797–1813

Palmer WC (1965) Meteorological drought. Research paper, 45. U.S. Weather Bureau, Washington, DC, p. 58

Pandey G, Lovejoy S, Schertzer D (1998) Multifractal analysis of daily river flows including extremes for basins five to two million square kilometers, one day to 75 years. J Hydrol 208:62–81

Peng CK, Buldyrev SV, Simons M, Stanley HE, Goldberger AL (1994) Mosaic organization of DNA nucleotides. Phys Rev E 49:1685–1689

Tessier Y, Lovejoy S, Hubert P, Schertzer D, Pecknold S (1996) Multifractal analysis and modeling of rainfall and river flows and scaling, causal transfer functions. J Geophys Res 101:26427–26440

Thomas J, Prasannakumar V (2016) Temporal analysis of rainfall (1871–2012) and drought characteristics over a tropical monsoon-dominated State (Kerala) of India. J Hydrol 534:266–280

Thomas T, Nayak PC, Ghosh NC (2015) Spatio-temporal analysis of drought characteristics in the Bundelkhand region of Central India using the standardized precipitation index. J Hydrol Eng. http://dx.doi.org/10.1061/(ASCE)HE.1943-5584.0001189

Van-Rooy MP (1965) A rainfall anomaly index (RAI) independent of time and space. Notos 14:43–48

Vicente-Serrano SM, Begueria S, Lopez-Moreno JI, Angulo M, El Kenawy A (2010) A new global 0.5 degrees gridded dataset (1901–2006) of a multiscalar drought index: comparison with current drought index datasets based on the palmer drought severity index. J Hydrometeorol 11:1033–1043

Wei H, Feng G, Yan P, Li S (2018) Multifractal analysis of the drought area in seven large regions of China from 1961 to 2012. Meteorol Atmos Phys 130:459–471

Wei H, Yan P-C, Li S-P, Tu G, Hu J-G (2015) Application of long-range correlation and multi-fractal analysis for the depiction of drought risk. Chin Phys B 25(1)

Yu ZG, Leung Y, Chen YD, Zhang Q, Anh V, Zhou Y (2014) Multifractal analyses of daily rainfall time series in Pearl River basin of China. Phys A 405:193–202

Zargar A, Sadiq R, Naser B, Khan FI (2011) A review of drought indices. Environ Rev 19:333–349

Zhang Q, Chong YX, Yu ZG, Liu CL, Chen DYQ (2009) Multifractal analysis of streamflow records of the East River basin (Pearl River), China. Phys A 388:927–934

Zhang Q, Lu W, Chen S, Liang X (2016) Using multifractal and wavelet analyses to determine drought characteristics: a case study of Jilin province, China. Theor Appl Climatol 125:829–840

Zhang Q, Xu CY, Chen DYQ, Gemmer M, Yu ZG (2008) Multifractal detrended fluctuation analysis of streamflow series of the Yangtze River basin, China. Hydrol Process 22:4997–5003

# Chapter 10
# Impacts of Climatic Variability and Extremes on Agriculture and Water in Odisha Coasts

**Devipriya Paikaroy, Anil Kumar Kar, and Kabir Mohan Sethy**

## 10.1 Introduction

Climate change brings serious negative impacts on environment. The extreme events such as drought, heatwaves, floods, cyclones lead to both human and livelihood losses. Agriculture crops were badly impacted by the changes in rainfall patterns, sea-level rising and decrease in soil fertility, etc. All these extremes could have various impacts on both water and agriculture sectors which lead to migration and population displacement.

About 36% of the Odisha population (Pashupalak 2009) live in eight coastal districts and a high range of temporary and permanent migration is already a widespread phenomenon in these areas. The people of the coasts are basically poor and forced to migrate from their homes because of sudden natural calamity/disasters. The livelihoods losses due to natural calamities have extreme effects on agriculture, fisheries, availability of drinking water impacting health and life expectancy in the coastal areas is now a crucial issue for Odisha State.

Demographically due to subtropical littoral location of Odisha, the state is prone to frequent floods, tropical cyclones and droughts. The coastal plains are densely populated with the alluvial deposits of its river systems. The river banks have heavy

D. Paikaroy (✉)
Technical Expert, Research Scholar, Department of Geography, Utkal University, Bhubaneswar, India
e-mail: d.paikaroy@gmail.com

A. K. Kar
Department of Civil Engineering, VSSUT, Burla, Sambalpur, India
e-mail: anilkarhy@gmail.com

K. M. Sethy
Department of Geography, Utkal University, Bhubaneswar, India
e-mail: kabirmohan2006@yahoo.com

© The Editor(s) (if applicable) and The Author(s), under exclusive license to Springer Nature Switzerland AG 2021
A. Pandey et al. (eds.), *Hydrological Extremes*, Water Science and Technology Library 97, https://doi.org/10.1007/978-3-030-59148-9_10

**Table 10.1** Natural disasters in Odisha since 1999

| Sl. No | Name of the event | Year | Fatalities |
|--------|-------------------|------|------------|
| 1 | Cyclone: Hud Hud | September, 2014 | Nil |
| 2 | Floods in Odisha | October, 2013 | 21 |
| 3 | Cyclone: Phailin | October, 2013 | 23 |
| 4 | Floods in Odisha | September, 2011 | 45 |
| 5 | Drought | 2009 | Nil |
| 6 | Super Cyclone | 1999 | 10,000 |

*Source* Disaster Data & Statistics, National Disaster Management Authority, Govt. of India

load of silt with lesser carrying capacity resulting in frequent floods in the state. A list of climate induces disasters since 1999 in the state is given below (Table 10.1).

An assessment study has been carried out by the (Lenesco and Traore 2015) Institute of Ocean Management, Anna University in the year 2011, which indicated that about more than a third of Odisha's coastline is prone to erosion, and 8% is vulnerable to severe erosion. The availability of freshwater gradually decreases as the salinity of water increases due to the incursion of salty water from the sea. The cyclone-associated floods increase the salinity of water, which leads to water-borne diseases, cholera epidemics, etc. The state has experiencing frequent flood and drought in many coastal areas and significantly the people were suffering a lot due to loss of their life and property and forced to migrate.

The main objective of the study would be to carry out national/state assessments of natural disaster risks and systematically account for disaster losses and impacts mainly on Water and Agriculture sectors of the impacted areas. The study also focuses on to improve understanding of investment needs and policy framework required for disaster risk management in the state and reducing the human migration.

The study needs to identify social protection and income-earning opportunities like skills training, alternative livelihood programs and community-driven development initiatives for the people affected by the climate change extremes and new adaptation strategies to control human life and property.

## 10.2   Rationale of the Study

Most of the people of the Odisha coats depend on Agriculture, Forestry and Fisheries which were climate sensitive. The states have been experiencing frequent natural calamities for more than a decade which resultantly affect the crop production, water availability and health issues.

Due to extreme climate events, there is high risk in food security as well as decrease local livelihood options in the coastal zones of Odisha. Most of the eligible farmer community either prefer or rather forced to migrate as laborers to neighboring states.

The vulnerability and socioeconomic risks were encountered in the Agriculture sector are illustrated below (Table 10.2).

The cyclone and its associated flood incidence have severe damage to agriculture production. (Dubash 2014) Odisha has about 1517.67 km of saline embankments covering eight districts. These embankments are the lifelines of the people living in the villages along the coast (Fig. 10.1).

A study conducted by Deltas, Vulnerability and Climate Change: Migration and Adaptation (DECCMA) in the year 2015, to map the climate change risks on the

**Table 10.2** Vulnerability and socioeconomic risks in agriculture sector

| Sector | Vulnerability | Socioeconomic risks |
|---|---|---|
| Agriculture | • Soil erosion and salinity<br>• Soil wetness and damage<br>• Drought condition<br>• Erratic precipitation<br>• Stagnation of water<br>• Reduced soil moisture | • Increase risk of land degradation<br>• Decrease in crop yields<br>• Food insecurity<br>• Malnutrition and other health-related diseases |

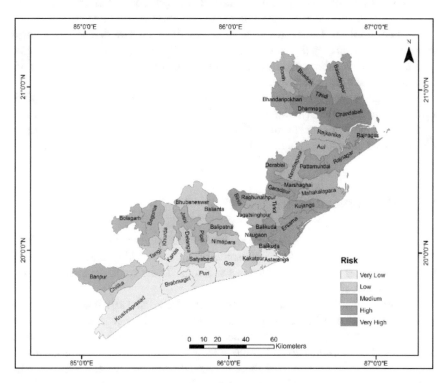

**Fig. 10.1** Block level climate change risk mapping of coastal zone of Odisha. *Source* Chilika Development Authority, Govt. of Odisha, Annual Report, 2015

**Table 10.3** Annual cropping
intensity of Odisha from 2011
to 2016

| Year | Net area sown (in 000 hect.) | Gross cropped area (in 000 hect.) | Cropping intensity (%) |
|---|---|---|---|
| 2011–12 | 5,292 | 8,928 | 157 |
| 2012–13 | 5,331 | 8,799 | 166 |
| 2013–14 | 5,422 | 9,054 | 167 |
| 2014–15 | 5,496 | 9,011 | 154 |
| 2015–16 | 5,608 | 8,180 | 146 |

*Source* Odisha Economic Survey Report, 2016–2017, p. 92

agriculture sector and clearly indicates that there would be huge agricultural losses every year.

The major impacts on agriculture sector are reducing yields of crops due to both irregular rainfall and drought in many regions of Odisha. It is also projected by Odisha University of Agriculture & Technology that the grain yield of rice will be decreased by 9% (Kumar and Nayak 2014) by 2020 due to change in variable weather patterns in the state. As per the latest Odisha Economic Survey report, the cropping intensity of the state gradually decreasing compared to previous trends (Table 10.3).

Frequent floods in Odisha resulting in soil damage, erosion due to soil wetness and water logging in crop fields substantially increase the crop losses. Due to extreme climate-induced weather conditions and warm nights during postflowering period, the quality of rice grain decreases.

Agriculture sector of the coastal areas severely affected by high speed wind, torrential rain and extensive flooding. The high tides may bring in saline water and sand mass to the fields makes unsuitable for agriculture. The crops are damaged by the stagnation of water inside the fields.

## 10.3 Recommendations

The natural disasters cannot be stopped, but we can improve the level of preparedness to tackle these issues. There are some best practices illustrated below, which can provide better resilient to cyclones and floods. These could potentially help in reducing the extent of crop damage.

Sub Surface Water Harvesting Structures (SSWHS) (Patel 2016) were developed by the Directorate of Water Management is highly useful for coastal cyclone prone and waterlogged areas. In this system, the freshwater floats above the saline water of ground could be tapped easily. It has the potential to enhance the water productivity of Ravi vegetable crops. Including these over-aged rice seedlings, techniques can be useful for the framers to coping with cropping losses due to invariable rainfall and floods.

Few flood-resistant aquatic crops (Odisha Climate Change Action Plan Report 2010) need to be cultivated in the agriculture farms, which is having high economic importance along with resistant to floods. Contingency crop planning in postcyclone and flood period is the need of the hour. (Gulati 2017) Modification of land and using sunken bed technique in low land regions would be highly effective in the utilization of available water, higher crop productivity, increase in cropping intensity and higher economic returns.

Better land management practices can be done by contour ploughing, contour planting, terracing, close spacing crops, soil conservation in sloppy lands to minimize soil erosion. Plantation of Casuarina and Eucalyptus (State Disaster Management Plan 2016) species in the flood-prone area would act as a viable flood resilient system. These plants potentially improve the soil drainage and operate better microclimate.

The farmers need to be trained to save their crops some instant from climate-induced disasters. The farmers would require initial orientation for seedling production and improved nursery management technologies. (Roy and Selvaranjan 2016) They need hands-on training on soil management, grafting, sowing, fertilization, raised beds, pest and disease management activities thereby to increasing their crop productivity.

## 10.4  Conclusion

The frequent occurrence of cyclones and floods in Odisha significantly reduces agricultural productivity, which affects the socioeconomic prospective of farmers and leads to food insecurity. These risks cannot be minimized successfully through application of isolated management measures. (Jena 2017) We have to implement the potential agricultural techniques and structural measures to enhance the accuracy of cyclone preparedness in coastal areas of the state.

Immediate attention must be paid to contingency crop planning and integrated cyclone and flood management strategies for reducing the extent of damage. The agricultural livelihoods of the coastal zones can be restored (Patnailk 2016) with suitable technological interventions compatible with the farming system.

## References

Dubash NK (2014) The centre for policy research. Research Gate Publications, 78

Gulati A (2017) Transforming agriculture in Odisha. Indian environmental portal, 34–35 & 65–67

Jena PP (2017) Impacts of climate change in Odisha, economic development–a comparative analysis. Coastline Predict Mahanadi River IOSR-JHSS 22:21–25

Kumar A, Nayak AK (2014) Management of cyclone disaster in agriculture sector in coastal areas. IIWM course book, 8–10 & 106–108

Lenesco D, Traore M (2015) Environment, IOM, The UN migration agency

Odisha Climate Change Action Plan Report (2010) 6.11: Water resources sector, 72

Pashupalak S (2009) Climate change and agriculture in Odisha. Orissa Rev 49–52

Patel SK (2016) Climate change and climate-induced disasters in Odisha, eastern India: impacts. Adapt Futur Policy Implic, Int J HumIties Soc Sci Inven 5:60–63

Patnailk BK (2016) Special issue on agriculture & farmer's welfare. SAMIKSHYA (DE&S J Socio-Econ Issue) 10, 1014–1015

Roy BC, Selvaranjan S (2016) Vulnerable to climate induced natural disaster. Draft for circulation 14–16

State Disaster Management Plan (2016) OSDMA, Government of Odisha 131–135

# Chapter 11
# Temporal and Spatial Variability of Daily Rainfall Extremes in Humid Northeast Assam State of India

**Maisnam Luwangleima and Salil K. Shrivastava**

## 11.1 Introduction

Climate change is one of the greatest environmental concerns facing mankind in the twenty-first century. It is likely to produce more extreme precipitation events (Allan and Soden 2008). One of the most substantial consequences of global warming would be an increase in the frequency and magnitude of extreme precipitation events carried by increased atmospheric moisture levels, large-scale storm activity and/or thunderstorm activity. According to the climatic model simulations made by Hennessey et al. (1997) and experimental evidences showed that warmer climates are responsible for increase in water vapor, which may result in more extreme precipitation events and consequently increased risks of floods (IPCC 2007).

Extreme rainfall leads to flash floods, landslides and crop damage that have major impacts on society, the economy and the environment. Although prediction of such extreme weather events is still fraught with uncertainties, a proper assessment of likely future trends would help in setting up infrastructure for disaster preparedness (Goswami et al. 2006; Swain et al. 2018). The most severe damages seem to result from extreme precipitation on a regional scale such as heavy precipitation and rainstorms events accounting for high percentages of the yearly total in a few rainy days are directly responsible for flood occurrence (Zhang et al. 2001).

Findings of extremes climate are also beneficial for the substantiation of climate model results and thus growing confidence in future climate projections (Kruger 2006; Shrestha et al. 2016). In different parts of the world, numerous studies (Haylock and Nicholls 2000; Manton et al. 2001; Salinger and Griffiths 2001; Rajeevan et al. 2008; Swain et al. 2017) have been made to identify trends in the climate extremes and

M. Luwangleima (✉) · S. K. Shrivastava
Department of Agricultural Engineering, North Eastern Regional Institute of Science and
Technology (Deemed to be University), Nirjuli, Itanagar 791109, Arunachal Pradesh, India
e-mail: luwangleimamaisnam@gmail.com

© The Editor(s) (if applicable) and The Author(s), under exclusive license
to Springer Nature Switzerland AG 2021
A. Pandey et al. (eds.), *Hydrological Extremes*, Water Science
and Technology Library 97, https://doi.org/10.1007/978-3-030-59148-9_11

149

increasing trends for extreme precipitation events were revealed. The most up-to-date and comprehensive global scenario of the observed trends in extremes in precipitation was presented (Alexander et al. 2006). They have reported a general increase in the heavy precipitation indices. However, compared with temperature variations, variations in extremes precipitation were found to be less spatially coherent and at a lower level of statistical significance. Their analysis showed that the major decreasing trends were found in the annual number of CDD for the Indian Region (Revadekar et al. 2011).

In India also many studies have been made of climate extremes. Bhaskaran et al. (1995) have observed a significant increase in moisture transport over India under doubled $CO_2$ environment that would lead to an increase in extreme rainfall events in the area. Soman et al. (1988) investigated annual extreme precipitation for stations in the state of Kerala, India and found declining trends, which were mainly observed in hilly terrain stations. Sen Roy and Balling (2004) studied extreme daily precipitation indices for 129 stations around India during the period 1910–2000 and observed that, 61 stations had a significantly decreasing trend and 114 stations had a signifi-cantly increasing trend from the 903 different time series (for seven variables); the percentage of the time series exhibited an increasing trend was found to be 61%. Most of the studies on extreme rainfall over India have used a limited number of stations.

The daily gridded rainfall data for the period 1951–2003 were used by the Goswami et al. (2006) to examine the trend of rainfall extreme over Central part of India and identified a rise in the magnitude and frequency of extreme rain events and significantly falling trend in the frequency of moderate events. Guhathakurta et al. (2011) have analyzed daily rainfall data for 1901–2005 period over 2599 stations spread all over the India and revealed an increase in the intensity of rainfall extreme over Orissa, Saurasutra and Kutch, West Bengal, coastal Andhra Pradesh and its adjoining areas, east Rajasthan and parts of northeast India in 1-day extreme rainfall. However, trends were significantly decreasing in rainy day, heavy rainfall days and rain days over Central and many portions of northern India.

For the preparation of disaster management and mitigation plan, it is a good practice to consider the variations in extreme weather events rather than the variations in mean (Guhathakurta et al. 2011). The use of climate extreme indices is now gaining popularity among researchers to investigate climate extremes. To facilitate research community, the Expert Team on Climate Change Detection, Monitoring and Indices (ETCCDMI) have developed set of climate extreme indices. Very few studies have been made of precipitation extremes over the northeastern part of India in spite of the fact that the region comes under high rainfall area.

In the present study, the districts of Assam in northeast region of India have been selected to investigate temporal and spatial variability and trends in ETCCDMI developed climate extreme indices.

## 11.2 Study Area and Data Used

### 11.2.1 Study Area

Assam in Northeast India covers an area of 78,438 km$^2$. The average annual rainfall of Assam varies between 1500 and 3750 mm and the temperature ranges from 35 to 38 °C in summer and 6 to 8 °C in winter. With the "Tropical Monsoon Rainforest Climate," the state experiences heavy rainfall and high humidity. During monsoon season, flood is a common feature in Assam. The details of the stations used are given in Table 11.1 and their locations are shown in Fig. 11.1.

**Table 11.1** Details of the districts and data period

| District | Latitude °N | Longitude °E | Elevation (m) | Data period |
|---|---|---|---|---|
| Tinsukia | 27.598 | 95.727 | 149.1 | 1901–2013 |
| Dhemaji | 27.624 | 94.912 | 108 | 1901–2013 |
| Dibrugarh | 27.393 | 95.156 | 111 | 1901–2013 |
| Lakhimpur | 27.218 | 94.250 | 87.3 | 1901–2013 |
| Sivasagar | 27.016 | 94.890 | 100.7 | 1901–2013 |
| Jorhat | 26.815 | 94.327 | 100.9 | 1901–2013 |
| Sonitpur | 26.860 | 93.003 | 92 | 1901–2013 |
| Kokrajhar | 26.668 | 90.220 | 111.9 | 1901–2013 |
| Udalguri | 26.816 | 92.151 | 133 | 1901–2013 |
| Chirang | 26.726 | 90.722 | 203.7 | 1901–2013 |
| Baksa | 26.738 | 91.417 | 112.2 | 1901–2013 |
| Golaghat | 26.485 | 93.910 | 95.6 | 1901–2013 |
| Nagaon | 26.303 | 92.923 | 390.3 | 1901–2013 |
| Darrang | 26.515 | 92.114 | 65.0 | 1901–2013 |
| Karbi Anglong | 26.244 | 93.494 | 839.8 | 1901–2013 |
| Barpeta | 26.375 | 91.063 | 41.0 | 1901–2013 |
| Nalbari | 26.421 | 91.459 | 51.3 | 1901–2013 |
| Bongaigaon | 26.428 | 90.646 | 207.2 | 1901–2013 |
| Dhubri | 26.139 | 90.053 | 46.4 | 1901–2013 |
| Morigaon | 26.345 | 92.364 | 57.7 | 1901–2013 |
| Kamrup | 26.197 | 91.653 | 159.4 | 1901–2013 |
| Goalpara | 26.142 | 90.698 | 41.9 | 1901–2013 |
| Cachar | 24.872 | 92.907 | 44.3 | 1901–2013 |
| Karimganj | 24.643 | 92.417 | 15.0 | 1901–2013 |
| Hailakandi | 24.545 | 92.635 | 31.0 | 1901–2013 |

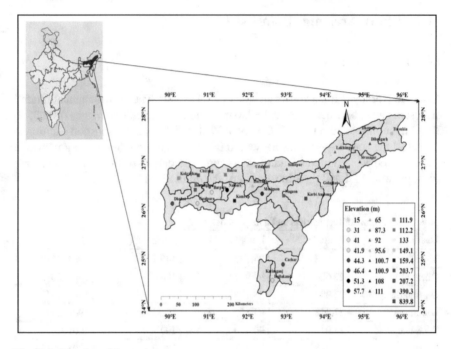

**Fig. 11.1** Location of the study area

## 11.2.2 Data Used

The gridded daily rainfall data for the period 1901–2013 for 25 districts of Assam were extracted from the $0.25° \times 0.25°$ India Meteorological Department (IMD), Pune grid dataset. The data were processed using the software, RClimDex, which executes data quality control, RH test, which performs homogeneity tests and calculates indices.

## 11.3 Methodology

The RClimdex1.0 software (available at http://cccma.seos.uvic.ca/ETCCDI/) was used in this study to obtain the extremes climatic indices (Zhang and Yang 2004). Five out of 11 extreme precipitation indices defined by ETCCDI were used in this present study to investigate the temporal and spatial variability of daily rainfall extremes in each district of Assam of humid northeast India. The slope of the trend in climate indices was obtained using Sen's slope estimator whereas the significance of upward (downward) trend of all the indices used was established using Mann-Kendall trend test (Jain et al. 2012). The details of indices used in the present study are given in Table 11.2.

**Table 11.2** Extreme precipitation indices with their definitions and units

| ID | Indicator name | Definition | Unit |
|---|---|---|---|
| CDD | Consecutive dry days | Maximum number of consecutive days with PRCP < 1 mm | days |
| CWD | Consecutive wet days | Maximum number of consecutive days with PRCP ≥ 1 mm | days |
| PRCPTOT | Annual total wet-day precipitation | Annual total PRCP in wet days (PRCP ≥ 1 mm) | mm |
| RX1day | Maximum 1-day precipitation | Annual Maximum 1-day precipitation | mm |
| R20 | Number of very heavy precipitation days | Annual count of days when rainfall ≥20 mm | days |

## 11.3.1 Magnitude of Trend

The magnitude of trend is tested using Sen's slope estimator (Jain et al. 2012). This test computes the slope (i.e., linear rate of change) in the time series. First, a set of linear slopes is calculated as follows:

$$T_i = \frac{x_j - x_k}{j - k} \tag{11.1}$$

for ($1 \leq j < k \leq N$), where $T_i$ is the slope, $x$ denotes the variable, $N$ is the number of data, and $j, k$ are indices.

Sen's slope is then calculated as follows:

$$\beta = \begin{cases} T_{\frac{N+1}{2}} & N \text{ is odd} \\ \frac{1}{2}\left(T_{\frac{N}{2}} + T_{\frac{N+2}{2}}\right) & N \text{ is even} \end{cases} \tag{11.2}$$

A positive value of $\beta$ indicates an upward (increasing) trend and a negative value indicates a downward (decreasing trend) in the time series.

## 11.3.2 Mann-Kendall Trend Test

The Mann-Kendall trend test is a nonparametric test most commonly employed to detect the presence of statistically significant trends in series of environmental data, climate data or hydrological data. In the present analysis of trend, the Mann-Kendall trend test (Jain et al. 2012) is used with the null hypothesis, $H_0$, that there is no trend in the data. The alternative hypothesis, $H_A$, is that the data follow a monotonic trend. The Mann-Kendall trend test statistics is calculated as follows:

$$S = \sum_{i=1}^{N-1} \sum_{j=i+1}^{N} \text{sgn}(x_j - x_i) \tag{11.3}$$

where $N$ is the number of data points. Assuming $(x_j - x_i) = \theta$, the value of $\text{sgn}(\theta)$ is computed as follows:

$$\text{sgn}(\theta) = \begin{cases} 1 & \text{if } \theta > 0 \\ 0 & \text{if } \theta = 0 \\ -1 & \text{if } \theta < 0 \end{cases} \tag{11.4}$$

For large samples ($N > 10$), the test is conducted using a normal distribution ($Z$ statistics) with the mean $S$ is $E[S] = 0$ and the variance is

$$\text{Var}(S) = \frac{N(N-1)(2N+5) - \sum_{k=1}^{n} t_k(t_k - 1)(2t_k + 5)}{18} \tag{11.5}$$

where $n$ is the number of the tied groups in the data set and $t_k$ is the number of data points in the $k$th tied group. The statistic $S$ is approximately normally distributed provided that the following $Z$-transformation is employed:

$$Z = \begin{cases} \frac{S-1}{\sqrt{\text{Var}(S)}} & \text{if } S > 0 \\ 0 & \text{if } S = 0 \\ \frac{S+1}{\sqrt{\text{Var}(S)}} & \text{if } S < 0 \end{cases} \tag{11.6}$$

If the computed value of $|Z| > z_{\alpha/2}$, the null hypothesis ($H_0$) is rejected at $\alpha$ level of significance in a two-sided test. In the analysis, the statistically significant trends were tested at 90% confidence level.

## 11.4 Results and Discussion

### 11.4.1 Analysis of Extreme Precipitation Indices for the Entire Assam

Five out of 11 extreme precipitation indices recommended by ETCCDI are used. The annual variation in the CDD, CWD, PRCPTOT, RX1day and R20 over Assam during 1902–2013 is shown in Fig. 11.2 whereas Table 11.3 is presented with maximum values of all the considered indices for different districts of Assam. It is clear from Fig. 11.2a that there is an insignificant increase in the values of CDD. Most of the values fluctuate around 40 days. In contradiction to CDD, a decreasing pattern is observed in CWD with a lowest value of 15 days in the year 2006 (Fig. 11.2b). The effect of decreasing pattern is seen, the PRCPTOT, which is also found to

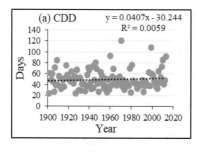

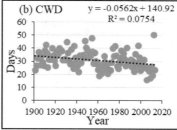

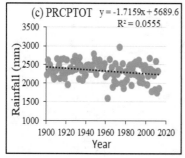

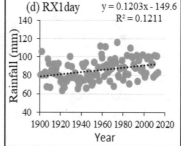

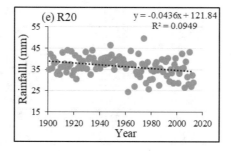

**Fig. 11.2** Annual variation in daily extreme precipitation indices **a** CDD, **b** CWD, **c** PRCPTOT, **d** RX1day and **e** R20 during 1901–2013 over Assam

be decreasing (Fig. 11.2c). The highest maximum 1-day precipitation, RX1day is detected in the year 1974 (116 mm) and the lowest in the year 1922 (Fig. 11.2d). On average, an increasing rate of 1.2 mm/decade during the study period was observed. The extreme heavy precipitation, R20 on average, shows a decrease in the value and it is observed to be highest in the year 1974 (50 days) and lowest in 1962 during the study period.

**Table 11.3** Maximum values of all the extreme precipitation indices

| Districts | CDD (days) | CWD (days) | PRCPTOT (mm) | RX1day (mm) | R20 (days) |
|---|---|---|---|---|---|
| Tinsukia | 81 | 69 | 5069.8 | 219.30 | 81 |
| Dhemaji | 81 | 73 | 4216.7 | 167.48 | 81 |
| Dibrugarh | 77 | 79 | 3874.8 | 148.08 | 75 |
| Lakhimpur | 104 | 74 | 3513.4 | 154.55 | 66 |
| Sivasagar | 138 | 69 | 3038.8 | 156.53 | 59 |
| Jorhat | 149 | 99 | 3101.4 | 136.00 | 54 |
| Sonitpur | 121 | 83 | 2979.4 | 125.42 | 57 |
| Kokrajhar | 153 | 60 | 4650 | 243.42 | 78 |
| Udalguri | 121 | 69 | 3073.6 | 374.15 | 51 |
| Chirang | 149 | 64 | 4337.6 | 339.40 | 70 |
| Baksa | 140 | 86 | 4425.3 | 168.08 | 78 |
| Golaghat | 140 | 82 | 2583.4 | 136.39 | 47 |
| Nagaon | 116 | 76 | 2334.7 | 117.54 | 39 |
| Darrang | 124 | 73 | 2698.9 | 302.15 | 47 |
| Karbi Anglong | 122 | 100 | 3984.1 | 116.98 | 67 |
| Barpeta | 176 | 57 | 3380.7 | 337.40 | 52 |
| Nalbari | 166 | 91 | 2879.3 | 175.50 | 51 |
| Bongaigaon | 148 | 71 | 4031.7 | 178.65 | 65 |
| Dhubri | 187 | 80 | 4039.7 | 280.88 | 69 |
| Morigaon | 144 | 82 | 3550.6 | 188.55 | 49 |
| Kamrup | 144 | 83 | 3300.6 | 165.23 | 50 |
| Goalpara | 148 | 92 | 3499 | 222.40 | 62 |
| Cachar | 133 | 95 | 4866.1 | 192.24 | 86 |
| Karimganj | 133 | 103 | 4113.1 | 224.05 | 73 |
| Hailakandi | 133 | 109 | 3732.9 | 267.30 | 70 |

## 11.4.2 District-Wise Analysis of Extreme Precipitation Indices

It is evident from Table 11.3 that the CDD had a variation from 77 days (minimum at Dibrugarh) to 187 days (maximum at Dhubri). Except Tinsukia, Dhemaji and Dibrugarh located at the north of Assam, the values of CDD are found to be more than 100 days indicating better moisture availability as compared with other districts. In case of CWD, maximum numbers of 109 days were observed in Hailakandi followed by 103 days in Karimganj at the south Assam exhibiting receiving of more precipitation. It is of significance to mention (from Table 11.3) that maximum amount of annual total precipitation was obtained in Tinsukia, which also supports the finding of less number of CDD. The RX1day that is important for designing of structures was 374.15 mm, the highest, observed in Udalguri indicating proper planning for control of runoff which may be generated through the 1 day maximum amount. As seen in that table, Cachar showed maximum number of days (86 days) for R20 as against 39 days for Nagaon (middle Assam).

## 11.4.3   Trend Analysis of Extreme Precipitation Indices

Table 11.4 shows the Mann-Kendall trend test results of the indices of extreme precipitation used in the present analysis. The significance of trend is tested at 90%

**Table 11.4**   Mann-Kendall trend test result of extreme precipitation indices

| District | CDD | CWD | PRCPTOT | RX1day | R20 |
|---|---|---|---|---|---|
| Tinsukia | 1.179 | −1.211 | **−2.484** | −0.109 | **−3.196** |
| Dhemaji | 1.489 | −1.303 | **−3.737** | −1.387 | **−3.211** |
| Dibrugarh | **2.412** | **−1.943** | **−3.382** | **2.427** | **−3.112** |
| Lakhimpur | 0.529 | 0.037 | **−3.330** | 0.045 | **−3.449** |
| Sivasagar | 0.851 | 1.546 | **−4.476** | **−2.536** | **−4.799** |
| Jorhat | **1.943** | −1.174 | **−4.528** | **−2.747** | **−3.866** |
| Sonitpur | 0.362 | −0.859 | **−2.548** | −0.839 | −1.429 |
| Kokrajhar | −1.598 | **−2.387** | −1.298 | 0.340 | **−2.181** |
| Udalguri | **1.963** | **−2.710** | **−2.094** | **3.677** | −1.189 |
| Chirang | **−1.714** | **−3.255** | **2.809** | **4.853** | 1.439 |
| Baksa | 1.642 | 0.266 | −0.300 | 0.933 | −0.017 |
| Golaghat | 0.325 | 1.625 | **−3.908** | **−3.084** | **−4.501** |
| Nagaon | 1.082 | **−3.151** | **−6.424** | **−1.978** | **−4.749** |
| Darrang | 0.466 | −1.012 | −0.266 | **3.030** | 0.459 |
| Karbi Anglong | 0.734 | **−2.558** | **−4.360** | 0.119 | **−2.402** |
| Barpeta | −0.462 | **−1.992** | −0.739 | **4.109** | −0.412 |
| Nalbari | 0.568 | **−4.612** | −0.891 | **2.769** | 0.159 |
| Bongaigaon | −1.104 | −1.186 | **2.042** | **4.072** | 0.859 |
| Dhubri | −1.203 | **−3.270** | −0.375 | 0.199 | −0.499 |
| Morigaon | 0.638 | **−2.913** | **−3.796** | −0.447 | **−2.308** |
| Kamrup | 0.506 | −1.571 | −0.960 | **3.119** | 0.829 |
| Goalpara | **−1.677** | −0.859 | **1.695** | **3.392** | 1.332 |
| Cachar | 1.030 | **−5.724** | −1.620 | **2.317** | **−1.881** |
| Karimganj | 1.112 | **−5.987** | **−1.861** | **2.762** | **−1.697** |
| Hailakandi | −0.104 | **−5.583** | **−2.662** | 1.506 | **−2.496** |

Values in boldface indicate significant at 90% confidence level

confidence level and the spatial variations of the trend of all the extreme precipitation indices used over Assam during the period 1901–2013 is depicted in Fig. 11.3a–e.

From the results of the trend analysis (Table 11.4), it can be observed that CDD is increasing in 18 districts and decreasing in 7 districts with significantly increasing in Dibrugarh, Jorhat and Udalguri districts while significantly decreasing in Chirang and Goalpara. The increase in the CDD may lead to drought-like situation in future over upper and central part of Assam (Fig. 11.3a). CWD is found to be increasing in 4 districts and decreasing in 21 districts, respectively, with significantly decreasing in 13 districts, i.e., Dibrugarh, Kokrajhar, Udalguri, Chirang, Nagaon, Karbi Anglong, Barpeta, Nalbari, Dhubri, Morigaon, Cachar, Karimganj and Hailakandi. Decrease in CWD is found to be more toward west of Assam (Fig. 11.3b). This finding of decrease in CWD also supports the reduction in the precipitation over Assam. A significant

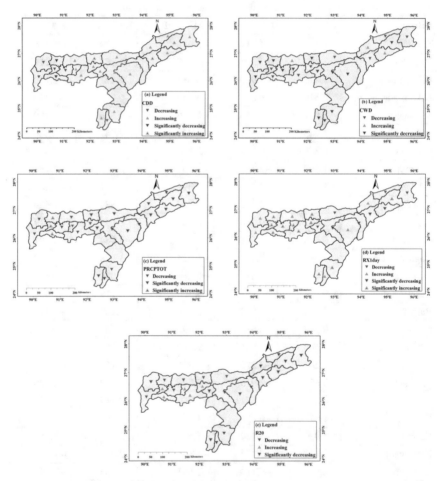

**Fig. 11.3** Spatial variations of the trend in extreme precipitation indices **a** CDD, **b** CWD, **c** PRCPTOT, **d** RX1day and **e** R20 over Assam during the period 1901–2013

upward trend is revealed for PRCPTOT in three districts, i.e., Chirang, Bongaigaon and Goalpara located at the west of Assam indicating better water availability over this part of Assam. The remaining districts showed downward trend of which 14 are significant and uniformly scattered over the entire Assam (Fig. 11.3c). It is evident from Table 11.4 and Fig. 11.3d that RX1day is increasing in 17 districts of which 11 are significant, i.e., Dibrugarh, Udalguri, Chirang, Darrang, Barpeta, Nalbari, Bongaigaon, Kamrup, Goalpara, Cachar and Karimganj while decreasing trend is found in 8 districts with significantly decreasing in Sivasagar, Jorhat, Golaghat and Nagaon. The R20 index is found to be increasing in 6 districts with no significantly increasing trend while decreasing in 19 districts with significantly decreasing in 14 districts namely Tinsukia, Dhemaji, Dibrugarh, Lakhimpur, Sivasagar, Jorhat, Kokrajhar, Golaghat, Nagaon, Karbi Anglong, Morigaon, Cachar, Karimganj and Hailakandi mainly cover the upper and lower Assam (Fig. 11.3e) during the study period, respectively.

## 11.5 Conclusions

The present study analyses the temporal and spatial variability of daily extreme precipitation indices in Assam of humid northeast India for the period 1901–2013. Five out of 11 extreme precipitation indices as recommended by ETCCDI were used for the analysis. The Mann-Kendall trend test was used to detect trend in the indices. The significance of upward (downward) trend is tested at 90% confidence level. From the analysis, the following inferences can be drawn:

1. The consecutive dry days (CDD), consecutive wet days (CWD), annual total wet days precipitation (PRCPTOT), maximum 1-day precipitation (RX1day) and extreme heavy precipitation (R20) were observed to be highest in Dhubri, Hailakandi, Tinsukia, Udalguri and Cachar districts, respectively.
2. Indicated a condition leading to dryness over the area as CDD is increasing, CWD, PRCPTOT and R20 are decreasing.
3. However, RX1day is increasing suggesting for proper conservation of runoff water by suitably designing water conservation structures to encounter the situation which may be faced by the area due to decrease in CWD, PRCPTOT and R20.

## References

Alexander LV, Zhang X, Peterson TC, Caesar J, Gleason B, Klein Tank AMG, Haylock M, Collins D, Trewin B, Rahimzadeh F, Tagipour A, Rupa Kumar K, Revadekar J, Griffiths G, Vincent L, Stephenson DB, Burn J, Aguilar E, Brunet M, Taylor M, New M, Zhai P, Rusticucci M, Vazquez-Aguirre JL (2006) Global observed changes in daily climate extremes of temperature and precipitation. J Geophys Res 111:D05109. https://doi.org/10.1029/2005JD006290

Allan RP, Soden BJ (2008) Atmospheric warming and the amplication of precipitation extremes. Science 321:1481–1483

Bhaskaran B, Mitchell JFB, Lavery JR, Lal M (1995) Climatic response of the Indian subcontinent to doubled CO2 concentrations. Int J Climatol 15:873–892

Goswami BN, Venugopal V, Sengupta D, Madhusoodanan MS, Xavier PK (2006) Increasing trend of extreme rain events over India in a warming environment. Science 314:1442–1445

Guhathakurta P, Sreejith OP, Menom PA (2011) Impact of climate change on extreme rainfall events and flood risk in India. J Earth Syst Sci 120(3):359–373

Haylock M, Nicholls N (2000) Trends in extreme rainfall indices for an updated high-quality data set for Australia, 1910–1998. Int J Climatol 20:1533–1541

Hennessey KJ, Gregory JM, Mitchell JFB (1997) Changes in daily precipitation under enhanced greenhouse conditions. Clim Dyn 13:667–680

IPCC (2007) Climate change 2007—the physical science basis. contribution of working group I to the fourth assessment report of the IPCC. Cambridge University Press, Cambridge, New York

Jain SK, Kumar V, Saharia M (2012) Analysis of rainfall and temperature trends in northeast India. Int J Climatol. https://doi.org/10.1002/joc.3483

Kruger AC (2006) Observed Trends in Daily Precipitation Indices in South Africa: 1910–2004. Int J Climatol 26:2275–2285

Manton MJ, Della-Marta PM, Haylock MR, Hennessy KJ, Nicholls N, Chambers LE, Collins DA, Daw G, Finet A, Gunawan D, Inape K, Isobe H, Kestin TS, Lefale P, Leyuk CH, Lwin T, Maitrepierre L, Ouprasitworng N, Page CM, Pahalad J, Plummer N, Salinger MJ, Suppiah R, Tran VL, Trewin B, Tibig I, Yee D (2001) Trends in extreme daily rainfall and temperature in Southeast Asia and the South Pacific: 1961–1998. Int J Climatol 21:269–284

Rajeevan M, Bhate J, Jaswal AK (2008) Analysis of variability and trends of extreme rainfall events over India using 104 years of gridded daily rainfall data. Geophys Res Lett 35:L18707

Revadekar JV, Patwardhan SK, Rupa Kumar K (2011) Characteristics features of precipitation extremes over India in the warming scenarios. Adv Meteorol 11. https://doi.org/10.1155/2011/138425

Salinger MJ, Griffiths GM (2001) Trends in New Zealand daily temperature and rainfall extremes. Int J Climatol 21:1437–1452

Sen Roy S, Balling RC Jr (2004) Trends in extreme daily precipitation indices in India. Int J Climatol 24:457–466

Shrestha AB, Bajracharya SR, Sharma AR, Duo C, Kulkarni A (2016) Observed trends and changes in daily temperature and precipitation extremes over the Koshi river basin 1975–2010. J Climatol, Int. https://doi.org/10.1002/joc.4761

Soman MK, Krishnakumar K, Singh N (1988) Decreasing trend in the rainfall of Kerala. Curr Sci 57:5–12

Swain S, Nandi S, Patel P (2018) Development of an ARIMA model for monthly rainfall forecasting over Khordha district, Odisha, India. In: Recent findings in intelligent computing techniques. Springer, Singapore, pp 325–331

Swain S, Patel P, Nandi S (2017) Application of SPI, EDI and PNPI using MSWEP precipitation data over Marathwada, India. In: 2017 IEEE international geoscience and remote sensing symposium (IGARSS). IEEE, pp 5505–5507

Zhang X, Yang F (2004) RClimDex (1.0) User manual. Climate Research Branch Environment Canada: Downsview, Ontario, Canada

Zhang X, Hogg WD, Mekis F (2001) Spatial and temporal characteristics of heavy precipitation events over Canada. J Clim 14:1923–1936

# Chapter 12
# Identification of Meteorological Extreme Years Over Central Division of Odisha Using an Index-Based Approach

**Sabyasachi Swain, S. K. Mishra, Ashish Pandey, and Deen Dayal**

## 12.1 Introduction

Climate change has become one of the most alarming issues for the present generation. The general impacts of climate change on hydrological cycle is well sought in hydrometeorology literature (Gosain et al. 2006; Himanshu et al. 2017; Palmate et al. 2017; Swain 2017; Ahani et al. 2018; Aadhar et al. 2019; Pandey and Palmate 2019; Veettil and Mishra 2020). The climate change leads to intensification of hydrological cycle and causes anomalous behavior of meteorological variables (Tsakiris 2014; Pandey et al. 2019). Rainfall is a key meteorological variable as its intensity, duration and frequency governs the extremes, that is, floods or droughts, based on its excessive surplus or deficit conditions (Chandrakar et al. 2017; Dabanlı et al. 2017; Pandey et al. 2017; Swain et al. 2017a; Tiwari and Pandey 2019). Due to the increase of anomalies in the rainfall pattern under the influence of climate change, these precipitation extremes are expected to increase (Wilhite et al. 2014; Garrote 2017; Bagirov and Mahmood 2018; Abbas et al. 2019; Amrit et al. 2019; Dayal et al. 2019). In simple words, the precipitation extreme can be defined as the remarkable alteration from the normal (averaged over a long period) conditions (Bellos et al. 2020). These extremes will cause severe consequences, especially in the developing

S. Swain (✉) · S. K. Mishra · A. Pandey · D. Dayal
Department of Water Resources Development and Management, Indian Institute of Technology Roorkee, Roorkee 247 667, Uttarakhand, India
e-mail: sabyasachiswain16@gmail.com

S. K. Mishra
e-mail: skm61fwt01@gmail.com

A. Pandey
e-mail: ashish.pandey@wr.iitr.ac.in

D. Dayal
e-mail: deemishra26@gmail.com

© The Editor(s) (if applicable) and The Author(s), under exclusive license
to Springer Nature Switzerland AG 2021
A. Pandey et al. (eds.), *Hydrological Extremes*, Water Science
and Technology Library 97, https://doi.org/10.1007/978-3-030-59148-9_12

countries like India due to poor adaptation and inadequate preparedness (Haguma and Leconte 2018; Swain et al. 2019a). These consequences also depend on the ability of water resource managers to respond to climate change in addition to pressures of increase in water demand due to rapid population growth, improved socioeconomic and legislative conditions (Gosain et al. 2006; Sethi et al. 2015; Tiwari et al. 2016; Pereira 2017; Swain et al. 2018a, b; Himanshu et al. 2019). This becomes a crucial issue for Odisha, a state in India, having an agro-based economy, and the agriculture of the state largely depends on rainfall during the south-west monsoon season (mid-June to mid-October). Therefore, the quantification of the extremes is necessary from the historical data and this will also be helpful to make predictions for the future.

Although the meteorological extreme years are commonly reported at large spatial scales corresponding to the administrative units (climate division, state, country), the extremes may vary significantly within such units (Mishra and Singh 2011; Kumar et al. 2017; Krishan et al. 2018). As precipitation possesses remarkably high spatial variation even within a small region, a better understanding of spatial dependency of extreme phenomena can be obtained, if assessed at sub-divisional units (e.g., district level).

The meteorological extremes have three important properties, viz., severity/intensity, return period and continuity (Byun and Wilhite 1999; Tsakiris and Vangelis 2004; Wilhite et al. 2007; Mishra and Singh 2010; Amrit et al. 2018a, b; Swain et al. 2019b). The knowledge of these properties for a particular region helps in developing proper planning and management strategies (Rossi et al. 1992; Myronidis et al. 2018; Zhu et al. 2019). In general, the meteorological extremes are assessed in several time steps, that is, hourly, daily, monthly, seasonal, annual or multi-annual (Swain et al. 2017b). However, identification of a dry or wet year based on historical data is commonly carried out in annual scales. In tropical regions, due to a clear seasonality in the pattern of rainfall, monsoon season contributes to a vast majority of the annual precipitation (Thornthwaite, 1948). Since the dry and wet seasons are well known for tropical regions, the assessment of only wet (monsoon) season may provide sufficient idea about whether the particular year can be regarded as an extreme year or not. This can be achieved by the use of suitable index that can represent both the wet and dry conditions effectively. Several studies exist in the literature to have used the indices for recognizing extreme conditions (Hayes et al. 1999; Seiler et al. 2002; Dubrovsky et al. 2009; Nkiaka et al. 2017; Swain et al. 2017b; Liu et al. 2018; Mukherjee et al. 2018; Marini et al. 2019; Mishra et al. 2019b; Hassan et al. 2020).

Considering all the above-mentioned issues, the objective of this study is to employ an index-based approach to identify the drought year or flood year using the monsoon rainfall records. The details of study area, data used, methodology, results and discussions, and the conclusions are presented in the subsequent sections of the chapter.

## 12.2 Materials and Methods

### 12.2.1 Study Area

The Central Division of Odisha is selected as the study area. The location of the study area is presented in Fig. 12.1. The constituent districts of the division are Baleshwar, Bhadrak, Cuttack, Dhenkanal, Jagatsinghpur, Jajpur, Kendrapada, Khordha, Mayurbhanj, Nayagarh and Puri. Six of these districts (Baleswar, Bhadrak, Jagatsinghpur, Kendrapara, Khordha and Puri) come under the Coastal Odisha region. The division is located between 19.450 to 22.570 North latitudes and 84.450 to 87.490 East longitudes. Both the largest and smallest districts of Odisha, that is, Mayurbhanj and Jagatsinghpur, respectively, are located within this division. The areal extent of the study area is 42500 km$^2$. According to the 2011 Census, the total population of the region is over 19 million with a density of 458 persons per km$^2$. Although the temperature remains moderate throughout the year, it can be very hot from March to June. In general, the humidity is high over the region. The region receives a good amount of rainfall, mostly during the monsoon season. Majority of the population have agriculture as their primary occupation, which is also the backbone of the economy of the state. There are several rivers flowing through the study area, with Mahanadi, Brahmani and Baitarani as the major ones.

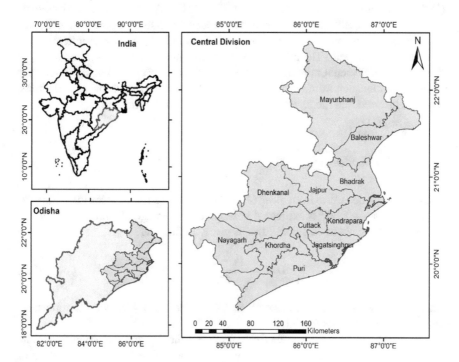

**Fig. 12.1** Location of the Central Division (Odisha), India

**Table 12.1** Summary of monsoon rainfall characteristics

| S. no. | Districts | Minimum monsoon rainfall | Maximum monsoon rainfall | Average monsoon rainfall | Standard deviation |
|---|---|---|---|---|---|
| 1 | Baleshwar | 658.2 | 2593.6 | 1375.0 | 303.1 |
| 2 | Bhadrak | 757.0 | 2015.1 | 1226.8 | 245.3 |
| 3 | Cuttack | 673.4 | 1894.0 | 1262.5 | 234.7 |
| 4 | Dhenkanal | 710.2 | 2106.0 | 1241.4 | 223.9 |
| 5 | Jagatsinghpur | 815.8 | 2113.3 | 1371.3 | 272.6 |
| 6 | Jajpur | 738.8 | 2079.8 | 1272.7 | 263.4 |
| 7 | Kendrapada | 784.2 | 2226.9 | 1271.6 | 269.3 |
| 8 | Khordha | 778.2 | 1767.2 | 1233.5 | 212.7 |
| 9 | Mayurbhanj | 787.3 | 1978.7 | 1354.5 | 270.5 |
| 10 | Nayagarh | 452.2 | 1885.0 | 1166.2 | 250.1 |
| 11 | Puri | 733.4 | 1854.0 | 1196.8 | 246.8 |

## 12.2.2 Data Used

The monthly rainfall records are collected from India Meteorological Department (IMD), Pune for a period of 113 years (1901–2013). As the region receives most of its annual rainfall during the monsoon season (mid-June to mid-October), this study uses sum of precipitation of 5 months (June, July, August, September and October) for each year. The range (maximum and minimum), average and standard deviation of monsoonal rainfall over all the districts are presented in Table 12.1. As the average monsoon rainfall varied between 1166 and 1375 mm over the districts, it can be inferred that the region receives a good amount of rainfall.

## 12.2.3 Methodology

The percentage departure from mean (*PDM*) has been used as an index to identify the extreme years. As *PDM* is expressed in terms of percentage, the ratio is multiplied by 100. The formula to calculate *PDM* is given by Eq. 12.1.

$$PDM = \frac{P_i - P_m}{P_m} \times 100 \qquad (12.1)$$

where $P_i$ represents monsoon rainfall in a particular (ith) year and $P_m$ represents long-term mean monsoon rainfall.

The *PDM* can be used to effectively represent the dry and wet conditions. The long-term mean rainfall is taken as the normal precipitation and a departure of

**Table 12.2** Classification of *PDM* into different categories

| Condition | Classes |
|---|---|
| *PDM* > 45 | Extreme wet |
| 35 < *PDM* < 45 | Severe wet |
| 20 < *PDM* < 35 | Moderate wet |
| −20 < *PDM* < 20 | (Near) normal |
| −35 < *PDM* < −20 | Moderate dry |
| −45 < *PDM* < −35 | Severe dry |
| *PDM* < −45 | Extreme dry |

less than 20% from it can be regarded as near normal condition. The detailed classifications of *PDM* into different categories are presented in Table 12.2.

## 12.3   Results and Discussions

The number of *PDM*-based events (or years) in different categories for both dry and wet conditions is presented in Table 12.3. It can be clearly observed that the number of moderate dry years vary from 9 over Cuttack district to 18 over Jagatsinghpur and Nayagarh districts. Similarly, the number of severe and extreme dry years varies from 1 to 3 and 0 to 2, respectively, over different districts. On the other hand, the number of moderate wet years is minimum over Dhenkanal and maximum over Mayurbhanj, that is, 6 and 20, respectively. Similarly, the number of severe and extreme wet years, respectively, varies from 2 to 6 and 0 to 3. It is to be noted that there have been at

**Table 12.3** Frequency of meteorological extremes in different classes over constituent districts in Central Division of Odisha

| Districts | Moderate dry | Severe dry | Extreme dry | Moderate wet | Severe wet | Extreme wet |
|---|---|---|---|---|---|---|
| Baleshwar | 17 | 2 | 2 | 10 | 6 | 2 |
| Bhadrak | 16 | 1 | 0 | 13 | 4 | 1 |
| Cuttack | 9 | 1 | 1 | 8 | 5 | 2 |
| Dhenkanal | 16 | 2 | 0 | 6 | 3 | 2 |
| Jagatsinghpur | 18 | 2 | 0 | 10 | 3 | 2 |
| Jajpur | 16 | 3 | 0 | 9 | 4 | 2 |
| Kendrapada | 13 | 3 | 0 | 12 | 3 | 3 |
| Khordha | 16 | 1 | 0 | 13 | 2 | 0 |
| Mayurbhanj | 17 | 2 | 0 | 20 | 2 | 2 |
| Nayagarh | 18 | 2 | 2 | 11 | 4 | 2 |
| Puri | 16 | 3 | 0 | 12 | 4 | 3 |

least three severe wet years over nine districts. Moreover, nine districts have faced at least two extreme wet years during 1901–2013. Hence, it can be inferred that the frequency of high-severity wet years is higher than that of dry years over Central Division of Odisha.

The total number of drought or flood years is the sum of the number of years in all the three categories. Hence, the total number of drought years vary from 11 to 22 over a span of 113 years, implying a return period of 5–10 years over the districts. Similarly, the total number of flood years vary from 11 to 24, implying a return period of 4–10 years. The district-wise return periods of drought years and flood years are presented in Figs. 12.2 and 12.3, respectively.

The return period is divided into three categories, that is, 4–5 years, 6–7 years, and 8 years or above. Regarding droughts, six districts (Baleshwar, Jagatsinghpur, Jajpur, Mayurbhanj, Nayagarh and Puri) have a return period of 4–5 years, whereas

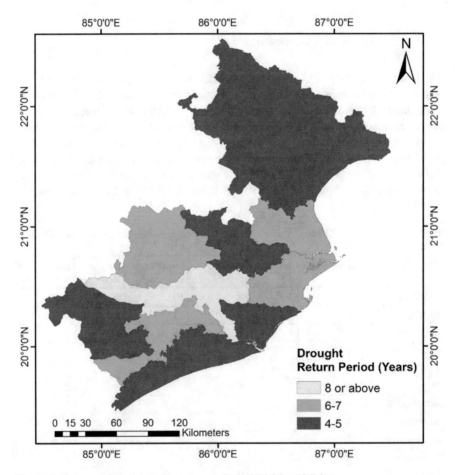

**Fig. 12.2** Return period of drought years over Central Division (Odisha)

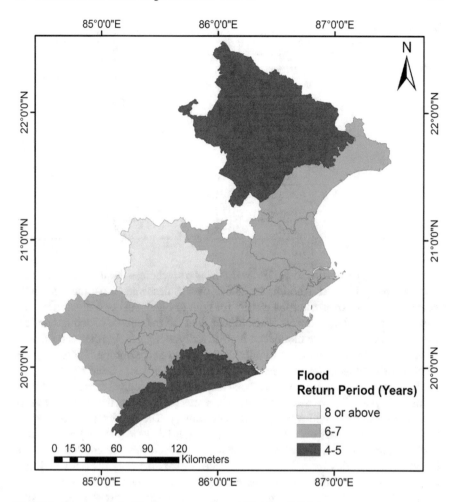

**Fig. 12.3**  Return period of flood years over Central Division (Odisha)

four districts (Bhadrak, Dhenkanal, Kendrapada and Khordha) have a return period of 6–7 years. Only the Cuttack district has a drought return period of 8 years or above. Similarly, regarding flood or wet years, two districts (Mayurbhanj and Puri) have a return period of 4–5 years, whereas eight districts (Baleshwar, Bhadrak, Cuttack, Jagatsinghpur, Jajpur, Kendrapada, Khordha and Nayagarh) have a return period of 6–7 years. Only the Dhenkanal district has a flood return period of 8 years or above. Therefore, it can be inferred that the region has undergone both drought and flood years frequently.

The intensity/severity of extremes is another crucial characteristic. The maximum intensity of drought or flood years are computed for each district using the maximum percentage of deficit or surplus precipitation with respect to the corresponding mean value, and the results are summarized in Table 12.4. The maximum deficit varies

**Table 12.4** *PDM*-based maximum intensity of dry or wet events over districts in Central Division (Odisha)

| Districts | Maximum deficit (%) | Maximum surplus (%) |
|---|---|---|
| Baleshwar | 52.1 | 88.6 |
| Bhadrak | 38.3 | 64.3 |
| Cuttack | 46.7 | 50.0 |
| Dhenkanal | 42.8 | 69.6 |
| Jagatsinghpur | 40.5 | 54.1 |
| Jajpur | 42.0 | 63.4 |
| Kendrapada | 38.3 | 75.1 |
| Khordha | 36.9 | 43.3 |
| Mayurbhanj | 41.9 | 46.1 |
| Nayagarh | 61.2 | 61.6 |
| Puri | 38.7 | 54.9 |

from 36.9% over Khordha to 61.2% over Nayagarh, whereas the maximum surplus varies from 43.3% over Khordha to 88.6% over Baleshwar. Thus, the maximum intensity was least over Khordha district for both dry and wet conditions. Also, the magnitude of rainfall surplus is much higher than the magnitude of rainfall deficit, which is reflected in more number of high-severity wet years. Further, the *PDM* is above 45 for surplus conditions for all the districts except Khordha. Therefore, it is the only district that has never been through extreme wet year (Table 12.3).

The continuity of an extreme is a crucial factor. If a hydrological extreme continues for multiple years, it will cause devastating impacts. In this study, the number of years a particular event continues is referred as continuity. The maximum continuity of drought and flood years over the districts of Central Division (Odisha) is presented in Figs. 12.4 and 12.5, respectively. It can be observed that the continuity of the drought events varies from 2 to 5 years. Baleshwar is the only district which has undergone a single drought event continuing for 5 years, that is, 1905–1909. Two districts (Jagatsinghpur and Nayagarh) have a maximum continuity of 3 years and the rest of eight districts have a maximum continuity of 2 years. Similarly, for flood/wet years, the maximum continuity is 5 years over Cuttack (2005–2009) and Mayurbhanj (1939–1943), followed by 3 years over Baleshwar district. All other districts have a maximum continuity of two years. However, for both the drought and flood years, there are multiple events over almost all the districts continuing for 2 years.

The assessment of properties of dry and wet years individually provides a clear idea that the region is subjected to frequent floods and droughts. However, there must be sufficient time to recover or alleviate the consequences of such extremes. Consecutive years of droughts or floods are even more dangerous and very difficult to manage. Therefore, it is required to assess the return period and maximum continuity of the meteorological extremes, irrespective of if its drought or flood year. The total number of meteorological extremes can be computed as the sum of the extremes in all the six categories presented in Table 12.3. The return period and maximum continuity

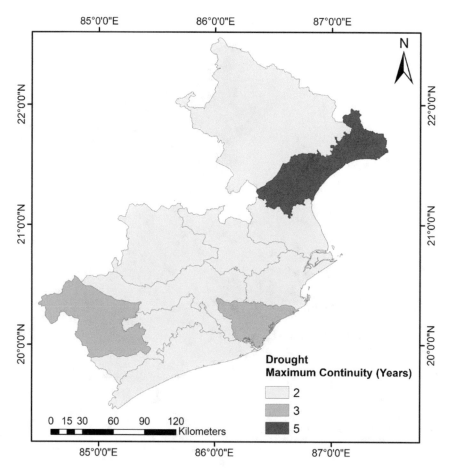

**Fig. 12.4** Maximum continuity of drought years over the districts of Central Division (Odisha)

of meteorological extremes over all the districts of Central Division (Odisha) is presented in Table 12.5.

The return periods of the meteorological extremes vary from 2.63 years over Mayurbhanj to 4.35 years over Cuttack. Therefore, the probability of occurrence of an extreme event over the region is very high. The maximum consecutive years of meteorological extremes (combining both dry and wet years) vary from 3 to 6 years over different districts. The continuity of 6 years is observed over Nayagarh from 1915 to 1920. During this six-year period, 1915, 1918 and 1920 were drought years, whereas 1916, 1917 and 1919 were the flood years. Baleshwar, Cuttack and Mayurbhanj have experienced five consecutive meteorological extreme years during 1905–1909 (all dry years), 2005–2009 (all wet years) and 1939–1943 (all wet years), respectively. Similarly, the maximum consecutive meteorological extreme years are four for Bhadrak, Kendrapada and Puri, and three for the rest of the districts.

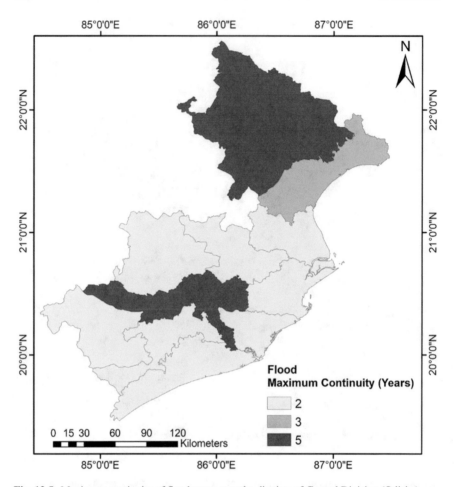

**Fig. 12.5** Maximum continuity of flood years over the districts of Central Division (Odisha)

Overall, it can be inferred that there have been a large number of meteorological extremes over the region during 1901–2013. A study on climate change impacts over hydrology of Indian river basins through projected scenarios revealed that severe flooding will occur in delta region of Mahanadi basin (Gosain et al. 2006). The increasing interventions with environmental changes in land use/land cover may cause deterioration of flooding conditions over coastal Odisha (Reddy et al. 2013). This study is about assessment of meteorological extremes based on monsoonal rainfall. However, apart from droughts and floods, the coastal districts of Odisha are also prone to several other issues like extreme cyclones, silt deposition, saline water intrusion, sea-level rise, and so on. It is reported that almost 200 cyclones have hit the coastal Odisha during 1970–2012 (Sahoo and Bhaskaran 2018). The depression formed in Bay of Bengal normally during summer and retreat monsoon seasons becomes detrimental for the coastal Odisha. Moreover, the problem of sediment

**Table 12.5** Return period and maximum continuity of meteorological extremes (dry or wet) over different districts in Central Division (Odisha)

| Districts | Return period (years) | Maximum continuity (years) |
|---|---|---|
| Baleshwar | 2.90 | 5 |
| Bhadrak | 3.23 | 4 |
| Cuttack | 4.35 | 5 |
| Dhenkanal | 3.90 | 3 |
| Jagatsinghpur | 3.23 | 3 |
| Jajpur | 3.32 | 3 |
| Kendrapada | 3.32 | 4 |
| Khordha | 3.53 | 3 |
| Mayurbhanj | 2.63 | 5 |
| Nayagarh | 2.90 | 6 |
| Puri | 2.97 | 4 |

deposition causes an advancement of delta region toward the sea, which may become fierce if the region is hit by a storm (Mohanti and Swain 2005; Swain 2014; Mishra et al. 2019a). The sea-level rise and consequent saline water intrusion pose another challenge for the water resources managers, which may further deteriorate under climate change. Therefore, development of proper adaptation strategies and adequate preparedness is necessary.

## 12.4   Conclusions

The *PDM*-based meteorological extreme years and their properties are assessed over Central Division (Odisha) considering the century-long monsoonal rainfall records over the constituent districts. It is found that the division has undergone both drought and flood years frequently. The return period of a meteorological extreme (either dry or wet) varies from 2.63 to 4.35 years over different districts. Further, the maximum consecutive years of either a drought or a flood varied from 3 to 6. The results presented in this study emphasize on development of proper planning and management strategies to reduce the ill effects of the extreme events.

## References

Aadhar S, Swain S, Rath DR (2019) Application and performance assessment of SWAT hydrological model over Kharun river basin, Chhattisgarh, India. World environmental and water resources congress 2019: watershed management, irrigation and drainage, and water resources planning and management. American Society of Civil Engineers, Reston, VA, pp 272–280

Abbas SA, Xuan Y, Song X (2019) Quantile regression based methods for investigating rainfall trends associated with flooding and drought conditions. Water Resour Manage 33(12):4249–4264

Ahani A, Shourian M, Rad PR (2018) Performance assessment of the linear, nonlinear and nonparametric data driven models in river flow forecasting. Water Resour Manage 32(2):383–399

Amrit K, Kumre SK, Mishra SK, Pandey RP (2019) Assessment of environmental flow Condition in Indian river Basin using SPI. World environmental and water resources congress 2019: watershed management, irrigation and drainage, and water resources planning and management. American Society of Civil Engineers, Reston, VA, pp 313–320

Amrit K, Pandey RP, Mishra SK (2018a) Assessment of meteorological drought characteristics over Central India. Sustain Water Resour Manage 4(4):999–1010

Amrit K, Pandey RP, Mishra SK, Kumre SK (2018b) Long-Term meteorological drought characteristics in Southern India. World environmental and water resources congress 2018: groundwater, Sustainability, and hydro-climate/climate change. American Society of Civil Engineers, Reston, VA, pp 207–215

Bagirov AM, Mahmood A (2018) A comparative assessment of models to predict monthly rainfall in Australia. Water Resour Manage 32(5):1777–1794

Bellos V, Papageorgaki I, Kourtis I, Vangelis H, Kalogiros I, Tsakiris G (2020) Reconstruction of a flash flood event using a 2D hydrodynamic model under spatial and temporal variability of storm. Nat Hazards 1–16

Byun HR, Wilhite DA (1999) Objective quantification of drought severity and duration. J Clim 12(9):2747–2756

Chandrakar A, Khare D, Krishan R (2017) Assessment of spatial and temporal trends of long term precipitation over Kharun watershed, Chhattisgarh, India. Environ Proc 4(4):959–974

Dabanlı İ, Mishra AK, Şen Z (2017) Long-term spatio-temporal drought variability in Turkey. J Hydrol 552:779–792

Dayal D, Swain S, Gautam AK, Palmate SS, Pandey A, Mishra SK (2019) Development of ARIMA model for monthly rainfall forecasting over an Indian river Basin. World environmental and water resources congress 2019: watershed management, irrigation and drainage, and water resources planning and management. American Society of Civil Engineers, Reston, VA, pp 264–271

Dubrovsky M, Svoboda MD, Trnka M, Hayes MJ, Wilhite DA, Zalud Z, Hlavinka P (2009) Application of relative drought indices in assessing climate-change impacts on drought conditions in Czechia. Theoret Appl Climatol 96(1–2):155–171

Garrote L (2017) Managing water resources to adapt to climate change: facing uncertainty and scarcity in a changing context. Water Resour Manage 31(10):2951–2963

Gosain AK, Rao S, Basuray D (2006) Climate change impact assessment on hydrology of Indian river basins. Curr Sci 346–353

Haguma D, Leconte R (2018) Long-term planning of water systems in the context of climate Non-stationarity with deterministic and stochastic optimization. Water Resour Manage 32(5):1725–1739

Hassan I, Kalin RM, Aladejana JA, White CJ (2020) Potential impacts of climate change on extreme weather events in the Niger Delta part of Nigeria. Hydrology 7(1):19

Hayes MJ, Svoboda MD, Wiihite DA, Vanyarkho OV (1999) Monitoring the 1996 drought using the standardized precipitation index. Bull Am Meteor Soc 80(3):429–438

Himanshu SK, Pandey A, Shrestha P (2017) Application of SWAT in an Indian river basin for modeling runoff, sediment and water balance. Environ Earth Sci 76(1):3

Himanshu SK, Pandey A, Yadav B, Gupta A (2019) Evaluation of best management practices for sediment and nutrient loss control using SWAT model. Soil and Tillage Res 192:42–58

Krishan R, Nikam BR, Pingale SM, Chandrakar A, Khare D (2018) Analysis of trends in rainfall and dry/wet years over a century in the Eastern Ganga Canal command. Meteorol Appl 25(4):561–574

Kumar D, Gautam AK, Palmate SS, Pandey A, Suryavanshi S, Rathore N, Sharma N (2017) Evaluation of TRMM multi-satellite precipitation analysis (TMPA) against terrestrial measurement over a humid sub-tropical basin, India. Theoret Appl Climatol 129(3–4):783–799

Liu D, You J, Xie Q, Huang Y, Tong H (2018) Spatial and temporal characteristics of drought and flood in Quanzhou based on standardized precipitation index (SPI) in recent 55 years. J Geosci Environ Prot 6(8):25–37

Marini G, Fontana N, Mishra AK (2019) Investigating drought in Apulia region, Italy using SPI and RDI. Theoret Appl Climatol 137(1–2):383–397

Mishra SK, Amrit K, Pandey RP (2019b) Correlation between tennant method and standardized precipitation index for predicting environmental flow condition using rainfall in Godavari Basin. Paddy Water Environ 17(3):515–521

Mishra M, Chand P, Pattnaik N, Kattel DB, Panda GK, Mohanti M, Baruah UD, Chandniha SK, Achary S, Mohanty T (2019a) Response of long-to short-term changes of the Puri coastline of Odisha (India) to natural and anthropogenic factors: a remote sensing and statistical assessment. Environ Earth Sci 78(11):338

Mishra AK, Singh VP (2010) A review of drought concepts. J Hydrol 391(1–2):202–216

Mishra AK, Singh VP (2011) Drought modeling—a review. J Hydrol 403(1–2):157–175

Mohanti M, Swain MR (2005) Mahanadi river delta, east coast of India: an overview on evolution and dynamic processes. Department of geology. Utakal University, Vani Vihar

Mukherjee S, Mishra A, Trenberth KE (2018) Climate change and drought: a perspective on drought indices. Curr Clim Change Rep 4(2):145–163

Myronidis D, Ioannou K, Fotakis D, Dörflinger G (2018) Streamflow and hydrological drought trend analysis and forecasting in Cyprus. Water Resour Manage 32(5):1759–1776

Nkiaka E, Nawaz NR, Lovett JC (2017) Using standardized indicators to analyse dry/wet conditions and their application for monitoring drought/floods: a study in the Logone catchment, Lake Chad basin. Hydrol Sci J 62(16):2720–2736

Palmate SS, Pandey A, Kumar D, Pandey RP, Mishra SK (2017) Climate change impact on forest cover and vegetation in Betwa Basin India. Appl Water Sci 7(1):103–114

Pandey BK, Khare D, Kawasaki A, Mishra PK (2019) Climate change impact assessment on blue and green water by coupling of representative CMIP5 climate models with physical based hydrological model. Water Resour Manage 33(1):141–158

Pandey A, Palmate SS (2019) Assessing future water–sediment interaction and critical area prioritization at sub-watershed level for sustainable management. Paddy Water Environ 17(3):373–382

Pandey BK, Tiwari H, Khare D (2017) Trend analysis using discrete wavelet transform (DWT) for long-term precipitation (1851–2006) over India. Hydrol Sci J 62(13):2187–2208

Pereira LS (2017) Water, agriculture and food: challenges and issues. Water Resour Manage 31(10):2985–2999

Reddy CS, Jha CS, Dadhwal VK (2013) Assessment and monitoring of long-term forest cover changes in Odisha, India using remote sensing and GIS. Environ Monit Assess 185(5):4399–4415

Rossi G, Benedini M, Tsakiris G, Giakoumakis S (1992) On regional drought estimation and analysis. Water Resour Manage 6(4):249–277

Sahoo B, Bhaskaran PK (2018) Multi-hazard risk assessment of coastal vulnerability from tropical cyclones—a GIS based approach for the Odisha coast. J Environ Manage 206:1166–1178

Seiler RA, Hayes M, Bressan L (2002) Using the standardized precipitation index for flood risk monitoring. Int J Climatol: J Roy Meteorol Soc 22(11):1365–1376

Sethi R, Pandey BK, Krishan R, Khare D, Nayak PC (2015) Performance evaluation and hydrological trend detection of a reservoir under climate change condition. Model Earth Syst Environ 1(4):33

Swain S (2014) Impact of climate variability over Mahanadi river basin. Int J Eng Res Technol 3(7):938–943

Swain S, Dayal D, Pandey A, Mishra SK (2019a) Trend analysis of precipitation and temperature for Bilaspur District, Chhattisgarh, India. World environmental and water resources congress 2019: groundwater, sustainability, Hydro-climate/climate change, and environmental engineering. American Society of Civil Engineers, Reston, VA, pp 193–204

Swain S, Mishra SK, Pandey A (2019b) Spatiotemporal characterization of meteorological droughts and its linkage with environmental flow conditions. AGUFM 2019:H13O–1959

Swain S, Patel, P, Nandi S (2017a) A multiple linear regression model for precipitation forecasting over Cuttack district, Odisha, India. In: 2017 2nd International conference for convergence in technology (I2CT), pp. 355–357. IEEE

Swain S (2017) Hydrological modeling through soil and water assessment toolin a climate change perspective a brief review. In: 2017 2nd International conference for convergence in technology (I2CT), pp 358–361. IEEE

Swain S, Patel P, Nandi S (2017b). Application of SPI, EDI and PNPI using MSWEP precipitation data over Marathwada, India. In: 2017 IEEE International geoscience and remote sensing symposium (IGARSS), pp 5505–5507. IEEE

Swain S, Nandi S, Patel P (2018a) Development of an ARIMA model for monthly rainfall forecasting over Khordha district, Odisha, India. Recent findings in intelligent computing techniques. Springer, Singapore, pp 325–331

Swain S, Verma MK, Verma MK (2018b). Streamflow estimation using SWAT model over Seonath river basin, Chhattisgarh, India. Hydrologic modeling. Springer, Singapore, pp 659–665

Thornthwaite CW (1948) An approach toward a rational classification of climate. Geogr Rev 38(1):55–94

Tiwari H, Pandey BK (2019) Non-parametric characterization of long-term rainfall time series. Meteorol Atmos Phys 131(3):627–637

Tiwari H, Rai SP, Shivangi K (2016) Bridging the gap or broadening the problem? Nat Hazards 84(1):351–366

Tsakiris G (2014) Flood risk assessment: concepts, modelling, applications. Nat Hazards Earth Syst Sci 14(5):1361

Tsakiris G, Vangelis H (2004) Towards a drought watch system based on spatial SPI. Water Resour Manage 18(1):1–12

Veettil AV, Mishra A (2020) Water security assessment for the contiguous United States using water footprint concepts. Geophys Res Lett

Wilhite DA, Sivakumar MV, Pulwarty R (2014) Managing drought risk in a changing climate: the role of national drought policy. Weather Clim Extremes 3:4–13

Wilhite DA, Svoboda MD, Hayes MJ (2007) Understanding the complex impacts of drought: a key to enhancing drought mitigation and preparedness. Water Resour Manage 21(5):763–774

Zhu S, Xu Z, Luo X, Wang C, Zhang H (2019) Quantifying the contributions of climate change and human activities to drought extremes, using an improved evaluation framework. Water Resour Manage 1–15

# Chapter 13
# An Approach Toward Mitigation of Cyclone Disaster: A Case Study of Odisha During Phailin

**Anil Kumar Kar, Krishna Kumar Gupta, Jaygopal Jena, and Dipti Ranjan Jena**

## 13.1  Introduction

The state Odisha is ranked as the fifth most flood-prone state of the country after UP, Bihar, Assam and West Bengal with a flood-prone area of 33,400 km$^2$. The southwest monsoon brings rain to the state from June to October every year. The state receives an average annual rainfall of 1500 mm and more than 80% of it occurs during monsoon period only. The coastal districts of the state are more vulnerable to frequent low pressure, cyclonic storms, depression and deep depression. The state has five major river basins, namely Mahanadi, Brahmani, Baitarani, Subarnarekha and Rushikulya which cause high floods in their respective delta. The rivers like Vamshadhara and Burhabalang also cause flash floods due to instant runoff from their hilly catchment. It is a fact that the three major river systems Mahanadi, Brahmani and Baitarani form a single delta during high flood, and in most of the cases the flood water of these three systems blend together causing considerable flood havoc. The state has 476.40 km of coastline on Bay of Bengal. The flood problem becomes more severe when its synchronies with high tides cause slow recede of flood. The silt deposited constantly

A. K. Kar (✉)
Civil Engineering Department, VSSUT, Burla, India
e-mail: anilkarhy@gmail.com

K. K. Gupta
Sikshya O'Anusandhan University, Bhubaneswar, India
e-mail: kkg3545@gmail.com

J. Jena
GITA, Bhubaneswar, India
e-mail: jenajoygopal@gmail.com

D. R. Jena
Utkal University, Bhubaneswar, India
e-mail: diptiluna@gmail.com

© The Editor(s) (if applicable) and The Author(s), under exclusive license
to Springer Nature Switzerland AG 2021
A. Pandey et al. (eds.), *Hydrological Extremes*, Water Science
and Technology Library 97, https://doi.org/10.1007/978-3-030-59148-9_13

by the waves in the delta area raises the bed level and the rivers often overflow their banks.

Among all the coastal states of India, Odisha is more prone to cyclone where nearly one-third of cyclones of east coast visit the state Odisha. The state is worst affected by tropical cyclones experiencing landfall of 260 cyclones within a time span of 100 years (Mishra and Kar 2016). Out of the total severe cyclonic storms of the Bay of Bengal, 15% affect Odisha, and districts like Balasore, Bhadrak, Jajpur, Cuttack, Puri, Ganjam, Kendrapara, Jagatsinghpur, Khordha, Gajapati are more prone to cyclone. Out of the total 260 cyclones that have confronted in last 100 years in Odisha coast, 180 were depression, 57 were storms and 23 were severe storms which accounted for 69, 22 and 9%, respectively, of the total disturbances. In Odisha cyclonic storm exposes people and landscape to the impact of three types of hazards, that is, high-speed wind, storm and tidal surge, heavy torrential rainfall which leads to physical destruction, saline inundation of low-lying area and flooding, respectively. The few severe storms that generated in Bay of Bengal during 1971, 1972, 1976, 1990, 1991, 1999-Super Cyclone and the recently occurred Phailin of 2013 have led to devastation of public life, property and great death toll. But the Super Cyclone of 1999 broke all the records of 100 years regarding intensity of the hazard and loss of life and property.

The majority of cyclonic storms occurred during October, when all the reservoirs are nearly in full condition due to end of monsoon and the filling schedules. The tracks of cyclones in the month of October (year 1951–2013) in Odisha coast are shown in Fig. 13.1. It shows that majority of cyclones are generated in Bay of Bengal around

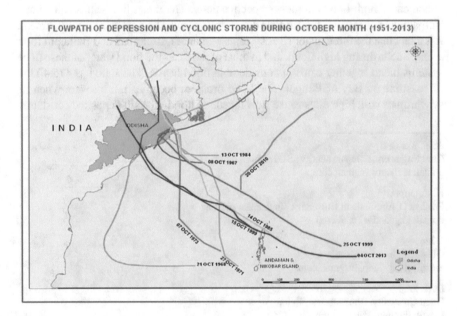

**Fig. 13.1** Cyclonic tracks during October month (1971–2013)

Andaman and Nicobar and travel toward Odisha and Andhra Pradesh. Sometimes these tracks may be deviated toward Bangladesh or Myanmar or go southwards. Other cyclones after land fall covered a distance along the land but during Super Cyclone 1999, it was encircled around Odisha coast near Paradip and Bhadrak, thus it continued for a longer period and intensified over a small area, resulting in a huge damage.

As these cyclones are associated with a huge rainfall and thus results in a flooding scenario, there are number of structural and non-structural techniques being applied in general for flood protection. Reservoir operation during extreme situations is a challenging task. Particularly, during cyclonic storm which carry a very heavy rainfall, it gives a very little time to react for taking an adequate structural measure. Further due to violent outside condition raising and strengthening of embankments, similar structural measures become impossible. The situation becomes further worsen when a cyclone approaches at the end of a good monsoon season and reservoirs are at full level. The management remains in a dilemma, how to create a flood space for accommodating the incoming flood.

There are number of suitable structural/non-structural measures adopted and advocated for different situation. In the face of increasing menace of various hazards, mitigation would remain the key and the most effective strategy to reduce the risks of cyclone (DDMP 2009). Two new concepts: (a) flood management virtual database and (b) flood management decision support system are presented by Simnovic (2001). Their benefits are demonstrated through the development of prototype systems for the Red River basin in Manitoba, Canada.

Gilbuena et al. (2013) were able to present a clear and rational approach in the examination of overall environmental effects of structural flood mitigation measures (SFMMs), which can be used by decision-makers and policy-makers to improve the Environmental Impact Assessment (EIA) practice and evaluation of projects in Philippines. Nquot and Kulatung (2014) have set to look into various mitigating measures meant to overcome the lapses and inadequacies of the present system and to encourage a better-informed approach in dealing with future occurrences of flood disasters, from the overflowing of the banks of the River Thames. Kundzewicz (2014) has observed that flood risk and flood preparedness became matters of widespread concern following the dramatic inundations in Poland in 1997 and 2010. Changes in flood risk are driven by change in the climatic system, in the hydrological/terrestrial system, and in the socioeconomic system. Wang (2013) has attempted to create a set of cultural heritage risk maps for New Taipei City in northern Taiwan as non-structural measures. This study analyzes the feasibility of using parks as water detention areas to reduce flood damage temporarily not only to cultural heritage areas but to human lives and property. Sahoo and Bhaskaran (2015) have mentioned about the semi-enclosed nature of Bay of Bengal (BoB) basin with its funnel shape that leads to high probability for cyclones to strike the land. With the wealth of cyclone track data available at present, it would be worthwhile to construct a synthetic track or the most probable track for practical applications and vulnerability studies for a coastline. There is an increasing trend noticed for post-monsoon cyclones in the past three decades, whereas for pre-monsoon season the trend is still unclear. Their study aims to construct synthetic tracks using IDW method for the pre- and post-monsoon seasons in the BoB basin using parameters such as cyclone eye, maximum sustained wind speed, and central pressure drop. Dewan (2015) has analyzed the vulnerability to floods, impacts and the coping strategies in Bangladesh and Nepal and focused on recommending a long-term mitigation policy. He has emphasized for application of traditional knowledge and indigenous practices for flood control to reduce the socioeconomic impacts and vulnerabilities.

As regards to non-structural measures, flood forecasting, flood plain zoning and so on are being employed. But cyclonic precipitation which brings flood is always an extreme case which disrupts the communication system and restricts the mobility.

## 13.2  Description of Cyclone PHAILIN and Action Taken

The cyclonic system was first noted at Gulf of Thailand on October 4 and around October 6. It was in Andaman sea and moved west–northwest with developed intensity. The IMD started monitoring the system as depression BoB 04 early on October 8. The system moved west–northwest and gradually intensified. The gradual developments are given in Table 13.1. With the declaration of onset of a cyclone, necessary instructions/warnings were transmitted by IMD. The current location of cyclone, probable tracks, position of landfall, wind speed and the quantitative precipitation forecast (QPF) given by IMD were of immense use for taking necessary decisions at the government level. The accuracy and frequency of these warnings were relevant at that time.

**Table 13.1**  PHAILIN movements time and wind speed

| S. No | Date | Time (hour) | Wind speed in kmph | Type |
|---|---|---|---|---|
| 1 | 9-Oct-2013 | | | Depression |
| 2 | 10-Oct-2013 | 1130 | 120 | Deep depression |
| 3 | 11-Oct-2013 | 1130 | 170 | Cyclone |
| 4 | 12-Oct-2013 | 0530 | 260 | Very severe cyclonic storm |
| 5 | 12-Oct-2013 | 1130 | 260 | Very severe cyclonic storm |
| 6 | 12-Oct-2013 | 1330 | 185 | Very severe cyclonic storm |
| 7 | 12-Oct-2013 | 1730 | 225 | Very severe cyclonic storm |
| 8 | 13-Oct-2013 | 1130 | 120 | Cyclonic storm |
| 9 | 13-Oct-2013 | 2330 | 56 | Deep depression |

## 13.3   PHAILIN-Associated Rainfall

As a warning the QPF during Phailin was received from time to time. From the probable track and the influence area of the track the basins of Odisha which will be affected most was being realized. The districts which will be mostly affected by rainfall, wind and storm surge were alerted accordingly. The probable track (Fig. 13.2) of cyclone was also provided by IMD which remains very helpful in tracking the cyclone.

At particular time interval IMD was giving the location of cyclone and track changes. The gradual refinement was clearly visible. These tracks obtained at different period of time are also overlaid (Fig. 13.3) with the basins of the state and the existing water resources storages falling on the influence area of the cyclone were verified, specifically their water storage capacity and current position. As that was the end of monsoon season the reservoirs were almost at full storage levels. Therefore, any kind of heavy rainfall that occur will be leading to a catastrophic flood as the existing reservoirs were not having flood space. The government decided to create flood spaces within the reservoirs of those which are within the influence zone of cyclone. The QPFs are judged accurately and the water volumes that will be generated in the upstream of reservoirs were estimated. Accordingly, the pre-depletion of the reservoir started before the arrival of the cyclone and necessary flood spaces were

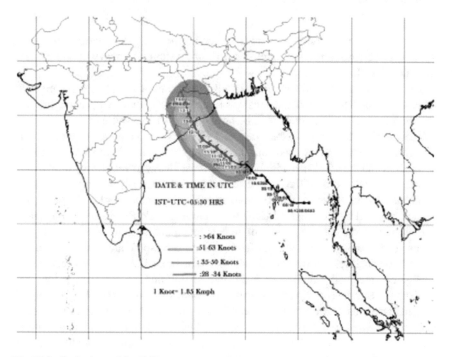

**Fig. 13.2** Cyclonic track by IMD

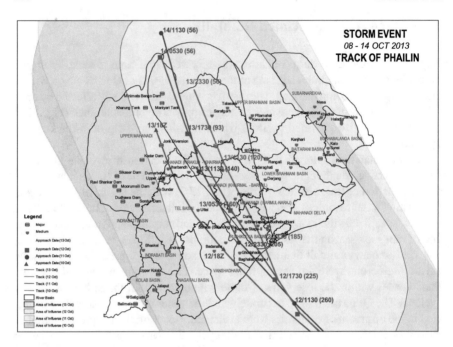

**Fig. 13.3** PHAILIN track of different periods overlaid on Odisha basins

generated. These water releases were so systematic and early that hardly any chance was left for downstream water logging due to storm surge and heavy flooding due to cyclonic rainfall.

The QPF versus actual rainfall over all the basins of the state were given in Table 13.2 from 8 to 14 October. This was remaining very helpful in calculating and assessing the real damage that could be possible due to an awaiting flood.

The districts that are affected due to PHAILIN are Ganjam, Gajapati, Puri, Khordha, Nayagarh, Kendrapada and Jagatsinghpur. The height of storm surge was around 3 m and the wind speed is above 225 kmph. Heavy to very heavy rainfall was recorded in many parts of the state. The heavy rainfall over Rushikulya, Baitarani, Burhabalanga and Subarnarekha basins lead to flood in deltaic part of the basin. The districts affected by the flood are Ganjam, Jajpur, Bhadrak, Balasore and Nayagada.

Following the Phailin rainfalls of higher magnitudes occurred in almost four basins. The one-day maximum rainfalls (in mm) corresponding to the station that occurred during these storm periods are shown in Table 13.3. It is seen that Burhabalanga has received a rainfall of 305.4 mm at Balimundali and then in next day 127.6 mm at Baripada just to add to the flood devastation. On 13 and 14 Keonjhargarh and Thakurmunda of Baitarani basin and Tiringi and Jamsholaghat of Subarnarekha received more than 100 mm rainfall. It was Madhabrida station of Rushikulya which received 184.2 mm on 13 October (i.e., on Phailin day). Although rainfall was heavy but a major flood situation was controlled.

**Table 13.2** The QPF versus actual rainfall values during Phailin

| S. No | Basin | 8-Oct QPF | 8-Oct OBS | 9-Oct QPF | 9-Oct OBS | 10-Oct QPF | 10-Oct OBS | 11-Oct QPF | 11-Oct OBS | 12-Oct QPF | 12-Oct OBS | 13-Oct QPF | 13-Oct OBS | 14-Oct QPF | 14-Oct OBS |
|---|---|---|---|---|---|---|---|---|---|---|---|---|---|---|---|
| | Date | 8-Oct | 8-Oct | 9-Oct | 9-Oct | 10-Oct | 10-Oct | 11-Oct | 11-Oct | 12-Oct | 12-Oct | 13-Oct | 13-Oct | 14-Oct | 14-Oct |
| 1 | Subarnarekha | 11–25 | 3.61 | 1–10 | 6.31 | 11–25 | 0.00 | 1–10 | 0.00 | 11–25 | 3.84 | 26–37 | 56.39 | >100 | 32.75 |
| 2 | Burhabalanga | 11–25 | 10.28 | 1–10 | 20.40 | 11–25 | 0.00 | 1–10 | 0.00 | 11–25 | 31.12 | 26–37 | 82.71 | 51–75 | 44.82 |
| 3 | Baitarani Basin | 1–10 | 8.53 | 1–10 | 14.32 | 11–25 | 0.14 | 11–25 | 0.14 | 26–37 | 15.56 | 51–75 | 87.75 | 51–75 | 16.93 |
| 4 | Upper Brahmani Basin | 26–37 | 2.01 | 1–10 | 13.40 | 1–10 | 0.00 | 1–10 | 0.00 | 11–25 | 0.27 | 26–37 | 28.24 | >100 | 19.19 |
| 5 | Lower Brahmani Basin | 11–25 | 5.48 | 1–10 | 22.94 | 11–25 | 0.57 | 1–10 | 0.57 | 11–25 | 15.50 | 51–75 | 70.46 | 38–50 | 10.52 |
| 6 | Upper Mahanadi | 26–37 | 18.59 | 11–25 | 6.10 | 1–10 | 0.56 | 1–10 | 0.56 | 11–25 | 0.00 | 51–75 | 4.47 | >100 | 10.34 |
| 7 | Mahanadi (Hirakud—Khairmal) | 1–10 | 39.20 | 1–10 | 6.56 | 1–10 | 0.01 | 26–37 | 0.01 | 51–75 | 0.11 | >100 | 28.47 | 38–50 | 10.22 |
| 8 | Mahanadi (Khairmal—Barmul) | 1–10 | 5.92 | 1–10 | 11.69 | 1–10 | 0.69 | 26–37 | 0.69 | 51–75 | 9.78 | >100 | 66.41 | 38–50 | 7.87 |
| 9 | Mahanadi (Barmul—Naraj) | 1–10 | 2.55 | 1–10 | 20.63 | 1–10 | 1.33 | 26–37 | 1.33 | 51–75 | 17.59 | >100 | 69.13 | 38–50 | 2.25 |
| 10 | Mahanadi Delta | 1–10 | 5.80 | 1–10 | 13.52 | 1–10 | 0.15 | 26–37 | 0.15 | 51–75 | 33.69 | >100 | 69.31 | 38–50 | 8.99 |
| 11 | Tel Basin | 1–10 | 9.11 | 1–10 | 3.02 | 1–10 | 3.09 | 26–37 | 3.09 | 51–75 | 3.04 | >100 | 36.62 | 38–50 | 4.78 |
| 12 | Rushikulya Basin | 1–10 | 1.86 | 1–10 | 16.29 | 1–10 | 9.12 | 26–37 | 9.12 | 51–75 | 21.19 | >100 | 67.53 | 26–37 | 1.07 |
| 13 | Vanshadhara | 1–10 | 4.11 | 11–25 | 5.26 | 11–25 | 2.67 | 11–25 | 2.67 | 51–75 | 7.72 | >100 | 62.98 | 26–37 | 0.12 |
| 14 | Indrabati Basin | | 7.74 | | 2.99 | | 2.48 | | 2.48 | | 0.00 | | 6.46 | | 1.95 |
| 15 | Nagavali Basin | | 5.70 | | 5.14 | | 1.20 | | 1.20 | | 2.62 | | 31.74 | | 0.03 |
| 16 | Kolab Basin | | 4.59 | | 3.12 | | 1.15 | | 1.15 | | 0.00 | | 6.89 | | 1.41 |

**Table 13.3** The maximum rainfall values at different basin with places during Phailin

| Date | Baitarani | Burhabalanga | Subarnarekha | Rushikulya |
|------|-----------|--------------|--------------|------------|
| 12-10-13 | 79.8 (Akhuapada) | 75.0 (Soro) | 25.0 (Jaleswar) | 56.1 (Gopalpur) |
| 13-10-13 | 150.8 (Keonjhargarh) | 305.4 (Balimundali) | 148.2 (Tiringi) | 184.2 (Madhabrida) |
| 14-10-13 | 104.0 (Thakurmunda) | 127.6 (Bariapada) | 135.2 (Jamsholaghat) | 30.4 (Chhatrapur) |

The fury of the cyclone with respect to damage was avoided due to the great preparation by the administration. All the concern departments like Revenue, Home, Water Resources, Panchayati Raj, Animal Husbandary, State Disaster Mitigation Authority, Special Relief Commissioner Energy, Civil Supplies and others worked together with a notion of achieving zero casualty. The evacuation from the low-lying areas was so prompt that around 6.5 lakh people were shifted to safer areas. All these were possible due to timely and accurate forecasting of cyclone and its magnitude by IMD.

## 13.4  Conclusion

Odisha remains the major cyclone-affected state of eastern coast of India. Many lessons were learnt from the most devastating Super Cyclone of 1999. So all the departments have remained alert when the cyclone Phailin approached in October at Odisha coast in 2013. The damage due to one of the high magnitudes of cyclone was averted by correct forecasting of IMD and timely preparation by all the departments. The evacuation of probable affected people and shifting them to cyclone shelter reduced the casualty. The pre-depletion of reservoir worked effectively as catastrophic floods have been avoided due to heavy cyclonic rainfall. The calculated zero-casualty target was achieved by the government whole heartedly.

## References

Dewan HT (2015) Societal impacts and vulnerability to floods in Bangladesh and Nepal. J Weather Clim Extrem 7:36–42

District Disaster Management Plan (DDMP), Kannur (2009). Chapter-VI

Gilbuena R Jr, Kawamura A, Medina R, Nakagawa N, Amaguchi H (2013) Environmental impact assessment using a utility based recursive evidential reasoning approach for structural flood mitigation measures in Metro Manila. Philipp, J Environ Manag 131:92–102

Kundzewicz ZW (2014) Adopting flood preparedness tool to changing flood risk conditions: the situation in Poland. OCEANOLOGIA 56(2):385–407

McMinn RW, Yang Q, Scholz WM (2010) Classification and Assessment of water bodies as adoptive structural measures for flood risk management planning. J Environ Manage 91(9):1855–1863

Mishra S, Kar D (2016) Cyclonic hazards in Odisha and its mitigation, pp. 37–41, Odisha Review, Jan

Nquot I, Kulatunga U (2014) Flood mitigation measures in the United Kingdom. In: 4th international conference on building resilience, Salford quays, United Kingdom, Procedia economics and finance, vol. 18, pp. 81–87, 8–11 Sept 2014

Sahoo B, Bhaskaran PK (2015) Assessment on historical cyclone tracks in the Bay of Bengal, east coast of India. Int J Climatol. https://doi.org/10.1002/joc.4331

Simonovic SP (2001) Two new non-structural measures for sustainable management of floods. In: International workshop on non-structural measures for water management problems, proceedings of the international workshop london. Ontario, Canada 18–20 Oct

Wang JJ (2013) Flood and debris flow risk maps to cultural heritage of New Taipei City. In: 2nd International conference on geological and environmental sciences IPCBEE, vol. 52 © (2013) IACSIT Press, Singapore. https://doi.org/10.7763/IPCBEE. V52

Wang J-J (2015) Flood risk maps to cultural heritage: Measures and process. J Cult Herit 16:210–220

# Chapter 14
# Research Needs for Stream Power Moderation in Hilly Torrents for Disaster Mitigation

Manoj Prasad Patel, Nayan Sharma, and Ashish Pandey

## 14.1 Introduction

According to India Meteorological Department (IMD), cloudburst is a natural phenomenon and is usually described by sudden, high-intensity and unforecastable rainfall with more than 100 mm/h rainfall, within a small period of time, over a small area (cloudburst over Leh, http://www.imd.gov.in). In the hilly regions, the sites where clouds are restricted in a closed valley are the ideal locations for the occurrence of cloudbursts. It happens when moisture-laden air lifts with adequate rapidity to form cumulonimbus clouds shedding water load with great strength and intensity (Kumar et al. 2012). In the Himalayan region, cloudburst is a recurring phenomenon. It has been noted that in recent years the intensity and the frequency of the cloudbursts have increased many folds. Often these cloudbursts, associated with flash floods and mass movement are the cause of concern, as these create havoc, particularly in the downstream region.

On Friday, August 3, 2012 at about 10 pm, the upper catchment areas of Assi Ganga and Bhagirathi rivers received very heavy (64.5–124.4 mm) rainfall/cloudburst accompanied with thunderstorms and lightning. The river observed a massive destructive flash flood on the night of August 3, 2012 due to the cloudburst that occurred at the Pandrasu ridge. Pandrasu ridge serves as a water divide between the Bhagirathi and Yamuna rivers. The flash flood took lives of around 500

M. P. Patel · N. Sharma · A. Pandey (✉)
Department of Water Resources Development and Management, Indian Institute of Technology Roorkee, Roorkee 247667, Uttarakhand, India
e-mail: ashish.pandey@wr.iitr.ac.in

M. P. Patel
e-mail: patelmanoj3681@gmail.com

N. Sharma
e-mail: nayanfwt@gmail.com

A. Pandey et al. (eds.), *Hydrological Extremes*, Water Science and Technology Library 97, https://doi.org/10.1007/978-3-030-59148-9_14

people, and affected approximately 12,000 people, damaging property costing about Rs 6.12 billion (Gupta et al. 2013). The reason for most of the damages in the area has been observed to be the physiographic disposition of the large boulders lying along the slopes of the valley. Many sections of the river, mainly the upper and the middle reaches got filled with about 1.5–3.0 m thick pebbles, cobbles and boulders brought by the flash flood (Gupta et al. 2013). This flash flood has destroyed three small hydro-power projects that are Assi Ganga-I, Assi Ganga-II and the one that was under construction in the Kaldi Gad.

In the present study, mathematical model investigation for study of different flow parameters was examined using HEC-RAS software. Stream power, flow velocity, shear stress and energy gradient of the channel for high flood event were analyzed, and an attempt was made to highlight the major causes of flash flood disaster that hit a part of the Uttarkashi district in the Upper Bhagirathi Valley of Garhwal Himalaya during the first week of August 2012. The plus point of mathematical modeling is that there are no scale distortions, since it involves actual size of a river channel. Thus, the applicability and accuracy with which the model works is dependent on the numerical methods and the physical foundation engaged in the model. It also focuses on area of interest which requires further research to moderate the disaster in Assi Ganga type hilly torrent river.

## 14.2 Materials and Methods

### 14.2.1 Study Area

The study was conducted for Assi Ganga river which is a tributary of Bhagirathi river and is located on the upstream side of Uttarkashi Township, extending from 30° 45′ N to 30° 48′ 31″ N latitudes and 78° 28′ 55″ E to 78° 31′ E longitudes. The study area comprised 237 pre-defined cross-sections, from Dandalka at the upstream side to Gangori at the downstream side, spanning over about 20.75 km length in Uttarkashi district of Uttarakhand, India.

The river located at an altitude of about 2755 m above the mean sea level originates from the lake called Dodi Tal. The river is known as Assi Ganga in the lower reaches, Kaldi Gad in the middle reach and Dodital Gad and Binsi Gad in its uppermost reaches. At an elevation of about 1170 m above the mean sea level, Assi Ganga river joins the Bhagirathi River near the area of Gangori. In its route from Dodi Tal to Gangori, it is nourished by many streams on either side, as shown in Fig. 14.1.

Geomorphologically, a highly unstable topography is depicted by the entire Assi Ganga valley as indicated by active erosional processes and high relief of the area. Assi Ganga river after flood event 2012 is shown in Fig. 14.2. There is a total relief of about 1300 m along the valley slopes, and a total relief of about 2800 m is observed

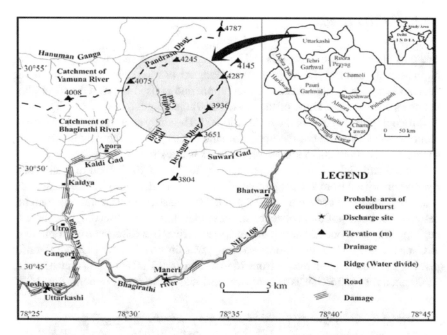

**Fig. 14.1** Location map of the area depicting the probable area of cloudburst along with the spatial location damages caused by it

**Fig. 14.2** Assi Ganga river after flood event 2012

for the entire area. Among the geomorphological processes, a vital role is played by the fluvial processes in shaping the present landscape of the area. Valley slope between 50° and 80° is observed, which reflects the steepness of the valley.

### 14.2.2  Data Used

The GIS image of Assi Ganga river, Uttarakhand, India is shown in Fig. 14.3. The digital elevation model (DEM) of the Assi Ganga river was downloaded from the site Bhuvan-Gateway to Indian Earth Observation, and is used in Arc-GIS for finding the cross-sections and slope of the river, for whole length of the river (Fig. 14.4). This topographical map is of the scale 1:50,000. The study area comprises 237 cross-sections, from Dandalka at the upstream side to Gangotri at the downstream side, spanning over about 20.75 km length in Uttarkashi district of India.

Normal and high flood discharge data for the August 2012 flood in Assi Ganga river is considered from a paper published by "Gupta et al. (2013)". According to the study, the cloudburst occurred on the top of the Pandrasu ridge at an elevation ranging between 4000 and 4500 m above the mean sea level. As was observed, discharge in huge quantity got concentrated in the Assi Ganga river that joins the Bhagirathi river at Gangori, and a high increase in discharge was observed in Assi Ganga river. In just 1 h, the discharge increased from 76 to 595 m³/s at 1:00 am on 3 August, and at 11:00 pm it rose to 2665 m³/s as compared to 135 m³/s the previous hour (Fig. 14.5).

**Fig. 14.3**  GIS image showing Assi Ganga river, Uttarakhand, India

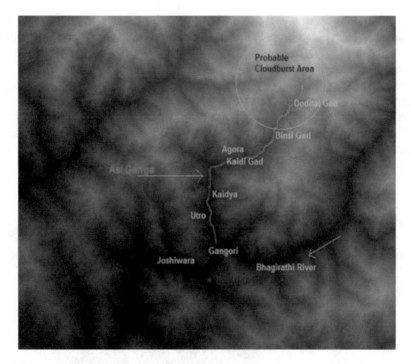

**Fig. 14.4**  DEM for study area of Assi Ganga river

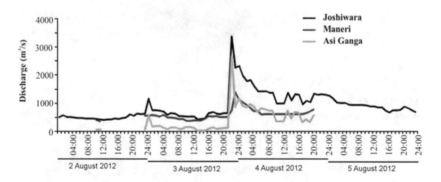

**Fig. 14.5**  Hourly discharge of Bhagirathi river at Joshiwara (depicting combined discharge of Assi Ganga and Suwari Gad), Maneri (depicting discharge of Suwari Gad) and Assi Ganga during August 2–5, 2012 (Gupta et al. 2013)

## 14.2.3   Methodology

In this study, mainly two software are used, that is, Arc-GIS for GIS applications and HEC-RAS for the hydraulic part of the work. The link between the two software was being performed by HEC-GeoRAS. Digital elevation model (DEM) consisting of entire Assi Ganga valley was downloaded from Bhuvan (Gateway to Indian Earth Observation) and was processed in Arc-GIS software and converted into triangular irregular network (TIN) and is presented in Fig. 14.6. With the help of HEC-GeoRAS software, river center line, bank line, flow path lines and cross-section lines were drawn and processed to export geometric data to HEC-RAS. In HEC-RAS geometric data was imported from HEC-GeoRAS (Fig. 14.7).

All the required modifications and editing were done at this stage. The flood discharge for the 2012 cloudburst event was entered in steady flow data. Reach boundary conditions were then entered in this window. Boundary condition was defined as normal depth for upstream, with upstream slope of 0.14. Super-critical

**Fig. 14.6** DEM processing in Arc-GIS for TIN

**Fig. 14.7** HEC-GeoRAS processing for geometric data

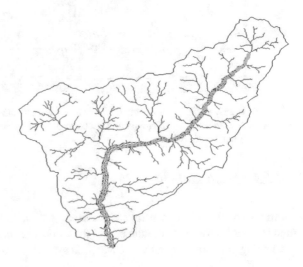

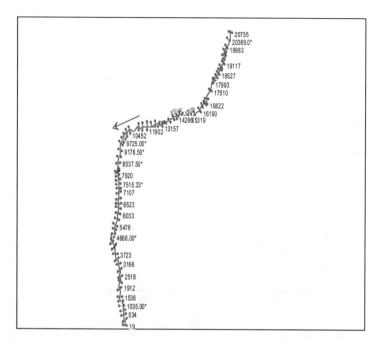

**Fig. 14.8** Assi Ganga river schematic in HEC-RAS

flow analysis was done in steady flow analysis. Then after, water surface profiles were computed. Slope of the river = $(2617.2 - 1133.6)/20736 = 0.0715 = 1{:}13.9$ was obtained from Fig. 14.6. The schematic of Assi Ganga river in HEC-RAS and L-sections are presented in Figs. 14.8 and 14.9, respectively. Some cross-sections of the river as obtained from HEC-RAS are presented in Figs. 14.10, 14.11, 14.12 and 14.13.

On reviewing and analyzing the profile plot and summary of errors and warnings and the notes from supercritical analysis, it was determined that additional cross-sectional information was required in most of the locations. Hence, interpolation was done for intermediate cross-section. Based on the table given in HEC-RAS Hydraulic Reference Manual 2010; G. J. Arcement, Manning's n chosen in this study is $n = 0.05$ as Assi Ganga is a mountain stream with cobbles and boulders at the bottom and all the calculations are made at flood discharge. HEC-RAS-based modeling was done for the analysis of the cloudburst devastated hilly river Assi Ganga, resulting in significant findings. Flow velocity, shear stress and stream power were calculated and the causes of disaster with respect to different flow parameters with the help of HEC-RAS were analyzed. Finally, core area was identified which requires further research to moderate the disaster potential of Assi Ganga type hilly streams. Detailed methodology flowchart depicting major activities is presented in Fig. 14.14.

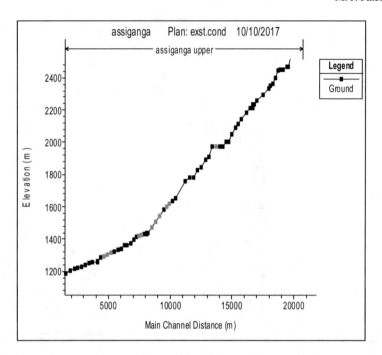

**Fig. 14.9** L-section of Assi Ganga river in HEC-RAS

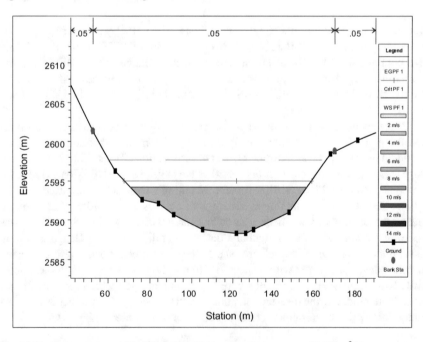

**Fig. 14.10** Cross-section at Ch. (19 + 117) (during cloudburst event of 2665 m³/s)

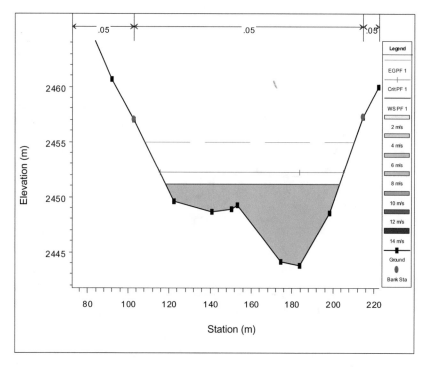

**Fig. 14.11** Cross-section at Ch. (18 + 572) (during cloudburst event of 2665 m$^3$/s)

## 14.3 Results and Discussion

### 14.3.1 Study of Flow Velocity

The flow velocity of the channel during high flood condition was computed with the help of HEC-RAS and is presented in Fig. 14.15, which uses Manning's formula for the calculation of flow velocity. Table 14.1 shows the flow velocity of the channel, and also shows that the flow velocity varies from very high value of 20–21 m/s at steep gradient to value as low as 3–4 m/s at flat gradient. The rapid change in flow velocity at certain points shows the rapid change of gradient from steep to flat and vice versa. This rapid change in gradient and thus high increase in flow velocity results in boulder movement. The steep bed gradients of this mountain stream impart very high magnitude of stream power, which poses very serious hazard of very large-sized bed load boulder movement causing devastation in the downstream areas. Thus, a strategy to minimize such hazard can be to reduce the excessive flow velocity by breaking up the steep bed gradients or by increasing bed roughness. Breaking slope of energy gradient or increasing bed roughness is necessary to reduce velocity at critical zones to reduce hazards due to boulder movement in downstream.

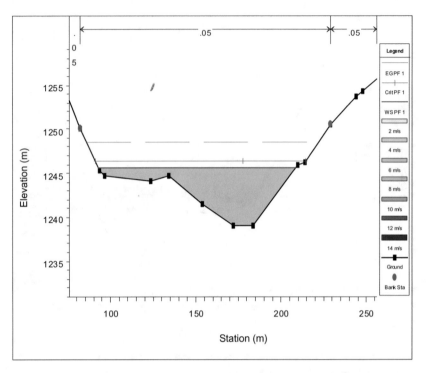

**Fig. 14.12** Cross-section at Ch. (1 + 912) (during cloudburst event of 2665 m³/s)

## 14.3.2 Study of Shear Stress

The key factor leading to instability of hilly streams is the mobilization of heavy bed load boulders along with significant suspended silt load for deformation of channel geometry. When these boulders of large size begin to move, they spin, roll and bounce along the bottom of the bed of the river, and it is also an important mechanism of transferring sediment from production to deposition areas. Bed load boulders movement occur when the drag forces (also known as shear stress) caused by water over the bed surpass the resisting forces of the banks and bed and the weight of the boulder (Schlunegger and Hinderer 2003). Shear stress on bank and bed particles is mainly a function of two factors: (1) mean flow depth and (2) the slope of the river (actually, energy grade line slope) (Dolores River Watershed Study: River Issues; Diplas 1987). In hilly streams, the slope of the river, that is, the energy gradient line of the river is very steep. These steep bed gradients of hilly streams results in very high velocity of flow and shear stress. Table 14.2 and Fig. 14.16 show variation in shear stress from very high value of 5000–6000 N/m² at steep gradient to value as low as 600–700 N/m² at flat gradient. Critical location of high shear stress is the major cause of heavy boulder movement as it is beyond critical shear stress where the

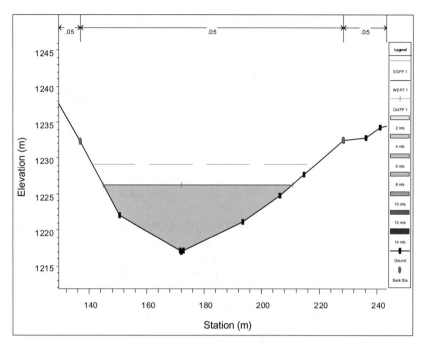

**Fig. 14.13**   Cross-section at Ch. (2 + 518) (during cloudburst event of 2665 m³/s)

incipient motion of bed particle starts. Therefore, on those locations it is necessary to break the energy gradient slope to reduce shear stress to safe limit.

### 14.3.3   Study of Stream Power

Stream power of the channel was computed with the help of HEC-RAS. Bagnold's stream power concept was used, which is defined as the product of shear stress along the bed and banks of stream, and average flow velocity. Thus, stream power has the dimensions of power per unit area of bed (Yang 1984; Yang et al. 1981)

$$\Omega = \rho gQS \qquad (14.1)$$

Table 14.3 and Fig. 14.17 show the stream power of the channel and shows variation in stream power from high value of 96,000 N/m s at steep gradient to a value as low as about 750 N/m s at flat gradient. The steep bed gradients of this mountain stream impart very high magnitude of stream power, which poses very serious hazard of very large-sized bed load boulder movement along with significant suspended silt load causing devastation in the downstream areas. Thus, it is identified that to minimize such hazard the stream power should be reduced to safe limit, which

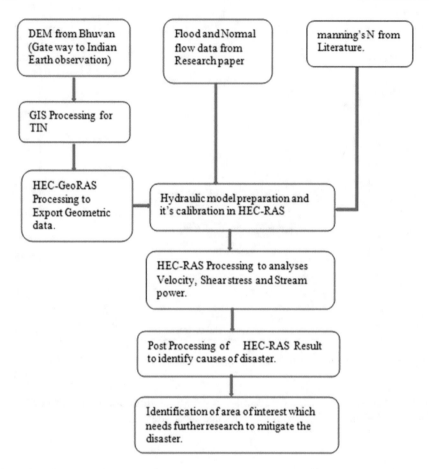

**Fig. 14.14** Flowchart depicting major activities

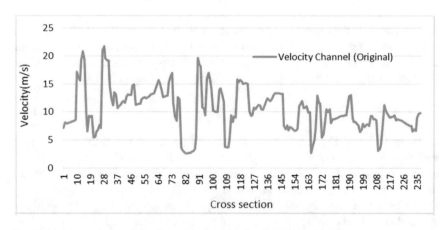

**Fig. 14.15** Flow velocity of the channel during high flood (HEC-RAS output)

**Table 14.1**  Flow velocity of the channel (HEC-RAS output)

| Velocity of channel (m/s) | | | | | |
|---|---|---|---|---|---|
| 7.07 | 11.99 | 2.85 | 15.07 | 10.79 | 7.74 |
| 8.09 | 11.66 | 2.55 | 15.07 | 11.1 | 7.8 |
| 7.89 | 12.8 | 2.58 | 15.01 | 9.92 | 7.48 |
| 8.03 | 13.12 | 2.64 | 10.12 | 9.98 | 9.32 |
| 8.11 | 13.03 | 2.72 | 9.25 | 2.7 | 9.27 |
| 8.2 | 13.05 | 2.84 | 9.41 | 3.98 | 8.71 |
| 8.31 | 14.75 | 3.02 | 10.73 | 5.02 | 8.72 |
| 8.44 | 14.89 | 3.33 | 10.43 | 7.04 | 8.49 |
| 8.58 | 11.22 | 5.45 | 10.87 | 12.97 | 3.04 |
| 17.24 | 11.33 | 19.63 | 11.27 | 11.59 | 3.41 |
| 16.16 | 11.43 | 18.36 | 11.13 | 11.58 | 4.05 |
| 15.58 | 11.48 | 18.13 | 10.44 | 5.42 | 7.39 |
| 19.32 | 12.27 | 10.75 | 10.38 | 6.03 | 11.23 |
| 20.82 | 12.54 | 10.53 | 11.17 | 7.71 | 10.51 |
| 19.29 | 12.62 | 9.42 | 11.75 | 10.52 | 9.82 |
| 9.96 | 12.48 | 15.87 | 12.41 | 9.94 | 9.26 |
| 6.51 | 12.64 | 16.95 | 12.19 | 10.46 | 8.95 |
| 9.29 | 12.87 | 15.54 | 12 | 8.03 | 9.13 |
| 9.07 | 13.1 | 14.12 | 12.38 | 8.69 | 9.22 |
| 9.28 | 13.29 | 10.18 | 13.08 | 8.72 | 9.36 |
| 5.43 | 13.31 | 10.05 | 13.36 | 8.84 | 8.55 |
| 5.53 | 13.97 | 9.96 | 13.34 | 8.94 | 8.69 |
| 6.43 | 14.85 | 9.99 | 13.32 | 9.05 | 8.59 |
| 6.79 | 15.75 | 13.92 | 13.3 | 9.15 | 8.5 |
| 7.71 | 15.19 | 14.23 | 13.27 | 9.25 | 8.38 |
| 7.15 | 14.07 | 13.18 | 13.25 | 9.34 | 8.24 |
| 21.02 | 12.62 | 11.72 | 7.51 | 9.41 | 8.04 |
| 21.73 | 12.77 | 3.79 | 6.94 | 9.43 | 7.85 |
| 19.52 | 12.8 | 3.68 | 7.57 | 11.56 | 7.69 |
| 19.24 | 12.91 | 3.67 | 6.72 | 12.82 | 7.56 |
| 19.13 | 15.21 | 5.16 | 7.3 | 13 | 7.47 |
| 14.67 | 16.29 | 9.36 | 7.1 | 8.87 | 6.5 |
| 12.31 | 16.97 | 8.19 | 6.91 | 8.24 | 6.97 |

(continued)

**Table 14.1** (continued)

| Velocity of channel (m/s) | | | | | |
|---|---|---|---|---|---|
| 11.13 | 11.53 | 9.15 | 6.61 | 8.19 | 6.67 |
| 13.56 | 9.16 | 9.01 | 6.71 | 7.82 | 9.04 |
| 13.13 | 8.43 | 15.85 | 6.99 | 7.64 | 9.65 |
| 10.64 | 12.68 | 15.31 | 11.09 | 6.44 | 9.76 |
| 10.95 | 12.26 | 15.68 | 11.57 | 6.95 | |
| 11.25 | 3.67 | 15.37 | 11.93 | 7.91 | |
| 11.77 | 3.21 | 15 | 10.63 | 7.29 | |

**Table 14.2** Shear stress of the channel (HEC-RAS output)

| Shear stress of channel (N/m$^2$) | | | | | |
|---|---|---|---|---|---|
| 817.58 | 3128.25 | 4077.46 | 3361.31 | 3117.17 | 1501.93 |
| 1098.43 | 2009.94 | 4604.55 | 2755.19 | 3110.01 | 1913.78 |
| 1033.58 | 2135.2 | 4935.95 | 210.54 | 3103.09 | 1034.93 |
| 1069.03 | 2175.79 | 2199.49 | 203.11 | 3096.3 | 1230.96 |
| 1083.25 | 2321.09 | 1358.91 | 207.57 | 3088.07 | 1229.11 |
| 1100.78 | 2422.14 | 1170.72 | 469.31 | 3080.53 | 1254.17 |
| 1121.64 | 2162.17 | 2919.14 | 1727.15 | 862.72 | 1278 |
| 1145.99 | 2592.05 | 2679.87 | 1178.04 | 744.15 | 1302.63 |
| 1171.88 | 2775.93 | 175.19 | 1380.13 | 1004.15 | 1326.67 |
| 5581.77 | 2739.16 | 132.99 | 1249.8 | 757.16 | 1350.16 |
| 4921.46 | 2741.18 | 103.82 | 4349 | 896.39 | 1374.54 |
| 4707.18 | 3533.62 | 83.06 | 3951.98 | 843.3 | 1397.49 |
| 6751.29 | 4025.41 | 85.87 | 4152.18 | 802.24 | 1411.79 |
| 8256.2 | 2110.4 | 90.79 | 4105.8 | 736.11 | 2128.43 |
| 7409.19 | 2145.53 | 98.4 | 3887.47 | 743.02 | 2667.61 |
| 1652.03 | 2180.4 | 110.01 | 3920.54 | 799.84 | 2812.2 |
| 642.62 | 2206.72 | 128.71 | 3926.82 | 2157.84 | 1209.39 |
| 1408.05 | 2528.26 | 164.18 | 3901.87 | 2306.49 | 1042.88 |
| 1306.93 | 2641.37 | 508.89 | 1616.85 | 2458.21 | 1052.73 |
| 1346.39 | 2687.5 | 7317 | 1458.48 | 1968.93 | 973.93 |
| 399.11 | 2671.62 | 6112.28 | 1553.87 | 1979.14 | 950.9 |
| 429.18 | 2713.4 | 6264.14 | 2096.24 | 2041.07 | 633.08 |
| 631.83 | 2787.31 | 1912.33 | 1966.23 | 1594.05 | 756.68 |
| 807.3 | 2870.47 | 1682.58 | 2066.68 | 1640.42 | 1045.14 |
| 1001.13 | 2945.76 | 1461.56 | 2214.78 | 94.33 | |
| 781.53 | 2998.76 | 3950.2 | 2205 | 211.08 | |

(continued)

**Table 14.2** (continued)

| Shear stress of channel (N/m$^2$) | | | | | |
|---------|---------|---------|---------|---------|---|
| 7247.9  | 3246.01 | 4859.75 | 2040.88 | 342.65  | |
| 7904.75 | 3655.03 | 4256.9  | 1992.27 | 725.42  | |
| 6048.33 | 4021.01 | 3648.56 | 2358.24 | 2894.74 | |
| 5904.01 | 3839.68 | 1734.27 | 2462.18 | 2235.95 | |
| 5416.94 | 3641.74 | 1679.77 | 2783.36 | 2285.35 | |
| 3208.54 | 2874.29 | 1670.23 | 2719.69 | 412.2   | |
| 2291.02 | 2954.19 | 1773.65 | 2646.57 | 495.97  | |
| 1910.63 | 2960.07 | 3520.22 | 2718.37 | 819.77  | |
| 3394.02 | 2973.53 | 3695.78 | 2972.43 | 1631.29 | |

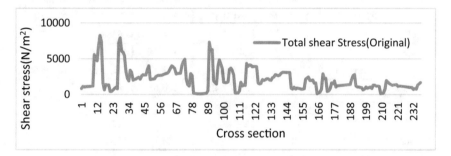

**Fig. 14.16**   Shear stress of the channel during high flood event (HEC-RAS output)

**Table 14.3**   Total stream power of the channel (HEC-RAS output)

| Total stream power of channel (N/m s) | | | | | | |
|----------|----------|----------|----------|----------|----------|----------|
| 5783.98  | 41088.2  | 62013.01 | 44292    | 41632.63 | 14932.84 | 868      |
| 8886.58  | 21389.89 | 75006.66 | 32292.87 | 41473.11 | 20017.52 | 5708.82  |
| 8150.82  | 23383.67 | 83768.96 | 797.47   | 41318.07 | 8305.34  | 22087.22 |
| 8587.86  | 24487.71 | 25357.52 | 747.59   | 41165.46 | 10702.71 | 18440    |
| 8780.57  | 27318.62 | 12450.25 | 761.57   | 40984.43 | 10722.24 | 15335.49 |
| 9024.52  | 29053.08 | 9869.13  | 2421.24  | 40815.6  | 11081.79 | 13089.97 |
| 9322.08  | 25208.09 | 37017.74 | 16170.15 | 6482.57  | 11429.73 | 11623.32 |
| 9673.6   | 33170.86 | 32842.8  | 9650.11  | 5168.09  | 11790.05 | 12297.23 |
| 10054.35 | 36432.02 | 643.64   | 12631.23 | 7597.7   | 12142.38 | 12628.69 |
| 96208.77 | 35691.35 | 427.21   | 11261.75 | 5086.87  | 12487.8  | 13101.14 |
| 79516.8  | 35772.84 | 295.71   | 68926.59 | 6542.55  | 12843.61 | 9839.78  |
| 73352.04 | 52110.22 | 212.18   | 60522.81 | 5985.2   | 13157.32 | 10419.27 |
| 130436.9 | 59933.76 | 221.95   | 65115.22 | 5541.44  | 13318.6  | 10119.28 |

(continued)

**Table 14.3** (continued)

| Total stream power of channel (N/m s) | | | | | | |
|---|---|---|---|---|---|---|
| 171858.2 | 23672.33 | 239.44 | 63093.19 | 4862.15 | 24607.17 | 9882.56 |
| 142944.9 | 24314.92 | 267.39 | 58307.87 | 4982.4 | 34189.54 | 9558.38 |
| 16460.56 | 24920.34 | 311.92 | 59075.97 | 5592.49 | 36571.2 | 9196.23 |
| 4181.7 | 25339.93 | 388.15 | 59190.61 | 23933.63 | 10731.24 | 8664.97 |
| 13085.04 | 31032.4 | 546.77 | 58554.86 | 26678.08 | 8592.14 | 8180.54 |
| 11857.08 | 33123.77 | 2773.46 | 16358.84 | 29336.75 | 8622.61 | 7785.34 |
| 12495.58 | 33928.35 | 143622.5 | 13496.65 | 20931.06 | 7611.86 | 7475.63 |
| 2166.16 | 33343.65 | 112197.5 | 14617.67 | 21353.53 | 7261.57 | 7265.46 |
| 2373.25 | 34288.12 | 113573.2 | 22483.62 | 22652.78 | 4076.46 | 4579.16 |
| 4061.13 | 35867.72 | 20564.16 | 20510.16 | 15813.51 | 5259.33 | 5762.16 |
| 5480.91 | 37604.71 | 17714.74 | 22474.17 | 16371.82 | 8271.73 | 4999.51 |
| 7723.55 | 39135.8 | 13774.27 | 24957.56 | 254.79 | 6258.38 | 12157.81 |
| 5584.38 | 39923.96 | 62670.8 | 24542.2 | 840.36 | 7417.65 | 14679.77 |
| 152363.8 | 45334.43 | 82357.63 | 21306.16 | 1718.47 | 8129.28 | 16496.56 |
| 171746.3 | 54264.12 | 66166.01 | 20681.42 | 5106.65 | 6276.71 | |
| 118085.4 | 63317.77 | 51503.02 | 26343.8 | 37537.68 | 12455.75 | |
| 113602.6 | 58309.08 | 17653.88 | 28919.49 | 25911.44 | 12271.19 | |
| 103646.9 | 51255.67 | 16887.02 | 34552.45 | 26454.84 | 10245.72 | |
| 47063.36 | 36259.68 | 16638.44 | 33142.77 | 2233.82 | 10493.32 | |
| 28192.15 | 37721.44 | 17726.41 | 31754.77 | 2992.06 | 9839.26 | |
| 21269.89 | 37898.84 | 49008.25 | 33652.6 | 6323.97 | 366.49 | |
| 46007.51 | 38377.48 | 52602.68 | 38877.55 | 17159.67 | 512.35 | |

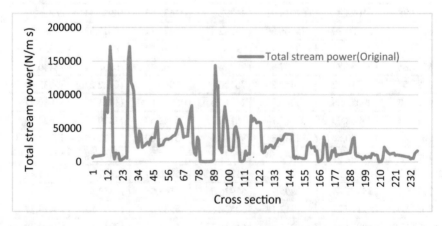

**Fig. 14.17** Stream power of the channel during high flood event (HEC-RAS output)

can be done by reducing velocity and shear stress. This required further detailed research for providing measures to reduce velocity and shear stress of the channel to the safe limit.

## 14.4   Conclusion

This study was carried out to identify the causes of disaster in Assi Ganga type steep slope hilly stream river. From the analysis of 20.75 km reach of study area, it was found that the major cause of disaster in downstream area is due to large boulder movement which rolls and bounce along bed due to high velocity, high shear stress and high stream power of channel during cloudburst or concentrated rainfall events. Study reveals that there is a need of further research in this type of river to reduce velocity, shear stress and stream power either by increasing roughness of bed or by flattering slope of energy gradient line which will reduce the stream power and protect the movement of heavy boulders to downstream.

**Acknowledgements** The authors wish to express deep sense of gratitude and sincere thanks to Department of Water Resources Development and Management (WRD&M), IIT Roorkee and Indian Water Resources Society (IWRS) for being helpful, providing opportunity and a great source of inspiration for presenting this innovative field of Research paper in "International Conference on Sustainable Technologies for Intelligent Water Management 2018". Also, the authors would like to thank all the teaching and non-teaching staff members of the department who contributed directly or indirectly to successful completion of this research work.

## References

Diplas P (1987) Bedload transport in gravel-bed streams. J Hydraul Eng 113(3):277–292
Gupta V, Dobhal DP, Vaideswaran SC (2013) August 2012 cloudburst and subsequent flash flood in the Assi Ganga, a tributary of the Bhagirathi river, Garhwal Himalaya, India. Curr Sci 105(2):249–253
HEC-RAS (2010) Hydraulic reference manual
Kumar MS, Shekhar MS, RamaKrishna SSVS, Bhutiyani MR, Ganju A (2012) Numerical simulation of cloud burst event on August 05, 2010, over Leh using WRF mesoscale model. Nat Hazards 62:1261–1271
Schlunegger F, Hinderer M (2003) Pleistocene/Holocene climate change, re-establishment of fluvial drainage network and increase in relief in the Swiss Alps. Terra Nova 15:88–95
Wang SSY, Langendoen EJ, Shields Jr FD (eds). ISBN 0-937099-05-8
Yang CT, Song CCS, Woldenberg MJ (1981) Hydraulic geometry and minimum rate of energy dissipation. Water Res Res 17(4):1014–1018
Yang CT (1984) Unit stream power equation for gravel. J Hydraul Eng 110(12):1783–1797

# Part II
# River Hydraulics

# Chapter 15
# Experimental Study and Calibration of Hydraulic Coefficients using Vertical Orifice

**R. Tejaswini and H. J. Surendra**

## 15.1 Introduction

The orifice is a device used to measure the discharge and is defined as an opening in a plate or wall or at the base of the vessel through which the fluid flows. When the water flows through the orifice the pressure developed at the upstream end gets decreased at the downstream end. The pipe present at the upstream end has a larger diameter when compared to the diameter of the orifice and this creates a pressure difference as it causes or restricts the flow of fluid (liquid or gaseous), with all these differential pressure produced by the flow meter called orifice plate which works on the theoretical principle of Bernoulli's equation.

Many methods have been used in determining the flow characteristics of a fluid flowing through the vertical orifice, and some of them are used for the analysis. Few researchers have studied and analyzed the characteristic effects under different significant conditions. The process of determining the effect of different geometrical design of orifice plates on the discharge coefficient is done by using the SST model on CFX as a solver (Akshay et al. 2016). Analytical and experimental considerations under different flow conditions for the sharp-crested circular orifices are explained (Ajmalam et al. 2016). Vertical sharp-edged orifice under different operating systems for the two different flow mechanisms and the mechanical energy lost in the process of flow are explained (Cao et al. 2011). For the incompressible fluid flow in pipes of geometries, the pressure distribution against the precise measurement scale and the relationship between the head losses and mass flow rate have been discussed

R. Tejaswini · H. J. Surendra (✉)
Department of Civil Engineering, Atria Institute of Technology, 560024, Karnataka Hebbal, Bangalore, India
e-mail: surendra@atria.edu

R. Tejaswini
e-mail: tejaswini9659@gmail.com

© The Editor(s) (if applicable) and The Author(s), under exclusive license to Springer Nature Switzerland AG 2021
A. Pandey et al. (eds.), *Hydrological Extremes*, Water Science and Technology Library 97, https://doi.org/10.1007/978-3-030-59148-9_15

(Farsirotoua et al. 2014). Numerical prescription of discharge coefficients of orifice plates and nozzles is done by using commercial finite-volume-based code FLUENT under all applicable conditions (Imada et al. 2013). CFD analysis of a fluid flowing through venturi under different flow parameters is explained (Jay et al. 2014). The application of laser Doppler measurement system used to analyze the orifice coefficients for different water orifices and the difficulty in measuring the water velocity at the exit of an orifice is explained (Massimiliano et al. 2008). The analysis of discharge coefficient for the stepped vertical orifice which contains non-coplanar vertical rectangular slots working under different relative heads is determined (Shesha Prakash et al. 2013). The geometric design of multi-hole orifices and the methodology in calculating the pressure loss coefficient is explained (Tianyi et al. 2011). The present experimental method in determining the calibration of orifice coefficients is very simple and constructible.

## 15.2   Principle

Bernoulli's equation states that "with the decrease in pressure of the fluid flowing along a streamline simultaneously increases the velocity of fluid". When the fluid flows from a larger region to a smaller region or smaller section, it gains velocity and kinetic energy and there is a loss of pressure to conserve the total energy. By using the law of conservation of energy principle, the Bernoulli's equation can be obtained. The principle of law of conservation of energy states that "the sum of all the forms of energy dissipated during the steady flow of fluid along a streamline remains same at all points" (Jay et al. 2014).

Newton's second law of motion is also used to establish the equation defining Bernoulli's principle. Due to the flow of fluid horizontally along a streamline from a region of higher pressure to a region of lower pressure, the velocity of fluid gets increased considerably and there is a decrease in the velocity of fluid that occurs due to the movement of fluid from a region of lower pressure to a region of higher pressure, that is, the velocity of the fluid flows along a streamline, and the pressure developed is inversely proportional (Jay et al. 2014).

Bernoulli's equation is given by integrating the Euler's equation for steady, incompressible fluid flow without friction.

$$\left(V^2/2g\right) + z + (p/\rho g) = \text{constant (H)} \tag{15.1}$$

where V is the velocity of the fluid flow at a point on a streamline; g is the acceleration due to gravity; p is the pressure at a point; $\rho$ is the density of the fluid at all points in the fluid; $(p/\rho g)$ is the flow work, that is, work done by the fluid (potential head); $(V^2/2\,g)$ is the kinetic energy of the flow per unit weight of fluid (the velocity head); z is the potential energy per unit weight (datum head); and H is the total energy.

Bernoulli's equation is also called an energy equation. If the flow is frictionless, the total energy between two points is given by $H_1 = H_2$.
That is,

$$(p_1/\rho g) + \left(V_1^2/2g\right) + z_1 = (p_2/\rho g) + \left(V_2^2/2g\right) + z_2 \qquad (15.2)$$

If the flow is frictional, then the total energy obtained is given by.

$$H1 + HE - HL = H2 \qquad (15.3)$$

where $H_E$ is the input energy between two different Sects. 15.1 and 15.2; $H_L$ is the loss of energy occurred between two sections due to friction; H1 is the total energy in Sect. 15.1; H2 is the total energy in Sect. 15.2.
Hydraulic gradient line (HGL) is the line joining piezometric head at different points in a flow.
That is

$$\text{Piezometric head } h = z + (p/\rho g) \qquad (15.4)$$

Energy line is a line joining the elevation of the total energy of flow measured above the datum (Fig. 15.1).
That is
Total energy $H = (V^2/2\,g) + [z + (p/\rho g)]$.

$$H = \left(V^2/2g\right) + h \qquad (15.5)$$

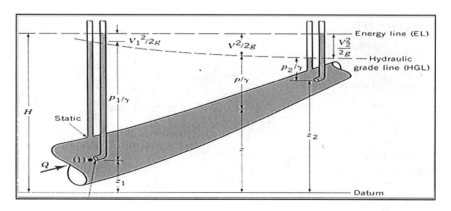

**Fig. 15.1** Bernoulli's equation energy line diagram (*Source* Google Earth: IIham Setiawan IIham)

## 15.3 Description

The oldest form of device used for measuring the discharge of water freely into the air is the vertical orifice which results in a significant loss of head. To overcome this head loss and for the flow to occur, the orifice is arranged in such a way that the inlet of water, that is, the upstream end level must always be placed above the top of the opening and the downstream end level must be placed at the bottom of the channel (Fig. 15.2).

The experimental studies are done by using circular sharp-edged orifices which are fully contracted so that the effect of contraction of the discharging jet can be controlled by control section by controlling the sides of the channel approach and the free water surface. Since the flow through this orifice must be submerged, it may be necessary to restrict the downstream channel to raise the water level above the top of the orifice.

### (a) Applications of orifice plate:

The most commonly used flow meter devices for industrial flow is the orifice plate, as it works based on the Bernoulli's principle.

In refrigeration and heat pump system, orifice plates are widely used as throttle fitting. At the bottom of the expansion valve orifices are installed. When the liquid refrigerant flows through the orifices because of the pressure drop created by the orifices it flashes to gas (Tianyi et al. 2011) (Fig. 15.3).

Orifice plates are used as restriction devices which are used to regulate the fluid flow or to reduce the pressure of the fluid flowing at the downstream of the orifice plate.

Orifice plates are used in many industries in extreme to calculate the discharge coefficient of a fluid flowing through a pipe corresponding to the measured pressure loss and the mass flow rate. This is an often-used process in several industries such as natural gas transport ducts and fluid feeding systems in the petrochemical, steel mill and ethanol distilling industries.

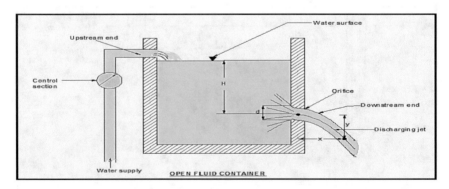

**Fig. 15.2** Open fluid orifice container

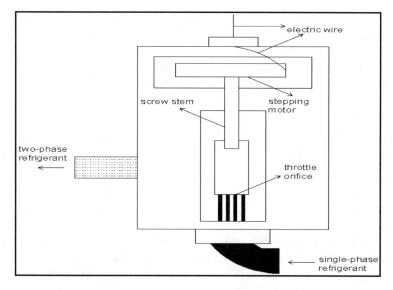

**Fig. 15.3** Typical electric expansion valve (*Source* Tianyi et al. 2011)

An orifice plate is also called a calibrated orifice since it has been calibrated with appropriate fluid flow and the trajectory path of fluid flow can be traced.

**(b) Merits and demerits of orifice plate**:

**Merits**:

Among the available flow metering methods, the pressure differential-based devices such as orifice plates, nozzles, venturi-tubes are used more due to their design and operational simplicity and low-cost maintenance.

- The construction of orifice plates is very simple due to the availability of the different standards of construction and calibration.
- Orifice plates are obtained at a cheap cost because the increase in the size of the orifice does not affect the cost of the plate, so they are inexpensive.
- They can be easily established and removed.
- These are expected to be good in quality and performance and are economical for many years.
- It is easy to measure a wide range of flow rates because it offers very little pressure drop across the orifice which can be recovered.
- The features of orifice plates are well known and predictable.
- They can be used in different types of sizes.

**Demerits**:

The disadvantages of orifice plates are: As it requires homogeneous and single-phase liquid and also axial velocity vector flow:

- Their accuracy is also affected by density, pressure and viscosity of a fluid.
- There is a comparatively high-pressure drop.
- Their distinctive features may change due to erosion and corrosion.
- The orifice plates get blocked due to the accumulation of wet matter in slurries. These are not good for the fluid having high viscosity.

**Limitations**:

To make sure full contraction and certainty in measurements, the restriction against the applications of the circular orifice should be considered (Source: Google Earth):

- The edge of the orifice should be keen and even.
- The distance between the slopes of the channel approach and stream water along with the edge of the plate to the bed should not be less than the radius of the orifice.
- The upstream water level should be placed at the top of the orifice to a height comparatively equal to the diameter of the orifice to prevent the entry of air.
- The orifice plate should be kept vertical and smooth at the upstream end.

## 15.4   Flow Through the Circular Sharp-Edged Vertical Orifice

The experimental studies are done by using sharp-edged orifice, as the device is simple, inexpensive and easy to install, which makes the device portable to measure the stream flow. Figure 15.4 shows a small circular orifice with sharp-edged in the sidewall of a tank discharging water freely into the atmosphere. The assumptions for the orifice flow were made by considering the Bernoulli's principle as follows:

- The flow properties at any point should be the same at all the time, that is, the flow should be steady.
- Along the streamline the density of the fluid should remain constant even if the pressure varies, that is, the flow should be incompressible.

**Fig. 15.4**  Discharge through the circular orifice (*Source* Google Earth: Dr. R K Bansal, 2005)

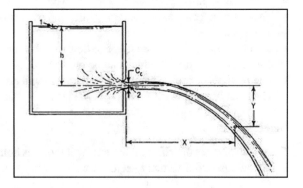

**Fig. 15.5** Diagram showing
vena-contracta

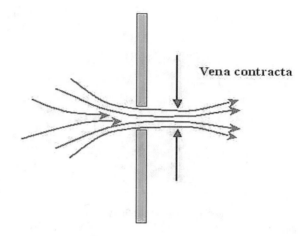

Vena contracta

- The friction developed by the viscous forces is considered to be negligible (Fig. 15.4).

Let the orifice is at a distance "H" below the free water surface at the upstream end. Due to the change in the direction of flow velocity when it approaches the orifice, the contraction of jet and the action of a lateral force on the jet are being compared.

Since the streamlines become parallel at a distance half of the diameter of the orifice from the plane of the orifice, the point at which the streamlines first become parallel with the increase in flow velocity by having the minimum diameter of the stream is termed as "vena-contracta", which is having a smallest cross-sectional area with utmost contraction taking place at the section approximately near the downstream of the orifice, where the jet is more or less horizontal. The pressure at this section is assumed to be uniform (Fig. 15.5).

When the pressure difference occurred because of surface tension is being neglected, the atmospheric pressure and pressure at vena-contracta become equal.

Considering the flow to be constant and by ignoring the frictional effects, apply the Bernoulli's theorem at the points 1 and 2 as shown in Fig. 15.2. We have

$$(p_1/w) + (V_1^2/2g) + z_1 = (p_2/w) + (V_2^2/2g) + z_2 \qquad (15.6)$$

But, $p_1 = p_2 = p_a$ ($p_a$ = atmospheric pressure).

If the area of the orifice is small when compared to the area of the tank, then the velocity at point 1 becomes negligibly small, that is, $V_1 = 0$.

$$\therefore V_2^2/2g = H \qquad (15.7)$$

$$V_2 = \sqrt{2gH} \qquad (15.8)$$

Equation (15.8) is known as *Torricelli's theorem*.

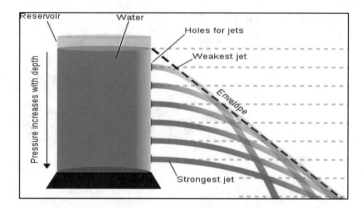

**Fig. 15.6** Typical diagram of Torricelli's theorem (*Source* Google Earth)

**Torricelli's theorem**:

*Torricelli's law* illustrates the departing speed of a water jet by assuming that there is no interruption for the fluid flow, and any resistance or viscosity which is explained is based on the distance at which the jet starts below the surface. This law itself is known as *Torricelli's theorem* in fluid dynamics related to the speed of fluid flowing out of an orifice at the height of fluid above the opening (Fig. 15.6).

The Torricelli's law states that "the velocity of fluid discharge in a sharp-edged hole at the bottom of a tank filled to a depth H is equal to the speed gained by the freely falling body from the same height H".

This diagram shows several jets arranged vertically in such a way that the jet is flowing out horizontally from the reservoir or takes a parabolic path from the surface of the water. When the two jet crosses each other at one particular depth, it is equal to their sum of depths at which the respective jet holes are present.

## 15.5 Hydraulic Coefficients

Hydraulic coefficients of the orifice which are briefly discussed below are:

- Coefficient of contraction
- Coefficient of velocity
- Coefficient of discharge
- Coefficient of resistance

Coefficient of contraction ($C_C$): It can be explained by comparing the area of the jet at the vena-contracta concerning the area of the orifice. It is denoted by $C_C$.

Let a = area of the orifice and

$a_C$ = area of the jet at vena-contracta

$$C_C = \frac{Area\,of\,jet\,at\,vena - contracta}{Area\,of\,the\,orifice}$$

$$C_C = \frac{a_C}{a} \qquad (15.9)$$

The value of $C_C$ varies from 0.61 to 0.69 depending on shape and size of the orifice and head of liquid under which flow takes place. In general, the value of $C_C$ may be considered as **0.64**.

Coefficient of velocity $(C_V)$: It can be defined by relating the actual velocity of a jet of liquid at vena-contracta to the theoretical velocity jet. It is denoted by $C_V$, and mathematically, it is given by the equation

$$C_V = \frac{The\,actual\,velocity\,of\,jet\,at\,vena - contracta}{Theoretical\,velocity}$$

$$C_V = \frac{V}{\sqrt{2gH}} \qquad (15.10)$$

where $V$ = actual velocity and $\sqrt{2gH}$ = theoretical velocity.

For various types of orifices, the value of $C_V$ ranges from 0.95 to 0.99, whose value depends on the shape and size of the orifice and the head under which the fluid flow takes place. Generally, the value of $C_V$ for sharp-edged orifices is very small and its value increases concerning the head of water but in common **0.98** is considered. It varies slightly with the different shapes of the edges of the orifice.

Coefficient of discharge $(C_d)$: It is defined as the rate at which actual discharge from orifice changes concerning the theoretical discharge from the orifice. It is denoted by $C_d$.

Let $Q$ = actual discharge and
$Q_{th}$ = theoretical discharge
Mathematically, it is given by

$$C_d = \frac{Q}{Q_{th}} = \frac{Actual\,discharge}{Theoretical\,discharge}$$

$$C_d = \frac{Actual\,velocity}{Theoretical\,velocity} \times \frac{Actual\,area}{Theoretical\,area}$$

$$C_d = C_V \times C_C \qquad (15.11)$$

The value of $C_d$ varies from 0.61 to 0.65. In general, the value of $C_d$ may be taken as **0.62**. The value of $C_d$ varies with the values of $C_V$ and $C_C$.

Coefficient of resistance $(C_r)$: It is defined as the proportion at which the head loss occurred in the orifice to the available head of water at the exit of the orifice.

$$C_r = \frac{Loss\ of\ head\ in\ the\ orifice}{Head\ of\ water} \qquad (15.12)$$

The head loss occurs in the orifice due to the resistance offered by the walls of the orifice against the fluid flow. For solving the numerical aspects its value is generally ignored.

## 15.6 Methodology

The vertical orifice instrument consists of a supply tank with overflow arrangement, and for the measurement of water in the tank, the manometer is attached to it. The inlet of the vertical orifice instrument is placed at the top of the opening, that is, the upstream end of the tank must be connected to the supply water in the device. An orifice of diameter 7 mm is placed at the downstream end of the tank where the outlet of the vertical orifice is placed. An overflow pipe was provided to permit changes in the head, and a flexible tube attached to the overflow pipe allows the return of the excess of water to the hydraulic bench. The arrangement is made in such a way that water comes out only through this attached opening in the form of a jet. The experiment is conducted by neglecting the head loss and the pipe losses in the instrument setup.

The material for the orifice is selected in such a way that it is a thin, plain circular plate with the circular hole aligned axially with the pipe and precaution should be taken so that because of the corrosion and erosion, the sharp edges and surface are not changed. A differential manometer is connected to both the supply tank and the collection tank (Fig. 15.7).

In the first place, the discharge is provided to the tank from the upstream end, and then discharge of water is controlled by the control valve to make the head of the water constant and also the flow of water through the orifice. The water discharged freely into the air forms a trajectory path. By using the measurement scale provided with a vertical needle, the jet called hook gauge is designed to plot the trajectory. With the help of screw to secure the positions of the needles, the "X" and "Y" coordinate values are noted at different points of the trajectory path. But initially, at the point of vena-contracta, the X and Y values are noted as zero. The actual velocity of the jet is determined by X and Y values.

The discharging water from the orifice is collected in the measuring tank of area 0.3 × 0.3 m. The time required for 10 cm rise in the piezometer fitted to measuring tank is noted and the actual discharge is calculated. By knowing the head of water in the tank we can calculate the theoretical velocity and theoretical discharge.

In the same way, again by the use of control valve make the head of water constant to some value; note the X and Y coordinates in the trajectory path obtained by the change in the flow of water. The same methodology is followed for 2–3 trials of the experiment (Fig. 15.8).

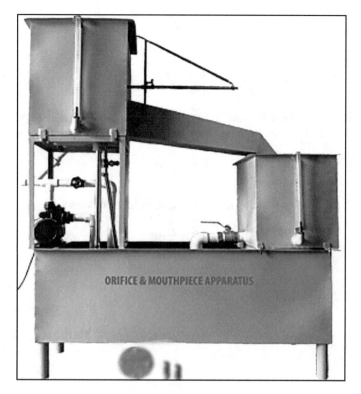

**Fig. 15.7**   Experimental setup of orifice apparatus

**Fig. 15.8**   Line diagram
showing the experimental
setup of the orifice apparatus

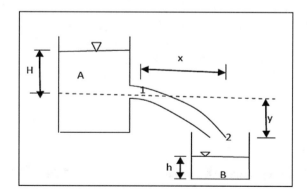

## 15.7   Observation

The diameter of the orifice fitted to the collecting tank, **d = 7 mm**
   Area of the orifice, **a = 3.8484X 10$^{-5}$m$^2$**
   Area of the measuring tank, **A = 0.09 m$^2$**

$$a = \frac{\pi d^2}{4} m^2 \tag{15.13}$$

$$Q_{act} = \frac{AR}{T} \frac{m^3}{s} \tag{15.14}$$

where A is the area of the measuring tank; R is the rise of water level in the piezometer; T is the time required for the 10 cm rise.

$$Q_{th} = a\sqrt{2gH} \tag{15.15}$$

where g is the acceleration due to gravity; H is the head of water.

$$C_d = \frac{Q_{act}}{Q_{th}} \tag{15.16}$$

$$V_{act} = \sqrt{\frac{gX^2}{2Y}} \tag{15.17}$$

where X and Y are the coordinates on the trajectory path; g is the acceleration due to gravity.

$$V_{th} = \sqrt{2gH} \tag{15.18}$$

where g is the acceleration due to gravity; H is the head of water.

$$C_V = \frac{V_{act}}{V_{th}} \tag{15.19}$$

$$C_C = \frac{C_d}{C_V} \tag{15.20}$$

$$V_{act} = \frac{1}{\text{Slope}} \times \sqrt{\frac{g}{2}} \tag{15.21}$$

where $\text{Slope} = \frac{\sqrt{Y}}{X}$ obtained by the graphs (Tables 15.1 and 15.2).

Table 15.1 Theoretical values at the different head of water

| Head of water "H" in m | 0.512 | 0.42 | 0.377 |
|---|---|---|---|
| Theoretical velocity $V_{th}$ in m/s | 3.1696 | 2.8706 | 2.7197 |
| Theoretical discharge $Q_{th}$ in m$^3$/s | $1.2197 \times 10^{-4}$ | $1.1047 \times 10^{-4}$ | $1.0466 \times 10^{-4}$ |

**Table 15.2** Result of $C_d$, $C_C$ and $C_V$ for different heads of water

| Trail no | Head of water "H" in m | Station | X | Y | Actual velocity $V_{act}$ in m/s | Coefficient of velocity $C_V$ | Avg. coefficient of velocity $C_V$ | Discharge measurements | | | Coefficient of discharge $C_d$ | Coefficient of contraction $C_C$ |
|---|---|---|---|---|---|---|---|---|---|---|---|---|
| | | | | | | | | Rise "R" in meters | Time "T" in seconds | Actual discharge $Q_{act}$ in m³/s | | |
| 01 | 0.512 | 1 | 0 | 0 | 0 | 0 | 0.717 | 0.1 | 157 | $5.732 \times 10^{-5}$ | 0.47 | 0.656 |
| | | 2 | 5.55 | 0.2 | 2.748 | 0.867 | | | | | | |
| | | 3 | 8.75 | 0.45 | 2.889 | 0.911 | | | | | | |
| | | 4 | 12.5 | 0.95 | 2.840 | 0.896 | | | | | | |
| | | 5 | 15.7 | 1.45 | 2.886 | 0.911 | | | | | | |
| 02 | 0.42 | 1 | 0 | 0 | 0 | 0 | 0.706 | 0.1 | 162 | $5.555 \times 10^{-5}$ | 0.503 | 0.712 |
| | | 2 | 6.5 | 0.35 | 2.433 | 0.848 | | | | | | |
| | | 3 | 9.5 | 0.65 | 2.609 | 0.909 | | | | | | |
| | | 4 | 12.5 | 1.2 | 2.527 | 0.88 | | | | | | |
| | | 5 | 13.7 | 1.4 | 2.564 | 0.893 | | | | | | |
| 03 | 0.377 | 1 | 0 | 0 | 0 | 0 | 0.76 | 0.1 | 168 | $5.357 \times 10^{-5}$ | 0.512 | 0.674 |
| | | 2 | 4.5 | 0.15 | 2.5732 | 0.946 | | | | | | |
| | | 3 | 5.6 | 0.25 | 2.481 | 0.912 | | | | | | |
| | | 4 | 7.3 | 0.4 | 2.556 | 0.94 | | | | | | |
| | | 5 | 10.3 | 0.7 | 2.727 | 1.00 | | | | | | |

## 15.8   Discussions

The experimental study gives the flow pattern through the vertical orifice whose diameter is less than the diameter of the inlet pipe. Since the orifice is placed at the downstream end of the tank, there is a contraction of the discharging flow at a distance half of the diameter of the orifice. It is observed that as the head of the water is decreasing the theoretical discharge value is also decreasing. Also, the theoretical value of velocity is changing concerning the head of water "H".
   i.e.,

$$V_{th} \propto H \tag{15.22}$$

$$Q_{th} \propto H \tag{15.23}$$

Figure 15.9 shows the dependence of coefficient of discharge on the head of water. The coefficient of discharge decreases with the increase in the head of water.
   The slope drawn to the graph in Fig. 15.10 gives the graphical value of coefficient of discharge.
   The discharge measurements of the measuring tank are taken, whose actual discharge depends on the time required for the 10 cm rise in the piezometer.
   The X and Y values which are noted along the discharging of water from the orifice defines the trajectory path of the jet formed by the orifice.
   Figure 15.11 gives the trajectory path formed by the orifice for the head of water 0.512 m to the X and Y values noted along with the discharge.

**Fig. 15.9**  Coefficient of discharge versus head of water

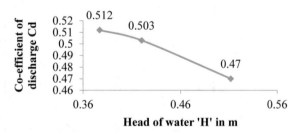

**Fig. 15.10**  Determination of coefficient of discharge

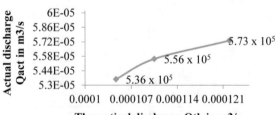

**Fig. 15.11**  Trajectory path for "H = 0.512 m"

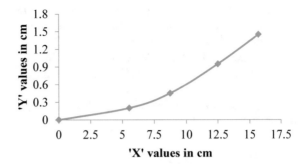

Figure 15.12 gives the trajectory path formed by the corresponding values of X and Y for the head of water 0.42 m noted along the discharging path.

Figure 15.13 gives the trajectory path followed by the curve for the values X and Y noted for the head of water 0.377 m.

Figures 15.11, 15.12 and 15.13 give the trajectory path for the different head of water. The graph of sqrt(Y) versus X gives the slope which is used to determine the graphical value for the theoretical velocity, and it is also used to determine the coefficient of velocity for the jet discharging from the orifice.

**Fig. 15.12**  Trajectory path for "H = 0.42 m"

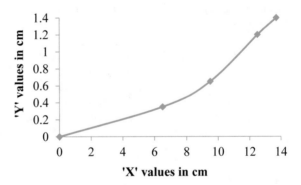

**Fig. 15.13**  Trajectory path for "H = 0.377 m"

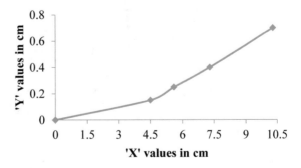

The slopes obtained by Figs. 15.14, 15.15 and 15.16 are used to determine the graphical value of the actual velocity of the jet and also the coefficient of velocity.

Figure 15.17 shows the dependence of coefficient of velocity on the head of water. The coefficient of velocity decreases with the increase in the head of water.

The slope drawn in Fig. 15.18 graph gives the graphical value of coefficient of velocity.

The slope drawn in Fig. 15.19 graph gives the graphical value of coefficient of contraction.

**Fig. 15.14** sqrt(Y) versus X for H = 0.512 m

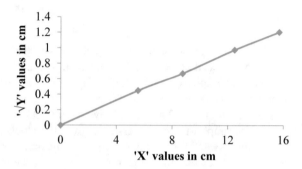

**Fig. 15.15** sqrt(Y) versus X for H = 0.42 m

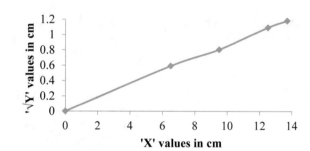

**Fig. 15.16** sqrt(Y) versus X for H = 0.377 m

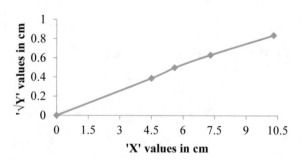

**Fig. 15.17**  Coefficient of velocity versus head of water

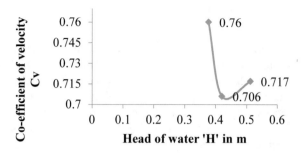

**Fig. 15.18**  Determination of coefficient of velocity

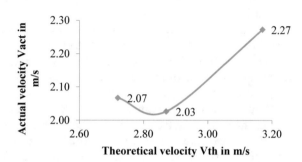

**Fig. 15.19**  Determination of coefficient of contraction

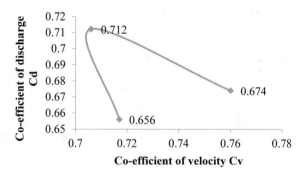

## Error Analysis

By the experimental determination, the values of hydraulic coefficients obtained are the coefficient of discharge $C_d = 0.495$; the error found is $-0.125$; coefficient of velocity $C_V = 0.728$; the error obtained is $-0.252$; and finally, the value of coefficient of contraction $C_C = 0.68$; the value is found within the range but for the general value the error obtained is $+0.041$. The errors are calculated based on the general values considered for the hydraulic coefficients.

## 15.9   Conclusion

The orifice is a small opening provided at the bottom of the tank or channel or the side of the wall. When the water flows from a region of higher pressure to lower pressure this simultaneously increases the speed of the jet discharging from the orifice. The trajectory path is plotted by the X and Y values for the different head of water discharging in the form of a jet from the orifice. An experimental study is made to know the dependence of the head of water on the hydraulic coefficients, that is, the coefficient of discharge Cd gives a direct representation of the effectiveness of the orifice that can be thought as its capability to provide the flow rate. The coefficient of velocity $C_V$ is the only orifice coefficient taking into account the energy losses. The coefficient of contraction $C_C$ explains how the actual mini section of the flow is smaller than the geometrical one. As the head of water increases, the theoretical discharge and theoretical velocity also increase. But the orifice coefficients $C_d$ and $C_V$ are inversely proportional to the head of water. The trajectory path obtained by the jet of water is always a parabolic curve in nature.

The point where the maximum contraction takes place called vena-contracta gives the maximum value of the discharge and the velocity. Among all the available flow metering devices, this is the one pressure differential device used in many industries. Throughout this experiment, the trajectory of the steady flow water condition is being measured and the characteristics of hydraulic coefficients are determined.

This experimental study is useful in the field of hydrology to know the variations in hydrological conditions which directly influence the quality of water. It is also useful for the interpretation of water quality data and water resource management by knowing the hydrological measurements which are essential. The discharge values estimated are also essential when calculating pollutant fluxes.

The calibration of orifice coefficients is essential to calculate the maximum flux of chemicals in water for which a series of discharge measurement is essential. The water level when correlated with stream discharge or with the storage volumes of reservoirs and lakes, water levels become the basis for computation of discharge or storage records.

## References

Ajmalam H, Ahmad Z, Ojha CSP (2016) Flow-through lateral circular orifice under free and submerged flow conditions. Flow Meas Instrum 52:57–66

Akshay D, Sagar M, Oshin U, Pallavi S, Rahul B (2016) Effect of orifice plate shape on performance characteristics. J Mech Civ Eng 13:50–55

Cao R, Liu Y, Yan C (2011) A criterion for flow mechanisms through vertical sharp-edged orifice and model for the orifice discharge co-efficient. Springer 8:108–113

Farsirotoua E, Kasiteropoulou D, Stamatopoulou D (2014) Experimental investigation of fluid flow in horizontal pipes system of various cross-section geometries. In: EPJ web of conferences, vol 67, p 02026

Imada FHJ, Saltara F, Baliño JL (2013) Numerical determination of discharge coefficients of orifice plates and nozzles. In: 22nd international congress of mechanical engineering (COBEM 2013), ISSN 2176-5480

Jay K, Jaspreet S, Harsh K, Gursimran SN, Prabhjot S (2014) CFD analysis of flow-through venturi. Int J Res Mech Eng Technol J Res Mech Eng Technol 4, 2249–5770

Massimiliano A, Loredana C, Marco F, Michele N (2008) Orifice coefficients evaluation for water jet applications, 16th IMEKO TC4 Symposium. Florence, Italy, p 174

Shesha Prakash MN, Ananthayya MB, Gicy K (2013) Flow-through non-coplanar vertical orifice. Int J Emerg Technol Comput Appl Sci 13–388:466–470

Tianyi Z, Jili Z, Liangdong M (2011) A general structural design methodology for multi-hole orifices and its experimental application. J Mech Sci Technol 25(9):2237–2246

# Chapter 16
# Discharge Prediction Approaches in Meandering Compound Channel

**Piyush Pritam Sahu, Kanhu Charan Patra, and Abinash Mohanta**

## 16.1 Introduction

Almost all the natural rivers meander. If the length of a river remains straight for 10–12 times of its channel width, then the river is considered as straight, which is quiet insignificant in its natural criteria. Sinuosity is defined as the ratio between the lengths of stream channel to the length of straight valley distance. River is stated to be meandering if its value is greater than 1.5. When heavy rainfall occurs, water from the main channel overtops its banks and spreads to the floodplains causing disasters. Accurate prediction of estimation of discharge is an important factor. Estimation of discharge prediction helps in various water resources engineering projects. A lot of researches are carried out to predict discharge for straight channels, but this paper comprises predictive approaches for discharge assessment in compound meandering channel.

Hooke (1974) measured bed elevation contour in a simple meandering flume with movable sand. For various discharges he calculated bed shear stress, sediment transport distribution and secondary flow. From his laboratory experiments, he concluded that secondary current increases with the increment of discharge. Toebes and Sooky (1967) conducted experiments on compound meandering channel. They analyzed

P. P. Sahu (✉) · K. C. Patra
Department of Civil Engineering, National Institute of Technology Rourkela, Rourkela 769008, Odisha, India
e-mail: pps14141@gmail.com

K. C. Patra
e-mail: kcpatra@nitrkl.ac.in

A. Mohanta
School of Mechanical Engineering, Vellore Institute of Technology, Vellore 632014, Tamil Nadu, India
e-mail: abinash.mohanta@vit.ac.in

A. Pandey et al. (eds.), *Hydrological Extremes*, Water Science and Technology Library 97, https://doi.org/10.1007/978-3-030-59148-9_16

the internal flow structure between floodplain and main channel. Their observational study signifies the existence of an imaginary horizontal fluid boundary between top edge of the main channel and floodplain. Ervine et al. (1993) analyzed various parameters, like bed roughness, sinuosity, aspect ratio, meander belt width and so on, and concluded that these parameters play an important role for conveyance in the compound meandering channel.

Classical discharge estimation follows basically three formulae for estimation of discharge prediction: Chezy's formula, Darcy-Weisbach formula and Manning's formula. This paper presents single-channel method (SCM), divided channel method (DCM) and coherence method for discharge prediction in compound meandering channel. Manning's "$n$" plays vital role as it has the influence of sinuosity ($s$), width ratio ($\alpha$), relative depth ($\beta$), roughness coefficient ($\gamma$), aspect ratio of main channel ($\delta$) and bed slope ($S_0$). Researches have proposed the value of Manning's roughness coefficient ($n$) in combination of above parameters based on some assumptions, which led to quiet efficient result of discharge prediction in compound meandering channel. Here all cross-sections have taken at the bend-apex of the compound meandering channel. Various predictive approaches have been applied through different datasets.

## 16.2 Methods of Estimation of Discharge in Compound Meandering Channel

### 16.2.1 Single-Channel Method (SCM)

SCM is one of the simplest methods for computation of discharge that considers a single entity of the channel and an average roughness coefficient. In this case, cross-sectional velocity is assumed uniform. Lambert and Myers (1998) concluded that single-channel method (SCM) underestimates the prediction of discharge when the floodplain depth is low. This is due to the assumption of uniform velocity at the cross-section. Discharge estimation can be carried out using SCM through Manning's, Chezy's or Darcy-Weisbach equation given by:

$$Q = \frac{1}{n} R^{\frac{2}{3}} S_0^{\frac{1}{2}} A \tag{16.1}$$

$$Q = AC\sqrt{RS_0} \tag{16.2}$$

$$Q = \sqrt{\frac{8g}{f}} A\sqrt{RS_0} \tag{16.3}$$

where $Q$ = total discharge of the compound channel, $n$ = Manning's roughness coefficient, $S_o$ = bed slope of the compound channel, $A$ = cross-sectional area of the compound channel, $R$ = hydraulic mean radius of the compound channel, $C$ = Chezy's constant, $g$ = acceleration due to gravity, $f$ = Darcy-Weisbach friction factor of the compound channel.

## 16.2.2 Divided Channel Method (DCM)

Divided channel method is extensively used for the calculation of discharge in the meandering compound channel. In this approach, the channel cross-section is divided into homogeneous subareas where uniform velocity is assumed in each subsection. Researchers follow three imaginary divisional interfaces while considering the subsections of the compound channel (horizontal, vertical and diagonal). These divisional methods are found to be more suitable than the single-channel method (SCM) for calculation of discharge. Again, horizontal division method (HDM) and vertical division method (VDM) are further divided into HDM-I, HDM-II, VDM-I and VDM-II. In HDM-I, the horizontal imaginary line is not considered for wetted perimeter of the main channel during calculation of discharge in compound meandering channel, while in HDM-II the horizontal imaginary line is considered as the wetted perimeter. In the similar way, an imaginary vertical division interface is considered from the top of the bank of the main channel to the top of the fluid layer. In VDM-I, the imaginary vertical lines are excluded for the wetted perimeter for the discharge calculation of the compound meandering channel, while in VDM-II the imaginary vertical lines are considered as the wetted perimeter. In the diagonal division method, an imaginary diagonal interface is assumed to have originated from the main channel—floodplain junction of the bank and extending to the central line of the water surface in the main channel. Such interface length is not considered as wetted perimeter for estimation of discharge. The sub-divisional discharges are added together to get total discharge of the compound channel. Mathematically, it can be represented as

$$Q_{\text{total}} = \sum_{i=1}^{i=N} Q_i = \left(\frac{1}{n_i}\right) A_i R_i^{\frac{2}{3}} S_0^{\frac{1}{2}} \tag{16.4}$$

where $Q_{\text{total}}$ = total discharge obtained through the compound channel, $Q_i$ = sectional discharge after applying divisional methods, $A_i$ = subsectional area, $R_i$ = subsectional hydraulic mean radius, $S_0$ = Bed slope of the compound channel, $n_i$ = Manning's roughness in each sub-division $i$ and $N$ = number of subdivisions (Fig. 16.1).

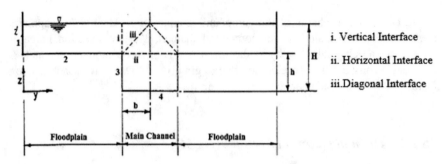

**Fig. 16.1** Preview of divisional methods in compound channel

## *16.2.3 Coherence Method (COHM)*

Coherence method (COHM) is a one-dimensional (1-D) method, which is a modification of traditional divided channel method. Ackers (1992–1993a, b) developed a new correction coefficient based on the previously experimental datasets and acknowledged it as "discharge adjustment factor" (DISADF). While considering the effect of momentum transfer, DISADF is being multiplied with QDCM. Coherence is expressed mathematically as the ratio of discharge obtained through single-channel method to discharge through divided channel method, given by

$$\text{COH} = \frac{Q_{\text{SCM}}}{Q_{\text{DCM}}} \tag{16.5}$$

If the value of COH is nearly equal to 1, then the channel is to be treated as single entity. For COH value significantly less than 1, DISADF is being multiplied with the QDCM. If the COH value is small, that is, less than 0.5, then DISADF remains insignificant. For this purpose, discharge deficit factor is being subtracted from the discharge obtained through divided channel method.

Figure 16.2 depicts the graph between DISADF with relative depth comprising four regions from the experimental study of FCF Phase A datasets. Acers formulated the discharge by consideration of the flow region based on the value of relative depth.

$$\text{For region 1, } Q_1 = Q_{\text{DCM}} \text{ DISDEF; } \beta < 0.2 \tag{16.6}$$

$$\text{For regions 2, 3, 4, } Q_i = Q_{\text{DCM}} \text{ DISADF}_i, i = 2, 3, 4; \beta \geq 0.2 \tag{16.7}$$

**Fig. 16.2** Discharge
adjustment factor (DISADF)
of FCF data (phase A)

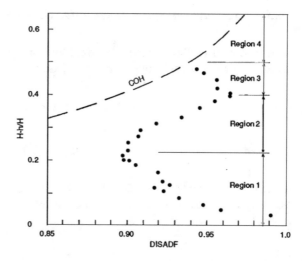

## 16.3  Roughness Coefficient Prediction Approaches for Meandering Channel

### 16.3.1  Soil Conservative Service (SCS)

The US soil conservative service (SCS) (1963) method was used for obtaining the
value of roughness coefficient based on the sinuosity of the meandering channel
given by

$$\frac{n'}{n} = \left(\frac{f'}{f}\right)^{\frac{1}{2}} = 1.0, \text{for } s < 1.2 \tag{16.8}$$

$$\frac{n'}{n} = \left(\frac{f'}{f}\right)^{\frac{1}{2}} = 1.15 \text{ for } 1.2 \leq s < 1.5 \tag{16.9}$$

$$\frac{n'}{n} = \left(\frac{f'}{f}\right)^{\frac{1}{2}} = 1.3 \text{ for } s \geq 1.5 \tag{16.10}$$

where $n$ = assumed Manning's $n$ value, $f$ = assumed friction factor, $n'$ = modified
Manning's $n$ value, $f'$ = modified friction factor, $s$ = sinuosity of the meandering
channel.

### 16.3.2 Linearized Soil Conservative Service (LSCS)

James and Wark (1992) proposed linearized soil conservative service (LSCS), which gives modified Manning's roughness values based on some sinuosity range given by

$$\frac{n'}{n} = 0.43s + 0.5, \text{ for } s < 1.7 \tag{16.11}$$

$$\frac{n'}{n} = 1.3, \text{ for } s > 1.7 \tag{16.12}$$

where $n =$ assumed Manning's $n$ value, $f =$ assumed friction factor, $n' =$ modified Manning's $n$ values, $f' =$ modified friction factor, $s =$ sinuosity of the meandering channel.

### 16.3.3 Shino, Al-Romaih and Knight (SAK)

Shiono et al. (1999) analyzed the bed slope and sinuosity of meandering channel for discharge estimation and formulated sinuosity $(s)$ in terms of friction factor $(f)$:

$$s = 10\left(\frac{f}{8}\right)^{\frac{1}{2}} \tag{16.13}$$

where $s =$ sinuosity of the meandering channel and $f =$ friction factor.

### 16.3.4 Dash and Khatua (DK)

Dash and Khatua (2016) proposed Manning's $n$ to be dependent on aspect ratio $(\delta)$, Reynolds number (Re), Froude number (Fr), bed slope $(S_o)$ and sinuosity $(s)$ which can be represented as

$$n = 0.013(1 - 0.015\delta^{-0.116} + 0.3021\ln(s) + 0.15\text{Re}^{0.0924}$$
$$- 0.3\ln(Fr) - 9.852S_o(1 - 374S_o) \tag{16.14}$$

### 16.3.5   Pradhan and Khatua (DK)

Pradhan and Khatua (2017) proposed Manning's $n$ which has the effect on relative depth, Reynold's number (Re), Froude's number (Fr), width ratio ($\alpha$), bed slope ($S_o$), sinuosity ($s$) and length scale factor ($m$), which is given by the following equation as

$$n = \frac{1}{250}\left(\frac{R^{0.84} S_o^{0.25} m^{0.08} s}{g^{0.25} \alpha \beta \nu^{0.5}}\right) \tag{16.15}$$

## 16.4   Meandering Channel Datasets for Analysis

To analyze the above discharge prediction methods, a large set of field and experimental data are collected and applied. The data comprises flood channel facility of HR Wallingford FCF B series, which is a large-scale data. Patra and Kar (2000) dataset was a doubly meandering channel data (both main channel and floodplain had sinuosity value of 1.043). Similarly, Khatua (2008) recorded data taken with variable flume size, sinuosity, bed slope and are marked as KII and KIII. Mohanty (2014) and Pradhan and Khatua (2017) data are also considered for meandering channel. Both these datasets had higher sinuosity. All these discharge results are recorded at the bend-apex of the meandering compound channel. For better presentation, FCF B series, Patra and Kar (2000), Khatua (2008) (two datasets), Mohanty (2014) and Pradhan and Khatua (2017) are named as B21, PIII, KII, KIII, MII and PKI, respectively. Three meandering channel datasets are also taken, where experiments are done at NIT Rourkela and the series are presented as NITR-I, NITR-II and NITR-III. The required parameters for calculation of discharge for various datasets are given in Table 16.1.

Department of Science and Technology, Government of India has provided funds for the setup of the large flume for various meandering channels, where experiments are conducted at the Hydraulic Engineering Laboratory, Department of Civil Engineering, National Institute of Technology, Rourkela, Odisha, India. Three flumes have been considered for the analysis—Mohanta et al. (2018; Mohanta) and Mohanta and Patra (2019). These are: (I) 60° meandering channel with straight floodplain, (II) doubly meandering compound channel (60° meandering at main channel and 30° meandering at floodplain) and (III) 30° meandering channel. All three meandering channels have rectangular smooth-bed cross-section that maintains a bed width of 0.28 m, main channel height of 0.12 m and a side slope of 1V:0H. All channels have same Manning's roughness coefficient value of 0.01. All three channels have bed slope of value 0.001. A movable bridge is provided across the meandering flume along the axes over the channel area so that measurements can be done at the meandering channel. For uninterrupted water supply to the overhead tank, a large underground sump was used. There is a measuring tank at the downstream end followed by a

Table 16.1 Experimental parameters for datasets of meandering compound channel

| Data series (No. of experimental runs) | Side slope of main channel (V:H) | Manning' n ($n_{mc} = n_{fp}$) | Sinuosity | | Height of the main channel, h(m) | Width of the main channel b(m) | Overall width of compound channel B(m) | Bed slope (So) | Relative depth ($\beta$) $= \frac{H-h}{H}$ | Actual discharge Q(m³/s) |
|---|---|---|---|---|---|---|---|---|---|---|
| | | | $S_{mc}$ | $S_{fp}$ | | | | | | |
| B21(16) | 1V:1H | 0.012 | 1.374 | 1.00 | 0.15 | 0.90 | 10.00 | 0.000996 | 0.08609–0.48048 | 0.0824–0.98939 |
| PIII(3) | 1V:0H | 0.026 | 1.043 | 1.043 | 0.25 | 0.44 | 1.38 | 0.0061 | 0.15254–0.20886 | 0.094535–0.108583 |
| KII(12) | 1V:0H | 0.01 | 1.44 | 1.00 | 0.12 | 0.12 | 0.577 | 0.0031 | 0.09909–0.33884 | 0.009006–0.031358 |
| KIII(12) | 1V:1H | 0.01 | 1.91 | 1.00 | 0.08 | 0.12 | 1.93 | 0.0053 | 0.08467–0.27992 | 0.012757–0.048474 |
| MII(5) | 1V:1H | 0.01 | 1.11 | 1.00 | 0.065 | 0.33 | 3.95 | 0.0011 | 0.19354–0.40909 | 0.017074–0.080667 |
| PKI(5) | 1V:1H | 0.01 | 4.11 | 1.00 | 0.065 | 0.33 | 3.95 | 0.00165 | 0.235–0.350 | 0.028–0.052 |
| NITR-I(4) | 1V:0H | 0.01 | 1.37 | 1.00 | 0.12 | 0.28 | 1.67 | 0.001 | 0.1367–0.3846 | 0.0248–0.0727 |
| NITR-II(4) | 1V:0H | 0.01 | 1.37 | 1.035 | 0.12 | 0.28 | 1.35 | 0.001 | 0.1304–0.3846 | 0.0184–0.0496 |
| NITR-III(4) | 1V:0H | 0.01 | 1.035 | 1.00 | 0.12 | 0.28 | 1.67 | 0.001 | 0.1429–0.3814 | 0.0236–0.0959 |

sump, which again supplies water to the overhead tank through pumping, thus establishing a complete recirculation path. The geometrical configuration of meandering channel is shown in Figs. 16.3, 16.4 and 16.5. Details of their geometric parameters

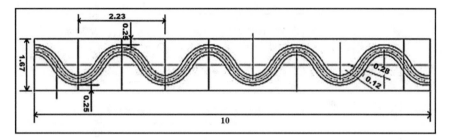

**Fig. 16.3**  Meandering compound channel (NITR type-I)

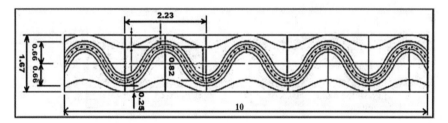

**Fig. 16.4**  Doubly meandering compound channel (NITR type-II)

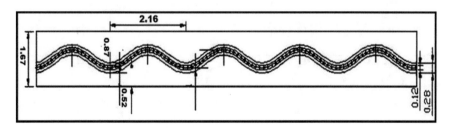

**Fig. 16.5**  Meandering compound channel (NITR type-III) (All dimensions in the above geometrical figures are in meters)

like type of channel, dimensions, bed slope, sinuosity, and wavelength and meander belt width are given in Table 16.2. The plan view of complete experimental setup is given in Fig. 16.6. Photographs of the three meandering channels are also shown in Figs. 16.7, 16.8 and 16.9.

**Table 16.2** Description of geometric parameters of NITR channel setup from Mohanta et al. (2018) and Mohanta and Patra (2019)

| S. No. | Parameters | Type I channel | Type II channel | Type III channel |
|---|---|---|---|---|
| 1 | Types of channel | Compound meandering channel with straight floodplain | Doubly meandering compound channel | Compound meandering channel with straight floodplain |
| 2 | Flume dimensions | (10 * 1.7 * 0.25) m | (10 * 1.7 * 0.25) m | (10 * 1.7 * 0.25) m |
| 3 | Side slope | 0 | 0 | 0 |
| 4 | Type of bed surface | Smooth | Smooth | Smooth |
| 5 | Bed slope of the channel ($S_b$) | 0.001 | 0.001 | 0.001 |
| 6 | Angle of arc of main channel ($\emptyset_m$) | 60° | 60° | 30° |
| 7 | Angle of arc of floodplain ($\emptyset_f$) | 0 | 30° | 0 |
| 8 | Sinuosity of the main channel ($Sr_m$) | 1.37 | 1.37 | 1.035 |
| 9 | Sinuosity of the floodplain ($Sr_f$) | 1 | 1.035 | 1 |
| 10 | Wavelength of the meandering channel ($\lambda$) | 2.23 | 2.23 | 2.16 |
| 11 | Width of the main channel (b) | 0.28 m | 0.28 m | 0.28 m |
| 12 | Total width of the channel (B) | 1.67 m | 1.35 m | 1.67 m |
| 13 | Bank full depth of main channel (h) | 0.12 m | 0.12 m | 0.12 m |
| 14 | Width of outer floodplain | 0.25 m | 0.25 m | 0.52 m |
| 15 | Width of inner floodplain | 1.14 m | 0.82 m | 0.87 m |
| 16 | Meander belt width (W) | 1.17 m | 1.17 m | 0.61 m |

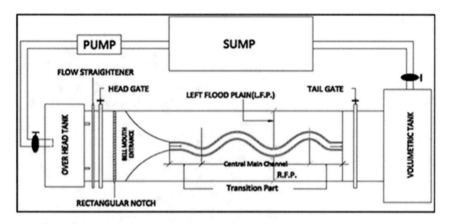

**Fig. 16.6**   Plain-metric view of experimental setup for the meandering channel

**Fig. 16.7**   NITR type-I
channel

**Fig. 16.8** NITR type-II
channel

**Fig. 16.9** NITR type-III
channel

## 16.5    Results and Discussions

Various discharge prediction approaches like SCM, HDM-I, HDM-II, VDM-I, VDM-II, DDM and COHM have been applied to the experimental datasets. Similarly, different formulae for determining Manning's roughness coefficient are evaluated and applied through the single channel method (SCM), while assessing the discharge calculation in meandering compound channel. To have a clear idea on the best method, graphs between actual discharge and predicted discharge are plotted, as can be seen in Figs. 16.10, 16.11, 16.12, 16.13, 16.14, 16.15, 16.16, 16.17 and 16.18.

For FCF B21 channel data, while all the methods gave better results when depth of water in meandering compound channel is less, it varies significantly when the depth

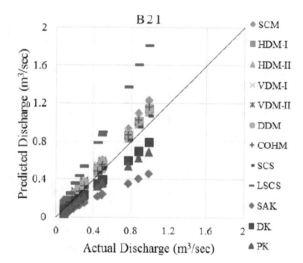

**Fig. 16.10** Graph between actual discharge and predicted discharge for B21 dataset

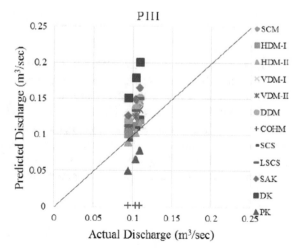

**Fig. 16.11** Graph between actual discharge and predicted discharge for PIII dataset

**Fig. 16.12** Graph between actual discharge and predicted discharge for KII dataset

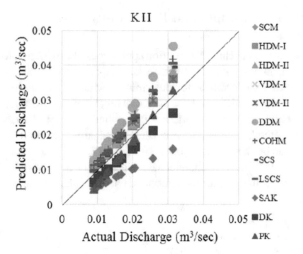

**Fig. 16.13** Graph between actual discharge and predicted discharge for KIII dataset

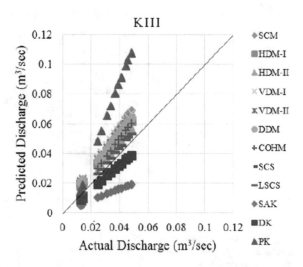

of flow increases. Among them, HDM-II, COHM and SCS give better predicted discharge. In PIII channel, HDM-II and SCS give better results. By considering KII and KIII channel datasets, HDM-II and LSCS are found to be more appropriate. In case of MII channel datasets, HDM-II, COHM and DK give better fitted curve between actual and predicted discharge. For PKI datasets, COHM and DK led to commendable results among other methods. In NITR-I and III datasets, HDM-II, COHM and PK methods give better results, while in NITR-II datasets VDM-II and DK method are more accurate while compared with actual discharge.

**Fig. 16.14** Graph between actual discharge and predicted discharge for MII dataset

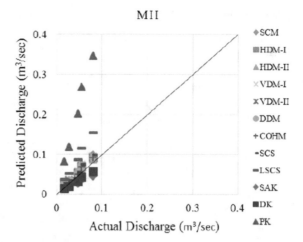

**Fig. 16.15** Graph between actual discharge and predicted discharge for PKI dataset

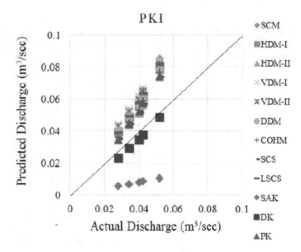

However, by plotting scattering points among actual and predicted discharge, definite conclusions cannot be made. For better analysis of each dataset, mean percentage error is calculated and plotted in line diagram for actual view of the accuracy, as shown in Fig. 16.19. From the divided channel methods (HDM-I, HDM-II, VDM-I, VDM-I and DDM), HDM-II agrees well with actual versus predicted discharge for all datasets. COHM method also found to be suitable for all datasets that consider the effect of momentum transfer. On the other hand, among the approaches, single-channel method (assumed "*n*" and different formulae for prediction of Manning's "*n*") and LSCS method give quiet significant result. Figure 16.20 signifies mean percentage errors with standard deviation of three best methods for the datasets. Values of standard deviation of their corresponding mean error gave a clear-cut idea about the range of the mean error and help to identify the appropriate method.

**Fig. 16.16** Graph between actual discharge and predicted discharge for NITR-I dataset

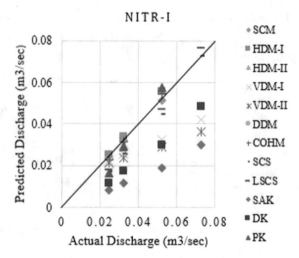

**Fig. 16.17** Graph between actual discharge and predicted discharge for NITR-II dataset

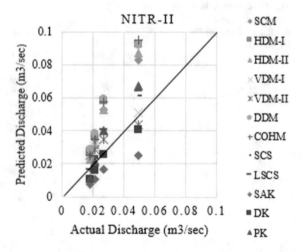

Similarly, for NIT Rourkela channel datasets, these approaches are applied, and the mean percentage of errors have been evaluated and plotted in Fig. 16.21. The same scenario has been encountered as previously taken datasets. The HDM-II, COHM and LSCS give the most appreciable results with suitable range of standard deviation, as shown in Fig. 16.22.

For better understanding of the behavior of flow mechanism at overbank stages, stage–discharge curves are plotted as shown in Figs. 16.23, 16.24, 16.25, 16.26, 16.27, 16.28, 16.29, 16.30 and 16.31. For B21 dataset it can be seen that when depth of water in over bank stage is low, all methods are found to be suitable, but when depth of flow increases, some try to over-predict and some tend to under-predict the actual discharge. HDM-II, COHM and LSCS method over-predict the actual discharge. COHM over-predicts slightly than HDM-II and LSCS. For PIII datasets,

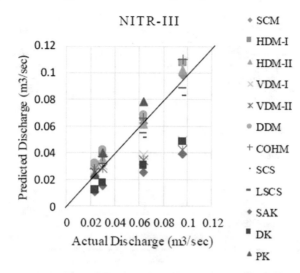

**Fig. 16.18**  Graph between actual discharge and predicted discharge for NITR-III dataset

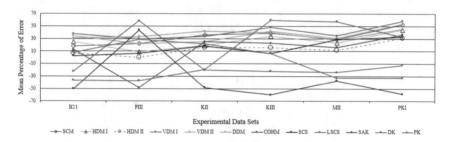

**Fig. 16.19**  Mean percentage errors of various approaches for different data series

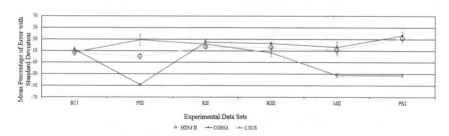

**Fig. 16.20**  Mean percentage of errors with standard deviation of best approaches for different data series

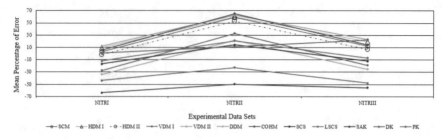

**Fig. 16.21** Mean percentage of errors of various approaches for NIT Rourkela data series

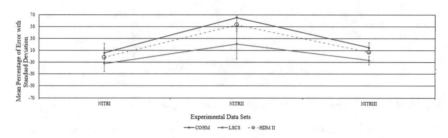

**Fig. 16.22** Mean percentage of errors with standard deviation of best approaches for NIT Rourkela data series

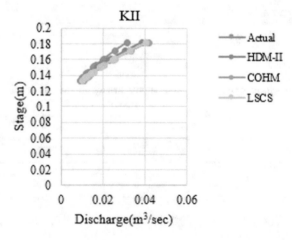

**Fig. 16.23** Stage-Discharge curve for KII dataset

HDM-II gives more accurate results than the COHM and DK method. HDM-II under-predicts the discharge when flow depth in overbank region is low and over-predicts the discharge, when flow depth increases. COHM under-predicts and LSCS overestimates the actual discharge value for PIII datasets. For the analysis of KII datasets, HDM-II, COHM, LSCS try to over-predict the actual discharge. COHM

**Fig. 16.24**  Stage-Discharge
curve for KIII dataset

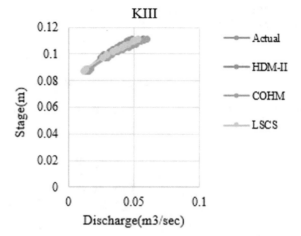

**Fig. 16.25**  Stage-Discharge
curve for B21 dataset

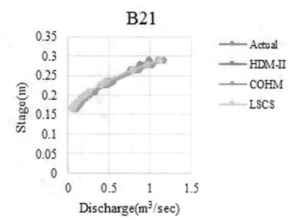

**Fig. 16.26**  Stage-Discharge
curve for PIII dataset

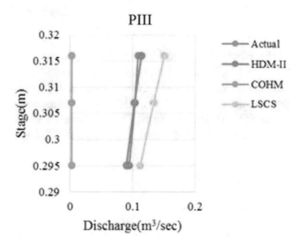

**Fig. 16.27** Stage-Discharge curve for MII dataset

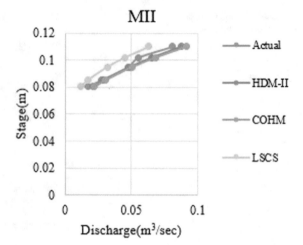

**Fig. 16.28** Stage-Discharge curve for PKI dataset

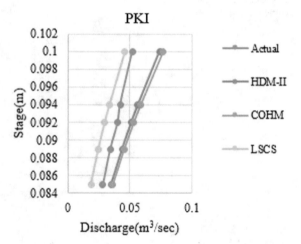

over-predicts more than HDM-II. The same scenario can be seen for KIII datasets too. The LSCS method was found to be more accurate with the actual discharge. For MII datasets HDM-II, COHM method over-predicts and LSCS method under-predicts the actual discharge. For PKI datasets, the same results have been found out as for MII datasets. In NITR-I datasets, HDM-II and LSCS under-predict, and COHM over-predicts the actual discharge. In NITRKL II and III datasets, LSCS gave

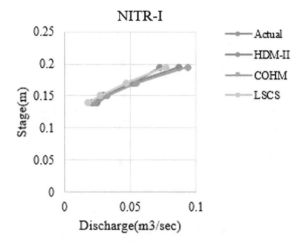

**Fig. 16.29** Stage-Discharge curve for NITR-I dataset

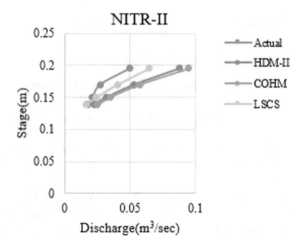

**Fig. 16.30** Stage-Discharge curve for NITR-II dataset

accurate results. In doubly meandering channel (NITR-II), LSCS over-predicts and in NITR-III channel datasets LSCS under-predicts.

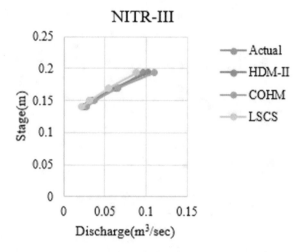

**Fig. 16.31** Stage-Discharge curve for NITR-III dataset

## 16.6  Conclusions

The above work led to the following conclusions:

1. Among all the divisional methods, HDM-II gives the most appropriate discharge results.
2. Predicted discharge by coherence method is also equally suitable for discharge estimation.
3. From the various Manning's roughness prediction approaches, the LSCS method gives more acceptable values that can be confidently used in SCM for better discharge prediction.
4. From the stage–discharge relationship, it is found that all the three best methods (HDM-II, COHM and LSCS) tend to converge with actual discharge, when flow depth in the overbank stages is low and they vary considerably when flow depth increases.
5. HDM-II over-predicts the actual discharge for most of the cases, while LSCS generally under-estimates the actual discharge. In doubly meandering channels, it over-estimates, while coherence method over-predicts the actual discharge.
6. However, the methods need to be tested with more channel data of various geometry and configurations before a last word can be said confidently.

**Acknowledgements**  The authors are thankful to the Department of Science and Technology, Government of India, for the financial support to carry out the research work smoothly during the period of study. The authors are thankful to Department of Civil Engineering, NIT Rourkela, for providing required facilities. The first author would like to give heart-felt thanks to second and third authors for guidance and assistance throughout the research work.

# References

Ackers P (1992) Hydraulic design of two stage channels. In: Proceedings of the Institution of Civil Engineers, Water, Maritime and Energy, December, Paper No. 9988, pp 247–257

Ackers P (1993a) Stage-discharge functions for two-stage channels. Impact New Res J Inst Water Environ Manage 7(1):52–61

Ackers P (1993b) Flow formulae for straight two-stage channels. J Hydra Res IAHR 31(4):509–531

Arcemen GJ Jr, Schneider VR (1963) Guide for selecting roughness coefficient n values for channels. Soil Conservation Services (SCS), U. S. Department of Agriculture, Washington, DC

Dash SS, Khatua KK (2016) Sinuosity dependency on stage discharge in meandering channels. J Irri Drain Eng ASCE. ISSN 0733-9437

Ervine DA, Willetts BB, Sellin RHJ, Lorena M (1993) Factors affecting conveyance in meandering compound flows. J Hydra Eng ASCE 119(12):383–1399

Hooke (1974) Distribution of sediment and shear stress in a meander bend. Uppsala Univ. Naturgeografiska Inst. Rapport 30, 58 pp

James CS, Wark JB (1992) Conveyance estimation for Meandering channels. Rep. SR 329, HR Wallingford, Wallingford, U.K., Dec

Khatua KK (2008) Interaction of flow and estimation of discharge in two stage meandering compound channels. Thesis Presented to the National Institute of Technology, Rourkela, in partial fulfilment of the requirements for the Degree of Doctor of Philosophy

Lambert MF, Myers WR (1998) Estimating the discharge capacity in straight compound channels. In: Proceedings of the Institution of Civil Engineers, Water, Maritime and Energy, No. 130, pp 84–94

Mohanta A, Patra KC (2019) MARS for prediction of shear force and discharge in two-stage meandering channel. J Irrigat Drain Eng 145(8):04019016

Mohanta A, Patra KC, Sahoo BB (2018) Anticipate Manning's coefficient in meandering compound channels. Hydrology 5(3):47

Mohanty PK (2014) Flow analysis of compound channels with wide flood plains. Thesis Presented to the National Institute of Technology, Rourkela, in partial fulfilment of the requirements for the Degree of Doctor of philosophy

Patra KC, Kar SK (2000) Flow interaction of meandering river with flood plains. J Hydra Eng ASCE 126(8):593–603

Pradhan A, Khatua KK (2017) Assessment of roughness coefficient for meandering compound channels. KSCE J Civil Eng 1–13

Shiono K, Al-Romaih JS, Knight DW (1999) Stage-discharge assessment in compound meandering channels. J Hydra Eng 125(1):66

Toebes GH, Sooky AA (1967) Hydraulics of meandering rivers with floodplains. Proc ASCE J Waterways Harbor Divi 93(WW2):213–236

# Chapter 17
# Performance Study of Cross Flow Hybrid Hydrokinetic Turbine

**Gaurav Saini and R. P. Saini**

## Nomenclature

| | |
|---|---|
| $C_P$ | Power coefficient |
| $\Lambda$ | Tip speed ratio |
| Dd | Darrieus rotor diameter |
| Hd | Darrieus rotor height |
| Ds | Savonius rotor diameter |
| Hs | Savonius rotor height |
| H | Clearance |
| H | Solidity |
| BR | Blockage ratio |
| $\Theta$ | Attachment angle |
| W | Width of channel |
| Hw | Height of channel |
| L | Length of channel |

G. Saini (✉) · R. P. Saini
Department of Hydro and Renewable Energy, Indian Institute of Technology Roorkee, Roorkee
247667, Uttarakhand, India
e-mail: gaurav161990@gmail.com

R. P. Saini
e-mail: saini.rajeshwer@gmail.com

A. Pandey et al. (eds.), *Hydrological Extremes*, Water Science
and Technology Library 97, https://doi.org/10.1007/978-3-030-59148-9_17

## 17.1  Introduction

The working principle of this hydro turbine technology is similar to wind energy technology (Kumar and Sarkar 2016). Various types of turbines can be used to harness the potential of hydrokinetic energy. These turbines are known as hydrokinetic turbines. Savonius and Darrieus are the most common type of rotor used for hydrokinetic turbines. The Savonius rotor works on the principle of drag force and have the excellent ability to start at low flow velocity of water but suffers with the low performance (Savonius 1931). On the other hand, Darrieus rotor works on the principle of lift force development and have the better performance but have the poor starting capabilities.

Various configurations of hybrid rotor are discussed in the literature. Zamani et al. (2016) developed a new model of hybrid hydrokinetic rotor having the characteristics of Savonius and Darrieus rotor. Gupta et al. (2008) carried out a comparative analysis to compare the performance of hybrid rotor with the single Savonius rotor. Another study survey was carried out by Chawla et al. (2016), which gave the parametric analysis of hybrid rotor in order to understand the performance under various operating conditions.

Gavalda et al. (1990) investigated experimentally a modified hybrid rotor and compared the results with conventional hybrid rotor under similar operating conditions. Abid et al. (2015) designed and tested a hybrid rotor on the base of self-starting capabilities. The performance and starting characteristics of a hybrid rotor with the use of deflector plate on the returning side blade of Savonius rotor was investigated. It has been found that the hybrid rotor with single deflector plate produces more torque and has better performance than solo Darrieus rotor. The placement of Savonius rotor also plays an important role in hybrid rotor, and it has been found that the arm of Savonius and Darrieus rotor should not be axial. Further, the performance of hybrid rotor is better if the Savonius part is near the central shaft of combined rotor (Sahim et al. 2013). In order to understand the placement of Savonius in hybrid rotor, Liang et al. (2017) investigated computationally, the parameters of hybrid rotor, that is, radius ratio and attachment angle and found that the best rotor is achieved with radius ratio of 0.25 and attachment angle of 0°.

The objective of the study in the present paper is to explore the performance and flow field of hybrid rotor under different operating conditions. The design of hybrid rotor has been conceptualized with the combination of Savonius and Darrieus rotor by placing the Savonius rotor inside the Darrieus rotor (Wakui et al. 2005). In order to carry out the analysis, a numerical approach is used, and the commercially available software ANSYS (v.18.1) is used to analyze the rotor.

## 17.2 Design Parameters of Hybrid Rotor

The hybrid rotor consists of Savonius and Darrieus rotor. The Savonius part of hybrid rotor is made of two semi-circular blades and the Darrieus rotor consists of three NACA0018 hydrofoils. The NACA0018 hydrofoil of symmetrical shape is considered. The rotor is designed for a water velocity of 1.5 m/s. Further, the blockage ratio and solidity for Darrieus rotor are considered as 30% and 0.42, respectively. The design of Savonius rotor of hybrid is based on the parameters of Darrieus rotor. The radius of Savonius rotor is considered as 0.4 times of the radius of Darrieus rotor to keep a radius ratio of 0.4. The Savonius rotor is attached to Darrieus rotor arm with an angle of 45° to maintain a fixed attachment angle optimized by Liang et al. (2017). For a fixed aspect ratio as 1.58, the height of the Savonius rotor is calculated as given by Kumar and Saini (2017). The details of the parameters for hybrid rotor and flow domain (channel) are given in Tables 17.1 and 17.2, respectively.

**Table 17.1** Parameters for hybrid rotor

| Parameters | Values |
|---|---|
| Hydrofoil | NACA0018 |
| Number of blades | 3 |
| Diameter of Darrieus rotor (Hd) | 225 mm |
| Height of Darrieus rotor (Hd) | 250 mm |
| Aspect ratio of Darrieus rotor | 1.1 |
| Solidity ($\eta$) | 0.42 |
| Diameter of Savonius rotor (Hd) | 90 mm |
| Height of Savonius rotor (Hs) | 142.2 mm |
| Aspect ratio of Savonius rotor | 1.58 |
| Number of Savonius blades | 2 |
| Types of blade | Semicircle |
| Radius ratio | 0.4 |
| Attachment angle ($\Theta$) | 45° |

**Table 17.2** Details for flow domain

| Parameters | Values |
|---|---|
| Width of channel (W) | 500 mm |
| Height of water in channel (Hc) | 375 mm |
| Length of channel (L) | 2500 mm |
| Clearance (h) | 75 mm |
| Blockage ratio (BR) | 30% |

## 17.3 Modeling and Grid Generation

The numerical simulation was done with the modeling of the problem domain. In the present case, the entire modeling of the rotor and its flow domain is carried out in the design module of ANSYS (v.18.1). A 3D model of hybrid rotor is created considering the dimensions given in Table 17.1. Simultaneously, the 3D flow domain is created along with the rotor as per the dimensions given in Table 17.2. The 3D hybrid rotor and rotor with its flow domain is shown in Figs. 17.1 and 17.2, respectively.

The model preparation is the first step for numerical simulation. Now, the 3D model of hybrid rotor is exported to the mesh module of ANSYS (v.18.1). Before

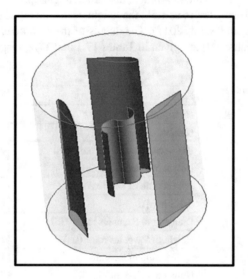

**Fig. 17.1** 3D model of hybrid rotor

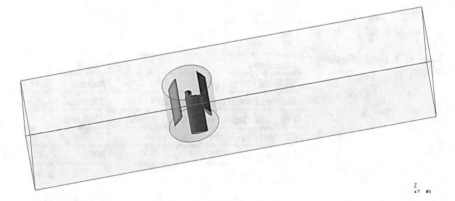

**Fig. 17.2** Hybrid rotor along with flow domain

exporting the model into the mesh module, the flow domain is selected and named for various parts, that is inlet, outlet, surface, wall and turbine. Further, the enclosure and the flow domain are set as fluid from the continuum media as fluid/solid. In the mesh generation the unstructured tetrahedron mesh is selected due to complex geometry of rotor. The inflation is applied by keeping the y + value less than 10 and providing the first layer thickness and growth ratio as 1.1 on the turbine and enclosure part of model. Elements (6.8 million) with aspect ratio (32), orthogonal quality (0.88) and skewness (0.88) are found to be satisfactory for mesh quality. The meshed rotor along with flow domain is shown in Figs. 17.3 and 17.4a and b, respectively.

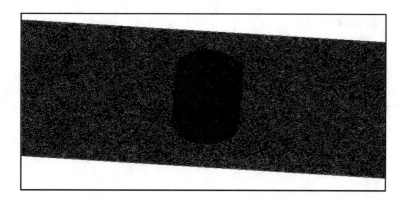

**Fig. 17.3**  3D mesh of flow domain along with rotor

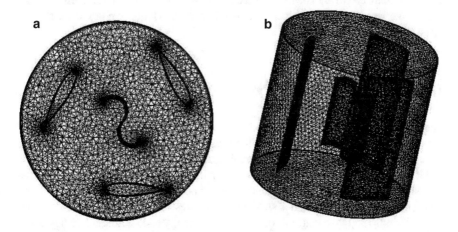

**Fig. 17.4  a** Boundary layer mesh at blades, **b** 3D mesh on hybrid rotor

**Table 17.3** Summary of boundary conditions

| Location | Boundary condition type |
|---|---|
| Inlet | Velocity inlet |
| Outlet | Pressure outlet |
| Upper surface | Symmetry |
| Other faces of channel | Wall (stationary) |
| Turbine | Wall (stationary) |
| Interfaces between channel and enclosure | Interface |

## 17.4  Solver Setup

The 3D meshed model has been solved in the Fluent module of ANSYS (v.18.1). For solution in Fluent, the msh file has been exported to the Fluent, and turbulence modeling was made by applying the desired boundary conditions over the named selection.

### 17.4.1  Boundary Conditions

In order to carry out the numerical simulation, boundary conditions are required to apply at the boundaries selected in named selections. The wrong selected boundary conditions can lead to unexpected results. The flow domain has been defined in Fluent and further exported to the solver. The detailed summary of boundary conditions is given in Table 17.3.

### 17.4.2  Turbulence Modeling

The numerical simulation of the model requires the selection of turbulence model for solution of RANS equations. For the present analysis, realizable k-ε turbulence model has been considered. The use of realizable k-ε turbulence model is quite simple and it solves the turbulence flow with high accuracy for complex geometry. Simultaneously, this turbulence model provides superior performance for the flows involving rotation, boundary layers under strong adverse pressure gradients, separation, and recirculation. The two equations, one for kinetic energy and another for energy dissipation rate, are required to solve in realizable k-ε turbulence model. For the present study multiple reference frame (MRF) approach has been applied to consider the rotating behavior of rotor.

## 17.5  Results and Discussion

The performance characteristics of hybrid rotor have been analyzed and a velocity of 1.5 m/s was considered for the analysis. Different angular velocity of rotor is considered with a tip speed ratio (TSR) range varying from 0.8 to 2.0. The performance of the hybrid rotor is analyzed in terms of coefficient of power and flow around the rotor (velocity, pressure).

### 17.5.1  Flow Contours

The contours for the variation in the pressure and velocity on the hybrid rotor is shown in Figs. 17.5 and 17.6, respectively. It has been observed from the pressure contours that the high-pressure zone is in the inner side of Darrieus blade and at the advancing side blade of Savonius rotor. The low-pressure zone has been observed on the outer side of Darrieus blade and on the returning side blade of Savonius rotor. The maximum and minimum pressure during the rotation of rotor are observed as 5973 Pa and −5700 Pa, respectively. Similarly, the velocity contours plot shows the maximum velocity at the outer side of Darrieus blade and the low velocity (wake zone) is found at the advancing side blade of Savonius rotor.

**Fig. 17.5** Pressure contour around the rotor

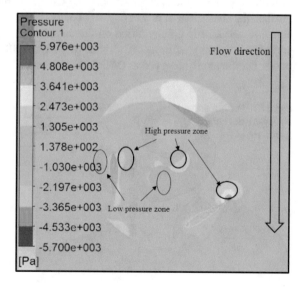

**Fig. 17.6** Velocity contour
around the rotor

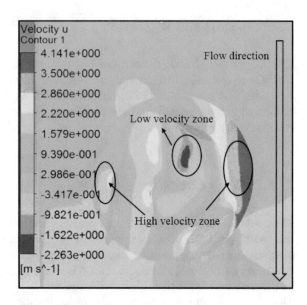

## 17.5.2  *Performance of Hybrid Rotor Turbine*

In order to analyze the performance of hybrid rotor, torque is directly calculated under
different angular velocity of rotor with the CFD post module of ANSYS (v.18.1).
Further, the coefficient of power ($C_P$) has been calculated by using the kinetic energy
equation and kinetic energy based on the calculated torque. Figure 17.7 shows the
variation in power coefficient ($C_P$) with respect to TSR. It is found that maximum
power coefficient is observed as 0.29 for a TSR value of 1.6.

**Fig. 17.7** Performance
characteristics curve for
hybrid rotor

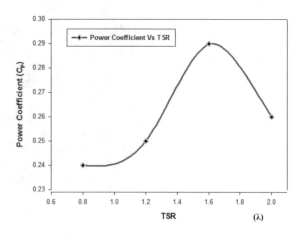

## 17.6 Conclusions

In the present paper, a numerical analysis carried out for hybrid rotor hydrokinetic turbine has been presented. The analysis of hybrid rotor is carried out under different angular velocity of rotor for a constant flow velocity of 1.5 m/s. The flow field around the rotor is discussed and it has been found that the net torque is the resultant of the lift and drag force produced due to the combination of Darrieus and Savonius rotor. Based on the analysis, it has been found that maximum power coefficient of rotor is observed as 0.29 for a TSR of 1.6. The increase in the angular velocity of rotor reduces the power coefficient, because the Savonius rotor tends to retard the rotation at higher peripheral velocity. The findings of the present study may be useful to develop further the hybrid rotor for different operating conditions.

## References

Abid M, Karimov KS, Wajid HA, Farooq F, Ahmed H, Khan OH (2015) Design, development and testing of a combined Savonius and Darrieus vertical axis wind turbine. Iranica J Energy Environ 6(1):1–4

Chawla S, Chauhan A, Bala S (2016) Parametric study of hybrid Savonius-Darrieus turbine. In: 2015 2nd International conference on recent advances in engineering and computational sciences, RAECS-2015 December 21–22, 2015 Chandigarh, India, pp 3–7

Gavalda J, Massons J, Diaz F (1990) Experimental study on a self-adapting Darrieus-Savonius wind machine. Solar Wind Technol 7:457–461

Gupta R, Biswas A, Sharma KK (2008) Comparative study of a three-bucket Savonius rotor with a combined three-bucket Savonius-three-bladed Darrieus rotor. Renew Energy 33:1974–1981

Kumar A, Saini RP (2017) Performance analysis of a single stage modified Savonius hydrokinetic turbine having twisted blades. Renew Energy 113:461–478

Kumar D, Sarkar S (2016) A review on the technology, performance, design optimization, reliability, techno-economics and environmental impacts of hydrokinetic energy conversion systems. Renew Sustain Energy Rev 58:796–813

Liang X, Fu S, Ou B, Wu C, Yh Chao C, Pi K (2017) A computational study of the effects of the radius ratio and attachment angle on the performance of a Darrieus-Savonius combined wind turbine. Renew Energy 113:329–334

Sahim K, Santoso D, Radentan A (2013) Performance of combined water turbine with semielliptic section of the savonius rotor. Int J Rotat Mach 985943:1–5

Savonius S (1931) The S-rotor and its applications. Mech Eng 53:333–338

Wakui T, Tanzawa Y, Hashizume T, Nagao T (2005) Hybrid configuration of darrieus and savonius rotors for stand-alone wind turbine-generator systems. Electr Eng Jpn 150:13–22

Zamani M, Nazari S, Moshizi SA, Maghrebi MJ (2016) Three dimensional simulation of J-shaped Darrieus vertical axis wind turbine. Energy 116:1243–1255

# Chapter 18
# Scaling of Open Channel Flow Velocities in Emergent, Sparse and Rigid Vegetation Patch with Rough Bed Interior of the Patch

**Chitrangini Sahu and Prashanth Reddy Hanmaiahgari**

## 18.1 Introduction

In this study, analysis of important turbulent features interior of a sparse vegetation patch in an open channel in turbulent flow condition is performed. The three-dimensional instantaneous velocities were measured with a down-looking ADV. The vegetation patch was placed in the middle cross-section of the channel. Uniform, rigid, acrylic cylindrical rods were used as vegetation stems. These rods are arranged in an array with regular spacing. Emergent vegetation canopy in OCF affects mass transfer, momentum transfer, hydraulic roughness, sediment transport, flow velocity, bottom shear stress, turbulence, fish, and species life. Further, flow through emergent canopy is characterized by sharp velocity gradients and discontinuity in drag at the interface producing shear layer formation between the interior and exterior flows. Discharge capacity also has been significantly affected due to the vegetation across the flow, so the presence of vegetation can be of major use for flood protection and flow management purposes. Various studies on the turbulence characteristics and mean fully developed flow through both submerged and emergent vegetation have been performed previously. It has been postulated in previous studies, where experiments were performed using cobbles and gravels as roughness particles, that the highly rough particles at the bed impact turbulence in inner and outer layers. This type of flow is typical in streams with high topography such as rivers in mountainous regions and upland first-order streams. Fox and Stewart (2014) carried out mixed

C. Sahu · P. R. Hanmaiahgari (✉)
Department of Civil Engineering, Indian Institute of Technology Kharagpur, Kharagpur 721302, West Bengal, India
e-mail: hpr@civil.iitkgp.ac.in

C. Sahu
e-mail: chitrangini.1702@gmail.com

© The Editor(s) (if applicable) and The Author(s), under exclusive license to Springer Nature Switzerland AG 2021
A. Pandey et al. (eds.), *Hydrological Extremes*, Water Science and Technology Library 97, https://doi.org/10.1007/978-3-030-59148-9_18

scaling for turbulent open-channel flow over gravel and cobble beds. Using asymptotic invariance principle (AIP), they performed a similarity analysis by means of flow as a balanced TBL. Equations developed using a combination of two scales for similarity analysis gave good results for logarithmic and outer layers. Fox and Stewart (2014) also performed statistical analysis which gave comparable results with that of George and Castillo (1997) and found that AIP equilibrium method is more suitable for the case of integral analysis. By using the lab scale experiments and field data, the analysis has demonstrated that goodness of fit is a good method. Tang and Hsieh (2015) carried out experiment related to dynamics analysis of vegetated water flow in a laboratory. Depending on its deformation, they classified vegetation as rigid or flexible. Most of the derived semi-empirical and empirical formulae for the flow through vegetation are related to drag coefficient and resistance force. Nepf and Vivoni (2000) added up more understanding about the impacts of vegetation on flow characteristics by performing experiments in open channel flume with model vegetation. They formulated the boundary-value problem by deriving dimensionless governing equations and boundary conditions. They also proposed a new parameter pertaining to permeability of a porous medium which shows the effects on the shear stress distribution. Neary et al. (2011) examined the effects of vegetation on turbulence sediment transport and stream morphology. Some of the key parameters like flow Reynolds number, canopy density, and submergence ratio are just a few of the many parameters that influence the spatial variability of the flow, momentum transfer, vortex shedding and dissipation, and instantaneous stresses that are known to affect sediment and morphological processes in river. Maji et al. (2016, 2017) observed that time averaged velocity profile becomes uniform after the half length of the patch.

## 18.2 Experimental Methodology

### 18.2.1 Experimental Flume Design

A recirculating flume of width 0.91 m, length 12 m, and depth 0.7 m with very mild longitudinal slope was used to conduct experiments. A tail gate was provided at the downstream end of the flume to control the flow depth.

### 18.2.2 Measuring Equipment

A three-dimensional acoustic Doppler velocimeter (ADV) with four down-looking probes was used to record the flow characteristics. It has an emitter at the center of the probes which emits acoustic pulses. Configuration of three acoustic sensors helps in detecting the reflected acoustic pulse and so measures the Doppler shift for a particle within the flow. The data are to be transmitted to the computer system via a conditioning module. These transmitted data are then interpreted to give the recorded flow characteristics such as velocity, turbulence intensities, and Reynolds shear stresses at a point. ADV measures the characteristics in three directions, that is, streamwise, transverse, and vertical directions. Data recording duration was 5 min at all points with measurement frequency of 100 Hz. Near to the channel bed and sidewall boundaries, the maximum number of measurements points were considered. Flow depth was measured using a point gauge.

### 18.2.3   Model of Vegetation Patch

The test section starts from a point 700 cm from the flume entrance and is 300 m long. The vegetation patch was located at the middle of the test section as shown in Fig. 18.1. To make the bed rough, circular marbles of diameter 2.5 cm were fixed at the patch base and over them vegetation rods were planted. The vegetation types adopted here were emergent, sparse, and rigid. These rigid vegetations were made from uniform acrylic cylindrical rods of diameter 0.64 cm and height 30 cm. The rods were fixed perpendicularly in an array arrangement of $7 \times 10$ on a perspex sheet which was fixed to channel bed. They were placed at a uniform streamwise gap of 9 cm and a lateral gap of 4 cm. The schematics of vegetation patch are shown in Figs. 18.2 and 18.3.

Velocity measurements were taken at streamwise distances $x = 0, 50, 80, 95.5$, 113.5, 131.5, 140.5, 149.5, 167.5, 185.5, 220, 240, 285.5, and 300 cm. At each streamwise section, lateral measuring points considered were 16.7, 35.4, 39.4, 43.4, 47.4, 51.4, 55.4, and 74.3 cm from the left wall of the flume. The point 47.4 cm in the mid-section and line passing through it was shown in Fig. 18.3.

**Fig. 18.1** Experimental setup

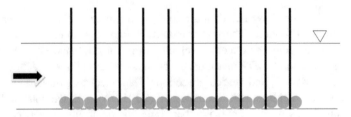

**Fig. 18.2** Side view of vegetation patch with marbles

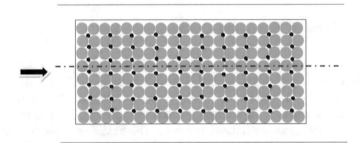

**Fig. 18.3** Top view of vegetation patch with marbles

## 18.2.4 Post-processing and Correction of Data

Three-dimensional coordinate system has been adopted where $x$, $y$, and $z$ denoted streamwise, transverse, and vertical directions, respectively, for the post-processing of data obtained from the ADV. To minimize the Doppler noise effect, the data, spike filtering, signal-to-noise ratio (SNR) threshold techniques, and correction were used. To further improve the accuracy of the turbulence measurements, low SNR and low correlations were removed. The spikes in velocity measurements removed from the raw time series were replaced by applying cubic polynomial interpolation. The present experimental conditions are given in Table 18.1. The average and maximum flow velocities refer to the approach flow.

## 18.2.5 Scaling Analysis

In this study, a length scale is proposed to normalize velocity profiles for vegetated open channel flow with rough bed. Gravels, submerged vegetation, dunes, and so on represent large-scale roughness in open channel flow and such flow differs considerably from those over smooth boundaries. Emergent vegetation with regular arrangement has been considered in this study. Flow is obstructed due to the presence of vegetation and so the velocity is reduced. Scaling is done to check the uniqueness

**Table 18.1** Experimental conditions

| Flow depth ($h$) | 15 cm |
|---|---|
| Average flow velocity $\left( \bar{u} \right)$ | 29 cm/s |
| Maximum flow velocity ($u_m$) | 30 cm/s |
| Reynolds number | 43,500 |
| Vegetation patch length | 81 cm |
| Vegetation patch width | 24 cm |
| Number of cylinders | 70 |
| Diameter of cylinders | 0.64 cm |
| Height of cylinders | 30 cm |
| c/c spacing of cylinders: $x$<br>$z$ | 9 cm<br>4 cm |
| Vegetation density | 0.0178/cm |

of the profiles that are obtained from data measured under different flow conditions. This operation helps to find out a standard distribution type of the profile. To normalize different profiles, some length scales are proposed and checked whether they normalize the profiles as desired.

## 18.3   Results and Discussion

Variation in shear velocity $u_*$ inside the vegetation patch is shown in Fig. 18.4. It is clear from the plot that shear velocity is maximum at the center of the patch in the streamwise direction. Figure 18.5 shows the plot of $z/h$ against $u/u_m$ with data collected at different streamwise points inside the patch along the midsection of the cross-section. The data are confined by a power law with two different values of m [power law: $u/u_m = (z/h)^{1/m}$]. The wide range of exponent (m-values) and the high degree of scattering imply that scales are inappropriate.

**Fig. 18.4** Shear velocity inside vegetation patch

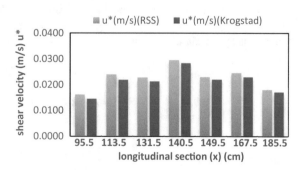

**Fig. 18.5** Scaling of
streamwise velocity in outer
layer variables

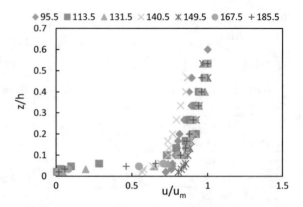

**Fig. 18.6** Scaling of
streamwise velocity using
shear velocity

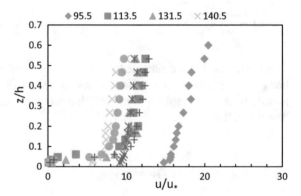

Figure 18.6 shows the plot of $z/h$ against $u/u_*$ with data collected at different streamwise points inside the patch along the midsection of the cross-section. The normalized profiles are unique, but they lie in a wide range and thus the scales are inappropriate. In Figs. 18.7 and 18.8, plots of $z/h$ against Reynolds shear stress scaled using $u_m^2$ and $u_*^2$, respectively, are shown. It is clear that the shear velocity scale is inappropriate as obtained profiles are not unique. Similarly, velocity defect was normalized with maximum velocity and shear velocity as presented in Figs. 18.9 and 18.10. Figure 18.10 shows that the spreading of the data points scaled with shear velocity is wide. Comparing the Figs. 18.9 and 18.10 shows that the maximum velocity is appropriate for normalizing velocity defect for such flows.

**Fig. 18.7** Scaling of Reynolds shear stress using maximum velocity

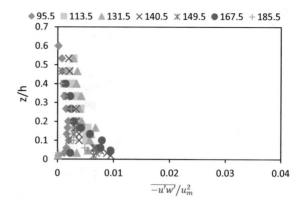

**Fig. 18.8** Scaling of Reynolds shear stress with shear velocity

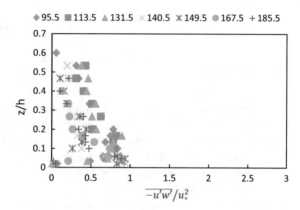

**Fig. 18.9** Scaling of velocity-defect using maximum velocity

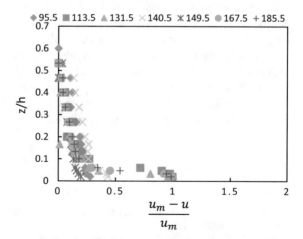

**Fig. 18.10** Scaling of velocity-defect using shear velocity

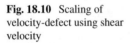

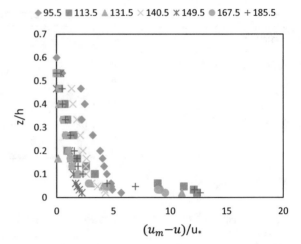

## 18.4 Conclusions

The laboratory experiments were carried out with emergent, sparse, and rigid vegetation patch in open channel flow. Three-dimensional velocities and depths were collected at different points inside the patch along the streamwise mid-section. The presence of emergent vegetation and roughness over the channel bed found to be obstructing the open channel flow and affecting the flow variables. For the scaling analysis, the flow depth is taken as the length scale and the maximum streamwise velocity and shear velocity are used as velocity scales. The scaling of streamwise velocity, turbulence intensities, and Reynolds shear stress were done with both the maximum streamwise velocity and shear velocity as velocity scales to normalize the profiles. The streamwise velocity was scaled well using the maximum streamwise velocity, but the range of exponent values of power law was found to be quite high and which is typical of vegetated flows. Velocity defect law scaled with shear velocity gives a good profile, but the spread of data points is still wide. Finally, it was concluded that the maximum velocity and flow depth are appropriate for scaling flow variables of vegetated flows with high roughness in the interior of the patch.

## References

Fox JF, Stewart RL (2014) Mixed scaling for open-channel flow over gravel and cobbles. J Eng Mech 140(10):06014010

George WK, Castillo L (1997) Zero-pressure-gradient turbulent boundary layer. Appl Mech Rev 50(12):689–729

Maji S, Pal D, Hanmaiahgari PR, Pu JH (2016) Phenomenological features of turbulent hydrodynamics in sparsely vegetated open channel flow. JAFM 9(6):2865–2875

Maji S, Pal D, Hanmaiahgari PR, Gupta UP (2017) Hydrodynamics and turbulence in emergent and sparsely vegetated open channel flow. Environ Fluid Mech 17:853–877

Neary VS, Constantinescu SG, Bennett SJ, Diplas P (2011) Effects of vegetation on turbulence, sediment transport, and stream morphology. J Hydraul Eng 138(9):765–776

Nepf HM, Vivoni ER (2000) Flow structure in depth-limited, vegetated flow. J Geophys Res 105(C12):28547–28557. https://doi.org/10.1029/2000JC900145

Tang CY, Hsieh PC (2015) Dynamic analysis of vegetated water flows. J Hydro Eng 21(2):04015064

# Chapter 19
# Energy Gain from Tehri PSP Due to Adoption of Variable Speed Technology

**L. P. Joshi and Nayan Raturi**

## 19.1 Introduction

Electrical energy plays a vital role in the development of country's economy. In order to ensure sustainable growth with minimum environmental damages, it is imperative that the provisions of Paris Agreement Convention for Climate Change that came into being from 4 November 2016 are followed in true sense. One of the vital conditions of the agreement for holding the increase in global average temperature to well below 2 °C above pre-industrial levels and to take further necessary measures to contain it to 1.5 °C above pre-industrial levels shifted the global attention toward renewable energy. However, renewable sources like wind, solar and so on have posed a great challenge for the grid stability by virtue of their highly variable and unpredictable nature. This has shifted the focus of the policy makers toward the development of various energy storage technologies, particularly the pumped storage energy storage plants.

The pumped storage plants having fixed speed synchronous machines contribute to about 140 GW in the global capacity. However, fixed speed synchronous machines suffer from the following limitations:

i. Power availability in grid for pumping operation
ii. Annual variation in dam water head level
iii. Grid demand-based variable power generation.

These factors severely impact the efficiency of the synchronous machines (Cavazzini and Pérez-Diaz 2014; Lefebvre et al. 2015). The impact on efficiency

L. P. Joshi · N. Raturi (✉)
Design Department, THDC India Ltd, Rishikesh, India
e-mail: nayanraturi@gmail.com

L. P. Joshi
e-mail: Joshilp3@gmail.com

has been overcome by the advancement of variable speed asynchronous machines wherein a mapping between speed of turbine and available water head can fetch maximum efficiency (Dean et al. 2010).

Synchronous machines equipped with variable speed feature for the rated capacity above 200 MW are not techno-economically viable as the power converters to be employed for such a system require high power rating. Also, the increased physical size of the inverter makes it difficult to accommodate them in underground power houses and in turn results in increased cost. Additionally, the impact on efficiency with respect to speed variation on full range of operation is same as for 10–15% speed of variation (Cavazzini and Pérez-Diaz 2014). These limitations have led to adoption of PSPs with variable speed asynchronous machines worldwide (e.g. some of the PSP based on this technology are employed in Linthal, Switzerland; 2X 400 MW PSP in Ohkawachi, Japan; 2X 390 MW PSP in Frades-II, Portugal). These variable speed machines with greater dynamic stability have the ability to perform the requisite operation without requirement of high-power converter. Tehri PSP is the first project in India to have 250 MW DFIM units having speed variation in the range of 10.73% to +8.33%. This project is under construction and is a part of Tehri Hydro Power Complex (Fig. 19.1).

## 19.2   Variable Speed Hydro Unit Generation Mode

The basic criterion for the selection of any hydro-generating unit is its maximum efficiency, which is based on the following parameters:

a. Available head
b. Discharge from the unit
c. Turbine speed.

Machine rotating at a fixed speed over the entire range of water head and discharge results in reduction in turbine performance (Singh et al. 2014). Therefore, the maximum efficiency is achieved by adjusting the turbine speed with respect to available water head and chosen output.

The study has been performed to analyze the impact of head variation over a wide range on a typical 250 MW Tehri PSP (India) doubly fed induction machine unit with fixed speed and variable speed. The parameters considered for the purpose are: (a) rated head: 188 m; (b) turbine speed: 230.77 rpm; (c) gross water head range: 127.5–224 m.

The rated output in generation mode is dependent on water head; that is lesser output electrical power at rated discharge will be generated for minimum water head. Hill curves of the turbine are used for the estimation of turbine efficiency for variable speed and fixed speed. Based on the hill curve the electrical power output with respect to water head level has been calculated and plotted graphically (shown in Fig. 19.3). It can be inferred from the graphical representation in Fig. 19.2 that at minimum water head level, that is, 127.5 m the gap in power generation for variable speed and fixed

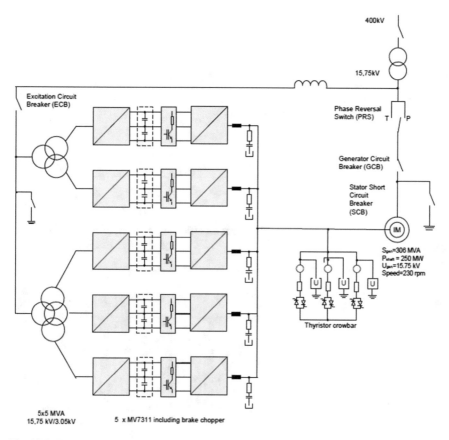

**Fig. 19.1** Tehri PSP 250 MW variable speed hydrogeneration

**Fig. 19.2** Additional power generation during variable speed mode

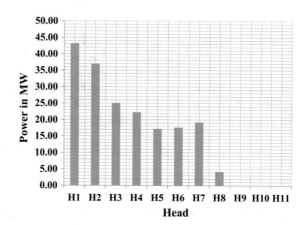

**Fig. 19.3** Comparative
graphs of a typical fixed
speed and variable speed
PSP schemes. **a** Impact on
efficiency at fixed power
output with varying water
head. **b** Power generation at
various water heads

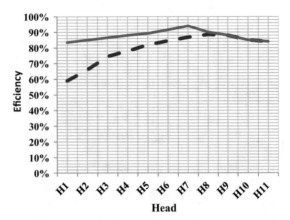

(a) **Impact on efficiency at fixed power output with
varying water head**

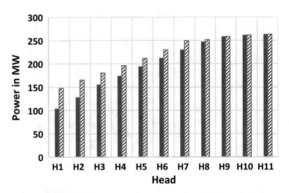

(b) **Power generation at various water heads**

speed is the highest, that is, 43.24 MW (147.37 − 104.13 = 43.24 MW) implying
that for minimum water level, the variable speed machines are more efficient than the
fixed speed machines. The analysis done for the complete year clearly indicates that
an additional gain of 8.3% will be achieved by employing a variable speed machine.
In terms of energy generation for a 1000 MW plant, this is equivalent to 127.09
MU/year. The detailed calculation is illustrated in Annexure-1.

## 19.3   Conclusion

The advantages of variable speed over fixed speed technology in terms of additional
power generation in generation mode and energy storage in pumping mode in case
of Tehri PSP have been showcased in this paper. It is evident that there could be 8.3%
additional power generation from a 250 MW machine by choosing variable speed
over the fixed speed.

Typical model of Tehri pumped storage plant for variable and fixed speed operation generation

| Head | Water head (m) | Head loss | Input power | Base input power | Input power p.u. | Rated speed | Peripheral velocity factor | Base peripheral velocity | Peripheral velocity factor p.u. | Efficiency of fixed speed | Efficiency of variable speed (%) | Output power for fixed speed | Output power for variable speed | Additional output power | Duration of generation (%) | Energy generation by fixed speed machine | Energy generation by variable speed machine | Additional generation |
|---|---|---|---|---|---|---|---|---|---|---|---|---|---|---|---|---|---|---|
| H1 | 127.5 | 3.8 | 176.49 | 268.24 | 0.66 | 234 | 108.98 | 88.40 | 1.23 | 59.0 | 83.5 | 104.13 | 147.37 | 43.24 | 11.78 | 23.18 | 32.81 | 9.63 |
| H2 | 140.0 | 3.7 | 194.47 | 268.24 | 0.72 | 234 | 103.82 | 88.40 | 1.17 | 66.0 | 85.0 | 128.35 | 165.30 | 36.95 | 11.40 | 27.65 | 35.62 | 7.96 |
| H3 | 150.0 | 3.7 | 208.74 | 268.24 | 0.78 | 234 | 100.21 | 88.40 | 1.13 | 74.5 | 86.5 | 155.51 | 180.56 | 25.05 | 7.11 | 20.90 | 24.26 | 3.37 |
| H4 | 160.0 | 3.8 | 222.86 | 268.24 | 0.83 | 234 | 96.99 | 88.40 | 1.10 | 78.0 | 88.0 | 173.83 | 196.12 | 22.29 | 7.11 | 23.36 | 26.35 | 2.99 |
| H5 | 170.0 | 3.9 | 236.99 | 268.24 | 0.88 | 234 | 94.05 | 88.40 | 1.06 | 82.0 | 89.3 | 194.33 | 211.51 | 17.18 | 7.11 | 26.11 | 28.42 | 2.31 |
| H6 | 180.0 | 3.8 | 251.40 | 268.24 | 0.94 | 234 | 91.32 | 88.40 | 1.03 | 84.5 | 91.5 | 212.43 | 230.03 | 17.60 | 7.11 | 28.55 | 30.91 | 2.36 |
| H7 | 190.0 | 3.9 | 265.52 | 268.24 | 0.99 | 234 | 88.85 | 88.40 | 1.01 | 86.8 | 94.0 | 230.34 | 249.59 | 19.25 | 7.11 | 30.95 | 33.54 | 2.59 |
| H8 | 200.0 | 3.8 | 279.93 | 268.24 | 1.04 | 234 | 86.54 | 88.40 | 0.98 | 88.5 | 90.0 | 247.74 | 251.94 | 4.20 | 7.11 | 33.29 | 33.86 | 0.56 |
| H9 | 210.0 | 4.0 | 293.91 | 268.24 | 1.10 | 234 | 84.45 | 88.40 | 0.96 | 88.0 | 88.0 | 258.64 | 258.64 | 0.00 | 7.11 | 34.76 | 34.76 | 0.00 |
| H10 | 220.0 | 4.2 | 307.90 | 268.24 | 1.15 | 234 | 82.51 | 88.40 | 0.93 | 85.0 | 85.0 | 261.71 | 261.71 | 0.00 | 7.46 | 36.90 | 36.90 | 0.00 |
| H11 | 224.0 | 4.3 | 313.46 | 268.24 | 1.17 | 234 | 81.78 | 88.40 | 0.93 | 84.0 | 84.0 | 263.31 | 263.31 | 0.00 | 19.59 | 97.49 | 97.49 | 0.00 |
| Sum | | | | | | | | | | | | | | | | 383.14 | 414.92 | 31.77 |
| 4 Units generation = | | | | | | | | | | | | | | | | 1532.57 | 1659.67 | 127.09 |
| Additional gain in % | | | | | | | | | | | | | | | | | | 8.3% |

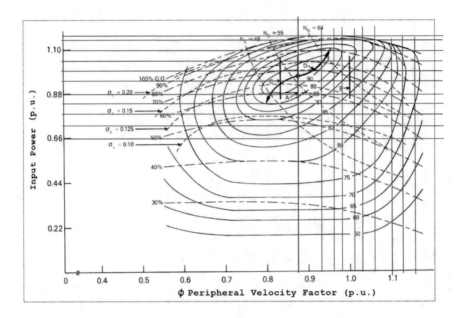

# References

Cavazzini G., Pérez-Diaz JI (2014) Technological developments for pumped hydro energy storage. European energy research alliance. Tech Rep 1–128

Deane JP, Gallachóir BÓ, McKeogh EJ (2010) Techno-economic review of existing and new pumped hydro energy storage plant. Renew Sustain Energy Rev 14(4):1293–1302

Lefebvre N, Tabarin M, Teller O (2015) A solution to intermittent renewables using pumped hydropower. Renew Energy World Mag 3:50–53

Singh RR, Chelliah TR, Agarwal P (2014) Power electronics in hydro electric energy systems–a review. Renew Sustain Energy Rev 32:944–959

# Chapter 20
# Rain Response Releases in Krishna Basin

**K. Venugopal and N. Srinivasu**

## 20.1  Issues Involved

Large basins are characterized by non-uniform distribution of rainfall. Catchment areas are large. There are dams in lower reaches. During the course of the time many dams come up in upper reaches rendering the flow uncertain in lower reaches, which in turn affect areas depending on inflows from upper reaches. This being the case, timely water availability is uncertain. During dry spells ayacut areas depending on receipts from upper reaches suffer (Smakhtin et al. 2008).

Large basins cross administrative boundaries, cutting across states. People become passionate about holding water received during monsoons in upper reaches. The releases take place only when flood is received at later stages. Skewed releases are taking place in time. All the water held at once and released continuously result in less optimal use of water during monsoon.

Nowadays, inter-basin transfer of water resources is taking place. Basins not only cut across administrative boundaries but also boundaries of basins while water is transferred from one basin to other. Water lying behind dams idle and crops withering due to dry spells in inter- and intra-basin is a regular occurrence.

Even when sufficient monsoon season is ahead, upper reaches hold water to the brim and wait until new inflows come. Water is released when they can no longer hold water. On the other hand, when monsoon fails and dry spells occur in other parts which otherwise would benefit if some releases happen in upper reaches. It saves crops from dry spells.

K. Venugopal (✉) · N. Srinivasu
Andhra Pradesh Ground Water and Water Audit Department, Amaravathi, India
e-mail: hydrologist321@gmail.com

N. Srinivasu
e-mail: srinalluri66@gmail.com

Contradictory situation happens wherein water is available in upper reaches of the basin and crops wither due to dry spell. It affects GDP of the nation because of less optimum use of available water in a basin (Water resources information system, http://www.apwrims.ap.gov.in). The basin water resources have to be viewed in totality to maximize benefits.

## 20.2   About Krishna Basin

The Krishna basin drains an area of 258,514 km² (that can be extended to 277,768 km² if large command areas located outside the hydrological boundaries of the Krishna basin, but receiving water from the Krishna River system are accounted for). Most of the basin lies on crystalline and basaltic rocks that create hard rock aquifers with low groundwater potential. The Krishna basin is subject to both the southwest and the northeast monsoon: rainfall decrease with distance inland from both coasts. This is particularly striking east of the Western Ghats where rainfall decreases from over 3,000 to 500 mm over less than 100 km. Precipitation decreases more gradually from 850–1000 mm in the Krishna delta in the east to 500–600 mm in the north-western part of the basin. The average rainfall in the basin is 840 mm, approximately 90% of which occurs during the monsoon from May to October. The climate of the Krishna basin is predominantly semi-arid to arid with potential evaporation (1457 mm a year on average) exceeding rainfall in all but three months thrice of the year during the peak of the monsoon. Irrigation is needed for agricultural development (Biggs et al. 2007; for further description of the physical setting of the Krishna basin).

## 20.3   Case Study of Krishna Basin

Krishna basin is one of the largest basins in India where water is used almost full during normal and below normal years (Gaur et al. 2007, 2008). Distribution of water is not timed properly. It is affecting water distribution in space though infrastructure is available to take water to the needy areas under dry spells. The line diagram of Krishna basin is shown in Fig. 20.1.

Krishna river runs for a length of 1295 km up to Prakasam barrage in Krishna district of Andhra Pradesh after originating at Mahadev hills in Maharashtra (NIH Roorkee, River basin information system). There are number of dams constructed across tributaries and on Krishna river. The water management becomes complex if water is held in dams for a longer period even if a monsoon period ahead is long and the likelihood of getting inflows into the dam is promising. The sub-basins of Krishna are shown in Fig. 20.1.

**Fig. 20.1**   Sub-basins of Krishna basin

River basin management for large basins like Krishna is so complex. Sub-basins receive different amounts of rainfall during different years. This year (2017) upper Bhima basin is running ahead of normal rainfall, whereas upper Krishna and Tungabhadra fell behind normal rainfall during most part of monsoon period. The inflows from Bhima basin is much more to reservoirs below Narayanpur. Different sub-basins contribute differently in different years in large basins like Krishna.

Intercepted catchments due to construction of dams across rivers and tributaries render water availability uncertain. The availability of water in time and space during dry spells becomes complex for planning water resources to meet various demands, which depends on inflows into the river during different periods of time (Bouwer et al. 2006). The inflows too dwindle over a period of time because of intercepted catchments in large basins like Krishna.

The inflows and storage in Almatti during southwest monsoon period are shown in the graphs. The daily average rainfall is taken from IMD website published daily basin-wise. The average rainfall for upper Krishna is taken into consideration to study the inflows vis-à-vis rainfall into Almatti dam. The daily rainfall basin-wise is considered as published in IMD website (sample of average rainfall for different basins as published by IMD for 6-09-2017 is shown) (Figs. 20.2, 20.3, 20.4, 20.5 and 20.6; Table 20.1).

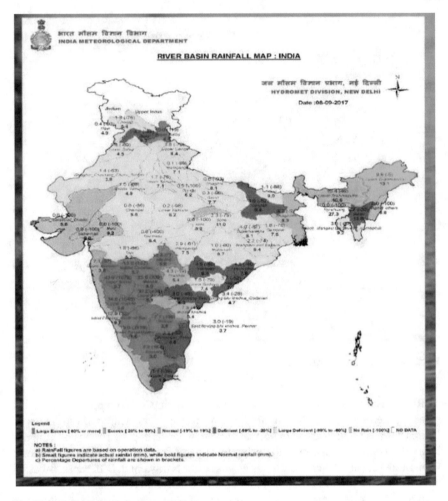

**Fig. 20.2** Rainfall in sub-basins of India

Outflows have not happened as to meet the downstream requirements, but the reservoir is kept full. There is no outflow from storage. After 19 September, heavy inflows into the river have been discharged downstream to fill big reservoirs like Srisailam and Nagarjunasagar. The worth of water during dry spell in Rayalaseema which the outflows from reservoirs would have served has been lost, as crops withered in many parts for the need of life-saving irrigation.

An exercise is done to know runoff from a spell of rainfall in upper Krishna catchment to compute approximate runoff for a cumulative rainfall from 8 September to 27 September (Table 20.2).

A cursory estimate shows that for a rainfall of 204.9 mm (240 tm.cuft) runoff of 63.25 tmc.ft has resulted in inflows, revealing rainfall runoff during later parts of monsoon has been reduced considerably from the past due to interception of natural

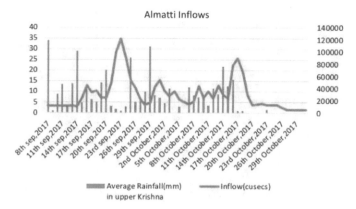

**Fig. 20.3**  Inflows into Almatti dam

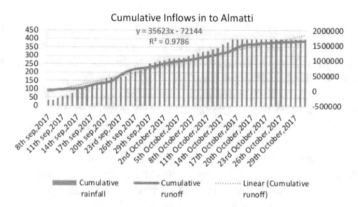

**Fig. 20.4**  Cumulative inflows into Almatti dam

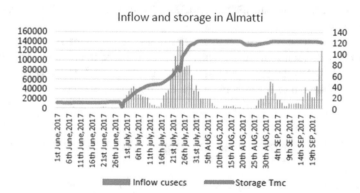

**Fig. 20.5**  Inflows and storage into Almatti

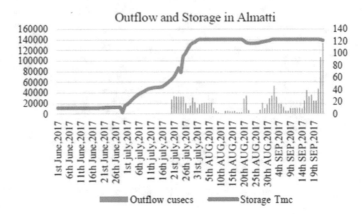

**Fig. 20.6** Outflow and storage in Almatti

catchment over a period of time, making it difficult to anticipate flow for planning as earlier. This is coming in the way of using water resources effectively.

Dwindling discharges to oceans from Krishna basin during different months from 1901 to 2007 (after Jean-Philippe Venot) (Fig. 20.7; Table 20.3).

As seen from above there is dwindling inflows into sea because of interception of upstream catchments (Venot et al. 2008). This is where intelligent water management comes into play. Simultaneous dry spells with water in upstream and full infrastructure in place give water to dry spell areas without any loss to the users under dam from releasing water, which any way made up during later rain spells from monsoons. This year about 700 mm rainfall in upper Krishna catchment filled up all the dams in upstream and releases of about 68 tmc took place. The releases mostly took place during later parts of the season (after September 18, 2017). By the time water is released irreparable damage has already been done because of dry spells in Anantapur and other Rayalaseema districts. Prakasam district too suffered due to dry spells in this season.

Conventional wisdom on water management resulted in untimely releases (though it meets quantity requirements annually) causing damage to crops and income loss to farmers GSDP to the state and GDP to the nation. A fresh thinking is needed to address this problem of releases to take care of dry spells in downstream.

## 20.4 Work Done so Far on Intelligent Water Management

Modeling and optimization techniques are done for river basin management. The approaches rarely addressed the real issues of water management at field level (Kumar et al. 2013). Many large-scale water resources systems, especially in transboundary contexts, are characterized by the presence of several and conflicting interests and managed by multiple, institutionally independent decision-makers. These systems

**Table 20.1** Inflows and storage in Almatti reservoir during dry spells in Rayalaseema

| Date | Inflow | Outflow | Storage |
|---|---|---|---|
| 25 July, 2017 | 142,593 | 33,100 | 94.92 |
| 26 July, 2017 | 86,530 | 20,322 | 100.64 |
| 27 July, 2017 | 89,092 | 10,738 | 107.41 |
| 28 July, 2017 | 89,742 | 16,266 | 113.76 |
| 29 July, 2017 | 69,658 | 31,638 | 117.04 |
| 30 July, 2017 | 37,322 | 21,683 | 118.39 |
| 31 July, 2017 | 48,906 | 11,681 | 121.61 |
| 1 August, 2017 | 35,672 | 18,388 | 123.08 |
| 2 August, 2017 | 20,849 | 20,849 | 123.08 |
| 3 August, 2017 | 21,001 | 21,001 | 123.08 |
| 4 August, 2017 | 21,001 | 21,001 | 123.08 |
| 5 August, 2017 | 21,001 | 21,001 | 123.08 |
| 6 August, 2017 | 21,003 | 21,003 | 123.08 |
| 7 August, 2017 | 11,863 | 11,863 | 123.08 |
| 8 August, 2017 | 6003 | 6003 | 123.08 |
| 9 August, 2017 | 3086 | 3086 | 123.08 |
| 10 August, 2017 | 1003 | 1003 | 123.08 |
| 11 August, 2017 | 1003 | 1003 | 123.08 |
| 12 August, 2017 | 6003 | 6003 | 123.08 |
| 13 August, 2017 | 6003 | 6003 | 123.08 |
| 14 August, 2017 | 5403 | 5403 | 123.08 |
| 15 August, 2017 | 5403 | 5403 | 123.08 |
| 16 August, 2017 | 6003 | 6003 | 123.08 |
| 17 August, 2017 | 3003 | 3003 | 123.08 |
| 18 August, 2017 | 3003 | 3003 | 123.08 |
| 19 August, 2017 | 3003 | 3003 | 123.08 |
| 20 August, 2017 | 0 | 29,499 | 120.42 |
| 21 August, 2017 | 0 | 35,000 | 117.4 |
| 22 August, 2017 | 0 | 7291 | 116.53 |
| 23 August, 2017 | 513 | 513 | 116.53 |
| 24 August, 2017 | 513 | 513 | 116.53 |
| 25 August, 2017 | 513 | 513 | 116.53 |
| 26 August, 2017 | 6871 | 1003 | 117.04 |
| 27 August, 2017 | 17,886 | 8106 | 117.87 |
| 28 August, 2017 | 20,792 | 21,023 | 118.73 |
| 29 August, 2017 | 20,803 | 11,023 | 119.57 |
| 30 August, 2017 | 27,885 | 18,105 | 120.42 |

(continued)

**Table 20.1** (continued)

| Date | Inflow | Outflow | Storage |
|---|---|---|---|
| 31 August, 2017 | 35,224 | 29,356 | 120.92 |
| 1 September, 2017 | 56,879 | 34,770 | 122.83 |
| 2 September, 2017 | 53,033 | 53,033 | 123.08 |
| 3 September, 2017 | 32,730 | 32,730 | 123.08 |
| 4 September, 2017 | 19,023 | 19,023 | 123.08 |
| 5 September, 2017 | 19,023 | 19,023 | 123.08 |
| 6 September, 2017 | 13,106 | 13,106 | 123.08 |
| 7 September, 2017 | 6023 | 6023 | 123.08 |
| 8 September, 2017 | 6023 | 6023 | 123.08 |
| 9 September, 2017 | 11,023 | 11,023 | 123.08 |
| 10 September, 2017 | 11,023 | 11,023 | 123.08 |
| 11 September, 2017 | 11,023 | 11,023 | 123.08 |
| 12 September, 2017 | 11,523 | 11,523 | 123.08 |
| 13 September, 2017 | 12,023 | 12,023 | 123.08 |
| 14 September, 2017 | 10,523 | 10,523 | 123.08 |
| 15 September, 2017 | 25,314 | 25,314 | 123.08 |
| 16 September, 2017 | 45,080 | 45,080 | 123.08 |
| 17 September, 2017 | 33,413 | 33,413 | 123.08 |
| 18 September, 2017 | 36,746 | 36,746 | 123.08 |
| 19 September, 2017 | 25,080 | 25,080 | 123.08 |
| 20 September, 2017 | 25,080 | 25,080 | 123.08 |
| 21 September, 2017 | 47,669 | 47,669 | 123.08 |
| 22 September, 2017 | 99,822 | 106,744 | 122.48 |
| 23 September, 2017 | 121,316 | 133,491 | 121.43 |

are often studied adopting a centralized approach based on the assumption of full cooperation and information exchange among the involved parties (Venugopal 2006). Such a perspective is conceptually interesting to quantify the best achievable performance but might have little practical impact given the real political and institutional setting (Giuliani and Castelletti 2013).

Given many constraints, a holistic approach is called to use water in real-time basis where it is badly needed from water available within reach on upstream basin.

**Table 20.2** Daily average rainfall inflows and average inflows into Almatti dam

| Date | Average rainfall (mm) in upper Krishna | Inflow (cusecs) in Almatti @ 8.00 am | Average inflows for 24 h | Quantity received up to 27/09/2017 |
|---|---|---|---|---|
| 8 Sep, 2017 | 34 | 11,023 | 11,023 | 0.9523872 |
| 9 Sep, 2017 | 0.9 | 11,023 | 11,023 | 0.9523872 |
| 10 Sep, 2017 | 8.9 | 11,023 | 11,023 | 0.9523872 |
| 11 Sep, 2017 | 13.5 | 11,023 | 11,273 | 0.9523872 |
| 12 Sep, 2017 | 3.2 | 11,523 | 11,773 | 1.0387872 |
| 13 Sep, 2017 | 13.7 | 12,023 | 11,273 | 0.9091872 |
| 14 Sep, 2017 | 29 | 10,523 | 17918.5 | 2.1871296 |
| 15 Sep, 2017 | 4.8 | 25,314 | 35168.5 | 3.8899872 |
| 16 Sep, 2017 | 11.3 | 45,023 | 39,218 | 2.8868832 |
| 17 Sep, 2017 | 6.2 | 33,413 | 35079.5 | 3.1748544 |
| 18 Sep, 2017 | 5.1 | 36,746 | 30,913 | 2.166912 |
| 19 Sep, 2017 | 14.4 | 25,080 | 25080 | 2.166912 |
| 20 Sep, 2017 | 20 | 47,669 | 36374.5 | 4.1186016 |
| 21 Sep, 2017 | 3.3 | 99,822 | 73745.5 | 8.6246208 |
| 22 Sep, 2017 | 1.8 | 121,316 | 110,569 | 10.4817024 |
| 23 Sep, 2017 | 0.9 | 88,497 | 104906.5 | 7.6461408 |
| 24 Sep, 2017 | 2.9 | 52,535 | 70,516 | 4.539024 |
| 25 Sep, 2017 | 25.9 | 41,250 | 46892.5 | 3.564 |
| 26 Sep, 2017 | 5.1 | 23,217 | 32233.5 | 2.0059488 |
| 27 Sep, 2017 | | | | 63.25 TMC |
| Rainfall | 204.9 mm | | | |
| Rainfall tm.cuft | 24.04 | | | |
| Runoff/Rainfall (63.25/235) | 27% | | | |

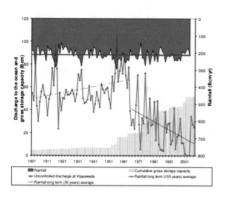
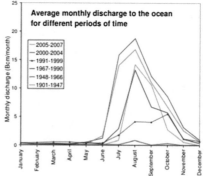

**Fig. 20.7** Average monthly discharges to the ocean

**Table 20.3** Number of dry spells in Anantapur district during southwest monsoon (http://core.ap.gov.in)

| District | Mandal | Dry spell count |
| --- | --- | --- |
| Anantapur | Uravakonda | 1 |
| Anantapur | Ramagiri | 2 |
| Anantapur | Somandepalle | 2 |
| Anantapur | Anantapurmandal | 2 |
| Anantapur | Agali | 2 |
| Anantapur | Lepakshi | 2 |
| Anantapur | Nallacheruvu | 2 |
| Anantapur | Bathalapalle | 3 |
| Anantapur | Rolla | 1 |
| Anantapur | Chennekothapalle | 2 |
| Anantapur | Rayadurg | 3 |
| Anantapur | Obuladevaracheruvu | 2 |
| Anantapur | Tadpatri | 1 |
| Anantapur | Brahmasamudram | 2 |
| Anantapur | Kambadur | 1 |
| Anantapur | Kanekal | 2 |
| Anantapur | Kadiri | 2 |
| Anantapur | Gooty | 1 |
| Anantapur | Tadimarri | 3 |
| Anantapur | Dhirchal | 1 |
| Anantapur | Chilamathur | 1 |
| Anantapur | Hindupur | 3 |
| Anantapur | Settur | 1 |
| Anantapur | Kundurpi | 1 |
| Anantapur | Parigi | 3 |
| Anantapur | Kudair | 1 |
| Anantapur | Pamidi | 1 |
| Anantapur | Kanaganapalle | 2 |
| Anantapur | Amadagur | 1 |
| Anantapur | Atmakur | 2 |
| Anantapur | Yellanur | 1 |
| Anantapur | Bukkapatnam | 2 |
| Anantapur | Roddam | 2 |
| Anantapur | Nallamada | 2 |
| Anantapur | Kalyandurg | 2 |

(continued)

**Table 20.3** (continued)

| District | Mandal | Dry spell count |
|---|---|---|
| Anantapur | Gorantla | 2 |
| Anantapur | Peddavadugur | 1 |
| Anantapur | Dharmavaram | 1 |
| Anantapur | Madakasira | 3 |
| Anantapur | Kothacheruvu | 2 |
| Anantapur | Narpala | 2 |
| Anantapur | Gummagatta | 2 |
| Anantapur | Mudigubba | 2 |
| Anantapur | Bukkaraya-samudram | 2 |
| Anantapur | Vidapanakal | 2 |
| Anantapur | Putlur | 1 |
| Anantapur | Amarapuram | 3 |
| Anantapur | Singanamala | 1 |
| Anantapur | Vajrakarur | 2 |
| Anantapur | Penukonda | 2 |
| Anantapur | Guntakal | 3 |
| Anantapur | Yadiki | 2 |
| Anantapur | Talupula | 3 |
| Anantapur | Puttaparthi | 3 |
| Anantapur | Raptadu | 2 |
| Anantapur | Garladinne | 1 |
| Anantapur | Nambulipulikunta | 3 |
| Anantapur | Tanakal | 2 |
| Anantapur | Gudibanda | 1 |
| Anantapur | Beluguppa | 2 |
| Anantapur | Gandlapenta | 2 |
| Anantapur | Bommanahal | 1 |
| Anantapur | Peddapappur | 2 |

## 20.5 Summary

Intelligent water releases

(1) With daily and seasonal forecasts available through latest technology, intelligent water management in large basins is a reality. Effects of El Niño and La Niña are better understood now to plan water resources with less uncertainty.

(2) Appropriate time for water releases in days and weeks to address dry spells in lower reaches is served by the water released from upper reaches.

(3) Total water resource inventory of rainfall, canal releases and groundwater will help in planning water resources better. Soil moisture is also one of the critical factors in intelligent water resource planning (http://www.apwrims.ap.gov.in).

(4) In basins where area of the catchments is large, different basins respond differently to rainfall in different years.

(5) With IOT it is now easy and economical to monitor basin water resources to address dry spells in time.

(6) With real-time rainfall, historical rainfall and gap in cumulative rainfall, a reasonable chance can be taken to manage releases without affecting upstream and downstream areas.

(7) Real-time groundwater levels and yields from wells will also help in planning available water resources more intelligently.

(8) Sowing and harvesting can be timed in coordination with water users to achieve efficient water utilization.

(9) At best of times water management is beset with uncertainties, but it can be overcome through proper evaluation of available water resources, demand and supply.

(10) It is not how much water you have from rain, surface and ground but how one resource intervenes to overcome shortages of other will decide how intelligently we manage water.

# References

Biggs TW, Gaur A, Scott CA, Thenkabail P, Rao PG, Krishna GM, Acharya S, Turral H (2007) Closing of the Krishna basin: irrigation, streamflow depletion and macroscale hydrology, vol 111. IWMI

Bouwer LM, Aerts J, Droogers P (2006) Detecting long-term impacts from climate variability and increasing water consumption on runoff in the Krishna basin (India), hydrology and earth

CM dash board, Government of Andhra Pradesh to access information on Rainfall, reservoir storages etc. on day to day basis. http://core.ap.gov.in/CMDashBoard/Index.aspx

Gaur A, McCornick PG, Turral H, Acharya S (2007) Implications of drought and water regulation in the Krishna basin, India. Water Resour Dev 23(4):583–594

Gaur A, Biggs TW, Gumma MK, Parthasaradhi G, Turral H (2008) Water scarcity effects on equitable water distribution and land use in a major irrigation project—case study in India. J Irrig Drain Eng 134(1):26–35

Giuliani M, Castelletti A (2013) Assessing the value of cooperation and information exchange in large water resources systems by agent-based optimization. Water Resour Res 49(7):3912–3926

Indian meteorological Department (IMD), information on daily rainfall accessed through. http://www.imd.gov.in/Welcome%20To%20IMD/Welcome.php

Kumar S, Pavelic P, George B, Venugopal K, Nawarathna B (2013) Integrated modeling framework to evaluate conjunctive use options in a canal irrigated area. J Irr Drain Eng 139(9):766–774

Relationship between basin development and downstream environmental degradation. IWMI research report, International Water Management Institute, Colombo, Sri Lanka

River basin information system of Krishna basin, NIH Roorkee, IWMI report

Smakhtin V, Gamage N, Bharati L (2008) Hydrological and environmental issues of inter-basin water transfers in India: a case study of the Krishna River Basin. Strategic analyses of the National River Linking Project (NRLP) of India Series 2, 79

Venot J-P. Drawing water for thirsty lands, Stories of the closing Krishna River basin in South India

Venot JP, Sharma BR, Rao KVGK (2008) The lower Krishna basin trajectory: water resource information system, water resource management system. http://www.apwrims.ap.gov.in/

Venugopal K (2006) Management of conjunctive use of water resources—a case study of SRBC command. PhD thesis, Sri Krishnadevaraya Univ., Anantapur, India

# Chapter 21
# Review of Flow Simulation Methods in Alluvial River

Deepak Dhakal, Nayan Sharma, and Ashish Pandey

## 21.1 Introduction

The river in the plain consisting of sand and silt and carrying the sediments which have the same characteristics as those of bed and bank of river is known as alluvial river. The most important characteristics of alluvial river is that their morphology (Church 2006), flow-resistance, and sediment-transport characteristics adjust in response to hydrodynamics of stream flow and sediment transport (Grade and Ranga Raju 2000). The primary driving force is the flow which governs the morphological behavior of an alluvial river (Trush et al. 2000).

Generally, in alluvial rivers the bed slope is flat, and consequently, the velocity is small (Arora 2002). The behavior of alluvial river also depends on the flood and sediment discharge. Alluvial river carries tremendously varying flow and silt load (Grade and Ranga Raju 2000). The alluvial river may be aggrading, degrading, or stable type depending on the surplus or deficit of sediment and sediment-carrying capacity of river (Arora 2002). The mechanism of morphological changes of alluvial rivers can be summarized as follows:

(a) Hydrodynamics, with conservation of mass and momentum,
(b) Bed changes, with conservation of sediment mass, and
(c) Sediment transport, with predictors for river sediment-carrying capacity.

D. Dhakal (✉) · N. Sharma · A. Pandey
Department of Water Resources Development and Management, Indian Institute of Technology Roorkee, Roorkee 247667, Uttarakhand, India
e-mail: deepakdhakal2006@gmail.com

N. Sharma
e-mail: nayanfwt@gmail.com

A. Pandey
e-mail: ashish.pandey@wr.iitr.ac.in

A. Pandey et al. (eds.), *Hydrological Extremes*, Water Science and Technology Library 97, https://doi.org/10.1007/978-3-030-59148-9_21

Previously, research approaches of river courses were mainly based on field observation and laboratory-scale modeling. Because of rapid development of computational fluid dynamics (CFD), mathematical modeling is growing in popularity. Presently, number of mathematical models have been developed and are in extensive use. All models are based on derivations of the basic principles, viz., conservation of mass, energy, and momentum (Toombes and Chanson 2011).

The characteristics of alluvial river, the available modeling programs, assumptions in the sediment modeling, and the governing equations of the model need to be identified as important aspects before simulation. This study focuses on the review of some of the flow simulation methods for alluvial rivers.

## 21.2   Major Issues of Flow Simulation for Alluvial Rivers

Recognition and understanding of the governing processes in the river system are essential for effective analysis of river problems.

Conceptualization of problems, formulation of mathematical model, solution, and interpretation are the major issues for simulation of alluvial river.

## 21.3   Existing Hydrodynamics and Sediment Transport Numerical Models

Computational hydrodynamic and sediment transport models, mostly, include numerical solution of one or more of governing differential equations of continuity, momentum, and energy of fluid, as well as differential equation for sediment continuity. For flow and sediment simulation there are various models available with varying level of complication.

Since early 1980s, one-dimensional (1-D) models have been successfully used in both research and engineering purposes. Basically, 1-D models are suitable for long duration and larger river reach conditions and has been widely used especially on a basin scale (large scale). But these models are less successful for local scale (Papanicolaou et al. 2008).

The mostly used 1-D models are: HEC-6 developed by Thomas and Prashum (1977), FLUVIAL 11 developed by Chang (1984), GSTAR developed by Molinas and Yang (1986), MIKE 11 by DHI (1993), HEC-RAS by Brunner (1995), and CCHE1-D by Vieira and Wu (2002). Most of the 1-D models can predict the basic parameters of a river channel such as average channel velocity, water surface elevation, bed-elevation variation, and total non-uniform sediment transport load in stream-wise direction only (Papanicolaou et al. 2008).

At early 1990s, computational research has been shifted toward 2-D models. 2-D models are depth-averaged models. They can provide spatially varied information about depth of water and bed elevation within rivers. Furthermore, these models can provide the magnitude of depth-averaged velocity component and sediment transport in principal and transverse directions (Papanicolaou et al. 2008). 2-D depth-average model is widely used for medium-scale domains and serves as a compromise between 1-D and 3-D models (Huybrechts et al. 2010).

Some of the 2-D models which were used for research and engineering purposes are: TAB-2 and RAM-2 developed by Thomas and McAnally (1985), MIKE 21C developed by DHI (1993), Delft 2-D developed by Walstra et al. (1998), and CCHE2D by Jia and Wang (1999) among others. Both 1-D and 2-D models are based on Saint Venant equations (Cao and Carling 2002).

3-D models typically solve continuity and the Navier–Stokes equations solving sediment continuity through finite difference, finite element, or finite volume approaches (Papanicolaou et al. 2008). EFDC3-D developed by Hamrick (1992), SSIIM developed by Olsen and Skoglund (1994), Delft 3-D developed by Delft-Hydraulic (1999), and TELEMAC by Hervouet and Bates (2000) are some 3-D models for flow and sediment simulation. 3-D models are used at a local scale.

Number of computer simulation methods have been developed for the simulation of alluvial river and are in extensive use. The selection of the model type is based on the scale and objective of the study.

## 21.4   HEC-RAS

The Hydrologic Engineering Center, River Analysis System (HEC-RAS) model was developed by US Army Corps of Engineers, in 1995 to calculate water surface profiles for steady, gradually varied flows in both prismatic and non-prismatic channels for 1-D flow consideration. HEC-RAS, a 1-D model, is an integrated package of hydraulic analysis programs, in which the user interacts with the system using a graphical user interface (GUI) (Brunner 1995).

One of the latest version of the HEC-RAS is version 5.0.3 for sediment transport and mobile bed modeling. The HEC-RAS system comprises steady flow water surface profile calculations, unsteady flow simulations, and movable boundary sediment transport calculations. The same geometric data and same geometric and hydraulic calculation routines will be utilized by all three components. Besides these, system also contains several hydraulic design features that can be calculated once the basic water surface profiles are computed. The main features include: ability to model full network of river channels, flood plain studies, channel dredging, several levee and encroachment alternatives, various equations for calculation of sediment transport, reservoir sedimentation, and long-term aggradation or degradation due to flow or channel geometry changes. HEC-RAS can import GIS data and HEC-2 data (Brunner 2016).

Model input requirement comprises the stream network, reach lengths, geometry, energy loss coefficients, flow, sediment, and hydraulic structures data. Boundary conditions are required to define the starting water depth at the stream system endpoints, that is, upstream and downstream.

1-D models are widely in practice might be because the model is simple to use, need lesser input data, and low computer power requirement, as well as the basic concept and programs have already around for long time.

### 21.4.1 Governing Equations

The equations that govern the hydrodynamic behavior of an incompressible fluid are based on the usual concepts of conservation of mass and momentum.

The HEC-RAS steady flow model is designed for calculating water surface profile for steady, gradually varied flow. Subcritical, supercritical, and mixed flow regime water surface profiles can be assessed using HEC-RAS. The effects of various obstructions such as bridges, culverts, weirs, and structures at over-bank region can be considered during calculation of water surface profile (Brunner 2016).

Energy equation is used to calculate the water surface profile. To solve the energy equation, an iterative process known as the standard step method is used. Empirical Manning equation is applied to calculate the friction loss (Dyhouse et al. 2003). The semi-empirical energy equation for water surface profile computation is Eq. (21.1) (Brunner 2016):

$$Z_1 + Y_1 + \alpha_1 \frac{V_1^2}{2g} = Z_2 + Y_2 + \alpha_2 \frac{V_2^2}{2g} + H_e \qquad (21.1)$$

where '$Z_1$' and '$Z_2$' represent the inverts elevations of the main river channel, '$Y_1$' and '$Y_2$' represent water depths in the channel, '$V_1$' and '$V_2$' represent the cross-section averaged velocities in the channel, '$\alpha_1$' and '$\alpha_2$' denote the energy correction factors, '$H_e$' denotes the energy loss between two sections, and '$g$' represents the acceleration due to gravity.

The quasi-unsteady flow simplifies the hydrodynamics by approximating the continuous flow hydrograph by series of steady flow profile. Hence, the principle of numerical solution that the quasi-unsteady model is based on is the standard step-solution method for solving energy equation.

The computation engine for the HEC-RAS 1-D unsteady flow simulator is based on the USACE's UNET model (USACE 2010). An unsteady model routes flow using the continuity and momentum equations, also referred as Saint-Venant shallow water equations. Equation (21.2) illustrates the continuity equation used for 1-D numerical solution:

$$\frac{\partial Q}{\partial x} + \frac{\partial A}{\partial t} - q_l = 0 \qquad (21.2)$$

where '$Q$' is flow at any time '$t$' and space '$x$', '$A$' is area of flow, and '$q_l$' denotes the lateral inflow per unit length. Similarly, the momentum equation in full dynamic wave form is given by Eq. (21.3):

$$\frac{1}{A}\frac{\partial Q}{\partial t} + \frac{1}{A}\frac{\partial(Q^2/A)}{\partial x} + g\frac{\partial y}{\partial x} - g(S_0 - S_f) = 0 \qquad (21.3)$$

where '$y$' is flow depth, and '$S_0$' and '$S_f$' are channel bed slope and friction slope, respectively. HEC-RAS mostly use the Manning's equation to calculate friction slope using Eq. (21.4):

$$S_f = \frac{Q^2 n^2}{2.208 A^2 R^{4/3}} \qquad (21.4)$$

where '$n$' is Manning's roughness coefficient.

HEC-RAS solves mass and momentum equations using implicit finite difference approximations and Preissman's second-order scheme (USACE 2010). To solve both Eqs. (21.2) and (21.3), an implicit finite difference approach is employed (Brunner 2016). It solves the equations using four-point implicit scheme (Fig. 21.1) with space derivatives and function values calculated at an interior point, $(j + \theta)\,\Delta t$, where '$j$' is any interior node number, '$\theta$' is the implicit partial derivative weighting factor, and '$\Delta t$' is the computational time step. The values of '$\theta$' and time step $\Delta t$ for the model under consideration should be determined based on a sensitivity analysis (Brunner 1995). The convective term discretization is done using the first-order upward upwind finite difference scheme. Upstream boundary condition at upstream of model area should be a flow hydrograph showing the variation of flow with time. Likewise, the boundary condition at the downstream could be either a stage or flow hydrograph, a rating curve, or normal depth specification.

The process of hydrodynamic and sediment solutions is inherently coupled with each other in the model building process. Whether it is a quasi-unsteady model or completely unsteady model, the river geometry, hydraulic parameters, and sediment transport capacity remain constant during the computational time step. Once hydraulic parameters are calculated from the hydrodynamic models during each time

**Fig. 21.1** Typical finite difference solution cell

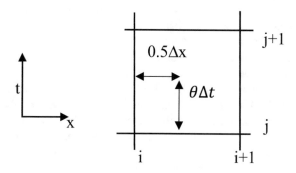

step, the sediment transport capacity is calculated using different transport functions (Brunner 2016). The total transport capacity for the corresponding control volume is calculated using Eq. (21.5), which is based on Einstein's (1950) classic assumption.

$$T_c = \sum_{j=1}^{n} \beta_j T_j \tag{21.5}$$

where '$T_c$' represents total sediment transport capacity, '$n$' represents number of grain size classes, and '$\beta_j$' denotes the percentage of active layer composed of material in grain size class '$j$'.

Sediment routing in HEC-RAS solves the Exner sediment continuity equation (Eq. 21.6) to determine the potential bed aggradation or erosion (Brunner 1995).

$$\frac{\partial z}{\partial t} + B \frac{1}{(1-\lambda)} \frac{\partial Q_s}{\partial \chi} = 0 \tag{21.6}$$

where '$B$' represents channel width, '$\lambda$' represents porosity of the bed, '$Q_s$' represents rate of sediment load in units of tons/day, and '$z$' is bed elevation.

During each time step, the continuity equation compares supply of sediment to the transport capacity for each grain class. If the supply is less than the capacity, the model calculates a sediment deficit resulting in erosion. If the supply is more than the capacity, the model calculates a sediment surplus resulting in deposition. HEC-RAS models typically consider three physical constraints, also referred to as continuity limiters that can decrease the transport capacity: temporal deposition limiter, temporal erosion limiter, and sorting and armoring effect (Brunner 1995). For instance, when armor layers are formed, surface particles that are available for transport will be coarser and more difficult to transport, implying that there will be less material available for transport (Brunner 1995).

### Fall Velocity and Transport Functions

The entrainment, sedimentation, and deposition of sediment depends on characteristics of flow and properties of sediment (Vanoni 1975). The most important parameter is the size of the sediment. Shape and density of the particle are also important parameters. Fall velocity of a sediment particle characterizes its response to flow and sediment studies (Vanoni 1975). Among others Stokes introduced the fall velocity calculation approach based on the static equilibrium of sediment particle for quiescent water column, typically for the laminar range of flows. However, fall velocities computation under natural flow conditions is complicated because of the irregularity of particle size and shape, relative densities of fluid and sediment, and strength of turbulence (Yang 1996). These variables were included in the propositions of (Rubey 1933; Toffaleti 1968; Van Rijan 1993), among others. Rubey (1933) came up with an analytical relationship for calculation of fall velocity of gravel, sand, and silt particles (Yang 1996). This formulation is a combination of Stokes' law (particles within laminar regime) and impact formula (particles outside Stokes' region), and the fall velocity is given by Eq. (21.7) (Brunner 1995):

$$\omega = F[gd(s - 1)]^{0.5} \tag{21.7}$$

where

$$F = \left[\frac{2}{3} + \frac{36\vartheta^2}{gd^3(s - 1)}\right]^{0.5} - \left[\frac{36\vartheta^2}{gd^3(s - 1)}\right]^{0.5} \tag{21.8}$$

where '$\omega$' is fall velocity, ft/s; '$d$' is particle size, ft; '$s$' is specific gravity of sediment; and '$\vartheta$' is kinematic viscosity, ft$^2$/s.

Toffaleti (1968) proposed a table for determining fall velocities using a shape factor of 0.9 and specific gravity of 2.65 (Brunner 2016). Fall velocities are referred for a range of temperatures and grain size classes based on the classification of American Geophysical Union (AGU). For non-spherical particles Van Rijn (1993) proposed a relation for fall velocity by considering shape factor of 0.7 in water at 20 °C as cited in Brunner (2016).

Sedimentation process in HEC-RAS is calculated using Exner sediment continuity equation and user-selected sediment transport function. Different sediment transport functions have been developed under different conditions of the flow and the sediment characteristic. Therefore, it is very important to understand the processes used in the development of the function for reasonable assessments of sediment transport in alluvial river channel (Brunner 2016). The range of particles sizes and sediment systems for which these transport functions could be employed is shown in Table 21.1.

All these transport functions are total-load predictor functions. Virtually all the sediment transport equations have been developed using cross-section averaged hydraulic variables, making the one-dimensional flow assumption appropriate for the sediment models (Copeland 1993).

**Table 21.1** Sediment transport functions (Brunner 2016)

| Transport function | Overall particle diameter (mm) | Mean particle size (mm) | Sediment range | Data source |
|---|---|---|---|---|
| Ackers-White (1973) | 0.04–7.0 | NA | Sand-Gravel | Flume |
| Engelund-Hansen (1967) | NA | 0.19–0.93 | Sand | Flume |
| Laursen (Copeland) | NA | 0.08–0.7 | Sand | Field |
| Laursen (Copeland) | NA | 0.011–29 | Sand-gravel | Flume |
| Meyer-Peter Muller1 (1948) | 0.4–29.0 | NA | Sand-gravel | Flume |
| Toffaleti (1968) | 0.062–4.0 | 0.095–0.76 | Sand | Field |
| Toffaleti (1968) | 0.062–4.0 | 0.45–0.91 | Sand | Flume |
| Yang (1973) | 0.15–1.7 | NA | Sand | Field |
| Yang (1973) | 2.5–7.0 | NA | Gravel | Field |

Even though the several supporting equations are required, the basic solution for the non-equilibrium condition of sediment transport is the Thomas's extension of the Einstein model of equilibrium condition.

### 21.4.2 Limitations of 1-D Modeling in HEC-RAS

HEC-RAS sediment transport model has the following limitations:

(a) The HEC-RAS 1-D model is based on one-dimensional gradually varied flow hydraulics and sediment transport theory. The equation of backwater is not valid for rapidly varied flow.
(b) The spatial and depth-wise variation of Manning's roughness could not be accounted for in 1-D hydraulic model.
(c) HEC-RAS could just measure the total sediment by reach only and could not differentiate the velocity and the condition in the curved channel. HEC-RAS can draw such as a meander river, but it still read the model as a straight channel. Therefore, HEC-RAS modeling system cannot calculate the curved channels on the model. Simulation of channel meanders needs better modeling system to know the distribution of sediment in each part of river.

### 21.4.3 2-D HEC-RAS Modeling

HEC-RAS version 5.0.3 can perform 1-D, 2-D, and combined 1-D and 2-D modeling. The ability to perform combined 1-D/2-D modeling within the same unsteady flow allows the user to work on large river system, utilizing the one whenever appropriate. The program solves either 2-D Saint-Venant equations or 2-D diffusion wave equations (user selectable option). 2-D unsteady flow equation solver uses the implicit finite volume algorithm. Mapping of inundation area and flooding animation can be done in HEC-RAS by RAS mapper feature. 2-D model can develop detailed flood mapping which is the main advantage of 2-D model (Brunner 2016).

### 21.4.4 Limitations of the 2-D Modeling Capabilities in HEC-RAS

The cell spacing is key in 2-D modeling. The value of cell spacing should be determined based on a sensitivity tests at each site. The computational time increases with the increase in resolution. Cell spacing value which develops a stable model and satisfies time step recommendation is based on Courant number (Brunner 2016).

Generally, 2-D models are limited to shorter time durations and smaller spatial extents than 1-D models. Moreover, 2-D models usually require more input data than 1-D models; hence, 1-D models are more commonly used than 2-D models. Currently, HEC-RAS cannot perform the sediment transport erosion/deposition in 2-D flow areas. HEC-RAS 2-D flow modeling software do not have sufficient tools for generating a detailed 2-D computational mesh. Several tools are still required to be added in HEC-RAS.

## 21.5 CCHE2D Model

CCHE2D models are the mathematical models developed by National Center for Computational Hydro-science and Engineering, University of Mississippi, USA, in 1999, for simulating flow and sedimentation processes in the natural environment. This model has been developed for 2-D flow schemes with sediment simulating capability (Zhang 2005).

### 21.5.1 The Governing Equations

The CCHE2D model uses the depth integrated momentum equations, continuity equations, and Reynolds equations. An efficient element method (special case of FEM) is used to discretize the governing equations. Dry and wet ability simplify the flow simulation of complicated topography. It can handle subcritical, supercritical flows, and their transitions (Zhang 2005). Suspended sediment is simulated by solving the advection/diffusion equation and the bed-load transport by empirical functions. Effect of bed slope and secondary flow in curved channels is also considered (Zhang 2005).

**Momentum Equations**
The Navier–Stokes momentum equations for depth-integrated 2-D turbulent flows in a Cartesian coordinate system are Eqs. (2.9) and (2.10) (Jia and Wang 1999):

$$\frac{\partial u}{\partial t} + \frac{u \partial u}{\partial x} + \frac{v \partial u}{\partial y} = -\frac{g \partial \eta}{\partial x} + \frac{1}{h} * \left( \frac{\partial h \tau_{xx}}{\partial x} + \frac{\partial h \tau_{xy}}{\partial y} \right) - \frac{\tau_{bx}}{\rho h} + f_{cor} * v \quad (2.9)$$

$$\frac{\partial v}{\partial t} + \frac{u \partial v}{\partial x} + \frac{v \partial v}{\partial y} = -\frac{g \partial \eta}{\partial y} + \frac{1}{h} * \left( \frac{\partial h \tau_{yx}}{\partial x} + \frac{\partial h \tau_{yy}}{\partial y} \right) - \frac{\tau_{by}}{\rho h} + f_{cor} * u \quad (2.10)$$

where depth-integrated velocity components in $x$ and $y$ directions are represented by '$u$' and '$v$', respectively. Similarly, $\rho$, $t$, $h$, and $f_{cor}$ representing the water density, time, local depth of water, and Coriolis parameter, respectively. Depth-integrated Reynolds stresses are represented by $\tau_{xx}$, $\tau_{xy}$, $\tau_{yx}$, and $\tau_{yy}$; and shear stresses on the bed and flow interface are represented by $\tau_{bx}$ and $\tau_{by}$ (Jia and Wang 1999).

### Continuity Equation

The depth-integrated continuity equation is used to compute the free surface elevation for the flow (Jia and Wang 1999). The continuity equation is solved on a staggered grid; especially the velocity correction method is applied to solve the system of equation. Implicit scheme by time marching is used for unsteady flow simulation. The continuity equation in Cartesian coordinate system is Eq. (21.11) (Jia and Wang 1999):

$$\frac{\partial h}{\partial t} + \frac{\partial uh}{\partial x} + \frac{\partial vh}{\partial y} = 0 \qquad (21.11)$$

Assuming bed elevation $\zeta$ would not change the flow simulation process. The simplified continuity equation is widely used which can be written as Eq. (21.12):

$$\frac{\partial \eta}{\partial t} + \frac{\partial uh}{\partial x} + \frac{\partial vh}{\partial y} = 0 \qquad (21.12)$$

where '$\eta$' and '$h$' represent free surface elevation and depth of water, respectively.

Equation (21.11) should be used if bed elevation is altered rapidly due to erosion or deposition (Jia and Wang 1999).

### Reynolds Stresses

Turbulence Reynolds stresses in Eqs. (2.9) and (2.10) are approximated according to the Boussinesq's assumption (Jia and Wang 1999). The equations are: Eqs. (21.13)–(21.17):

$$\tau_{ij} = (v_t)(ui, j + uj, i) \qquad (21.13)$$

$$\tau_{xx} = \frac{2v_t \partial u}{\partial x} \qquad (21.14)$$

$$\tau_{xy} = v_t \left( \frac{\partial u}{\partial y} + \frac{\partial v}{\partial x} \right) \qquad (21.15)$$

$$\tau_{yy} = \frac{2v_t \partial v}{\partial y} \qquad (21.16)$$

$$\tau_{yx} = v_t \left( \frac{\partial u}{\partial y} + \frac{\partial v}{\partial x} \right) \qquad (21.17)$$

### Bed Shear Stresses

CCHE2D model uses the depth-integrated logarithmic law and Manning's coefficient method to calculate bed shear stresses. From the logarithmic law the bed shear stresses are obtained by Eqs. (21.18) and (21.19) (Jia and Wang 1999):

$$\tau_{bx} = \frac{1}{8}\rho(f_c)uU \qquad (21.18)$$

$$\tau_{by} = \frac{1}{8}\rho(f_c)vU \qquad (21.19)$$

where shear stresses on the bed and flow interface are represented by $\tau_{bx}$ and $\tau_{by}$, respectively. Similarly, '$f_c$' represents the Darcy-Weisbach coefficient, '$U$' is the resultant velocity component, $u$ and $v$ are depth-integrated velocity components in $x$ and $y$ directions, and '$\rho$' is the density of water. From the Manning's coefficient method, the bed shear stresses are calculated using Eqs. (21.20) and (21.21):

$$\tau_{bx} = \frac{1}{\sqrt[3]{h}}\rho gn^2 uU \qquad (21.20)$$

$$\tau_{by} = \frac{1}{\sqrt[3]{h}}\rho gn^2 vU \qquad (21.21)$$

where '$n$' and '$h$' are Manning's roughness coefficient and depth of water, respectively.

### Suspended Sediment Transport Equation

The CCHE2D model uses depth-averaged sediment transport equation for suspended sediment (Jia and Wang 1999), that is, Eq. (21.22):

$$\frac{\partial c}{\partial t} + \frac{u\partial c}{\partial x} + \frac{v\partial c}{\partial y} - \frac{\partial}{\partial x}\left[\beta v_t * \frac{\partial c}{\partial x}\right] - \frac{\partial}{\partial y}\left[\beta v_t * \frac{\partial c}{\partial y}\right] = S \qquad (21.22)$$

where '$c$' represents depth-integrated sediment concentration, '$u$' and '$v$' represent velocity components. The coefficient to convert turbulence eddy viscosity to eddy diffusivity for sediment is represented by '$\beta$'. Local balance of suspension and deposition is denoted by the term '$S$'.

### Bed Load Transport Equation

The bed load transport ($q_b$) formula developed by van Rijn is adopted in CCHE2D model (Jia and Wang 1999). The formula is presented in Eq. (21.23):

$$q_b = 0.053\left[\left(\frac{\rho_s}{\rho} - 1\right)g\right]^{0.5} d_{50}^{1.5} D^{-0.3} T^{2.1} \qquad (21.23)$$

where

$$D = d_{50}\left[(s-1)g/v^2\right]^{0.3} \qquad (21.24)$$

$$T = \frac{\tau - \tau_{cr}}{\tau_{cr}} \qquad (21.25)$$

where '$T$' denotes the bed shear stress parameter, '$\tau_{cr}$' denotes the critical shear stress according to Shields, and '$s$' represents the ratio of density of sediment to density of water.

**Bed Deformation Equation**

Bed morphological change in CCHE2D model is computed with the sediment continuity equation, that is, Eq. (21.26) (Jia and Wang 1999):

$$\frac{\partial \xi}{\partial t} = \frac{1}{1 - p}\left(\frac{\partial q_x}{\partial x} + \frac{\partial q_y}{\partial y}\right) \tag{21.26}$$

where '$p$' denotes porosity of bed material and '$\zeta$' is elevation of bed.

**Total Load**

As per the conventional classification, moving sediment is divided into suspended load and bed load. Generally, in CCHE2D, total load sediment transport is simulated (Jia and Wang 1999).

### 21.5.2  Limitations of CCHE2D Model

CCHE2D is a highly data descriptive model that needs more data than other models like HEC-RAS and MIKE 21C. Since it is a depth-averaged model, it could not predict velocity accurately on sharp changing zone.

## 21.6  Mike 21C Model

MIKE 21C is a 2-D numerical modeling system developed by Danish Hydraulic Institute (DHI) in 1990, for the simulation of hydrodynamics of vertically homogeneous flows and sediment transport. This model has the ability of utilizing both the rectilinear and curvilinear computational grid. This model is suitable for simulating two-dimensional free surface flow and sediment transport in alluvial rivers where accurate description of the flow along the bank line is important (DHI 2017). This model is basically applicable to protection scheme against bank erosion and bed scour measures to minimize shoaling, structures including weirs, spurs, barrages, and sedimentation of water intake, outlets, locks, harbors, and reservoirs (DHI 2017).

The numerical grid in MIKE 21C is created by means of a user-friendly grid generator. Areas of special interest can be resolved using a higher density of grid lines at these locations. The MIKE 21C is particularly suited for river morphological studies and includes modules to describe flow hydrodynamics, sediment transport, river bed changes, and river bank erosion (DHI 2017).

Flow hydrodynamics (velocity and water level) are calculated over a curvilinear or a rectangular computational grid covering the study area by solving the vertically integrated Saint Venant equations of continuity and momentum. Helical flow (secondary currents) developing in channel bends due to curved streamlines.

Sediment transport based on various model types viz. Van Rijan (1993), Meyer-Peter and Müller (1948), Engelund and Hansen (1967), Yang (1973), or user-defined empirical formulas) can be computed. The effect of helical flow, gravity on a sloping river bed, shapes of velocity, and concentration profiles are considered in separate bed load and suspended load sub-models based on the theories by Galappatti (1983). Graded sediment descriptions can be applied as well by defining several different sediment fractions, which are treated separately by the sediment transport module.

For scour and deposition, changes in bed elevations at each grid cell at every time-step is computed by solving continuity equation for sediment. Effect of supply-limited sediment layers can be incorporated as well. Bank erosion is calculated from a formula relating near-bank conditions to bank erosion rates. At every time step, the accumulated bank erosion can be used for updating the bank lines, and for updating the curvilinear grid (extent of the modeling area). By defining a silt factor, the bank erosion products can be included in the sediment budget for the adjacent riverbed, or it can be disregarded depending on the composition of the bank material.

Mike 21C is one of the most comprehensive and well-established tools for simulating alluvial river. Mike 21C is approximated by using FDM in curved coordinates (Ahmadi et al. 2009; DHI 2017; Dang and Tran 2016). Structurally, Mike 21C has three main modules, namely hydrodynamic, sediment transport, and river morphology module.

## 21.6.1 Hydrodynamic Module

For simulating on long time scale, three-dimensional flow model is very complicated. To overcome this complexity a hydrodynamics module is developed by the scientists which convert the flow model into two-dimensional equations (Ye and McCorquodale 1997). The hydrodynamics model solves the Saint-Venant equations in curvilinear coordinates (Ahmadi et al. 2009). These equations are given in Eqs. (21.27)–(21.29):

$$\frac{\partial H}{\partial t} + \frac{\partial P}{\partial s} + \frac{\partial q}{R_s} + \frac{P}{R_n} = 0 \qquad (21.27)$$

$$\frac{\partial p}{\partial T} + \frac{\partial}{\partial s}\frac{(p^2)}{h} + \frac{\partial}{\partial h}\frac{(pq)}{h} + \frac{2pq}{hR_h} + \frac{p^2 - q^2}{hR_s} + gh\frac{\partial H}{\partial S} + \frac{y}{c^2}\frac{P\sqrt{P^2+q^2}}{h^2} = \text{RHS} \qquad (21.28)$$

$$\frac{\partial q}{\partial T} + \frac{\partial}{\partial s}\frac{(q^2)}{h} + \frac{\partial}{\partial h}\frac{(pq)}{h} + \frac{2pq}{hR_h} + \frac{p^2 - q^2}{hR_n} + gh\frac{\partial H}{\partial n} + \frac{y}{c^2}\frac{P\sqrt{P^2+q^2}}{h^2} = \text{RHS} \qquad (21.29)$$

where '$s$' and '$n$' represent the coordinates in the curvilinear system, '$h$' represents water depth, '$p$' is the mass flux in $s$-direction and '$q$' is the mass flux in $n$-direction. Water level is represented by '$H$', Chezy roughness coefficient is denoted by '$C$', and gravitational acceleration is represented by '$g$'. Similarly, '$R_s$' and '$R_n$' are radius of curvatures of $s$- and $n$-line, respectively. Right-hand side Reynolds stress is denoted by RHS (Dang and Tran 2016).

### 21.6.2 Sediment Transport Module

Suspended load transport equations in sediment transport module are as follows (Eq. 21.30) (Galappatti and Vreugdenhil 1985):

$$\frac{\partial c}{\partial t} + u\frac{\partial c}{\partial x} + v\frac{\partial c}{\partial y} + w\frac{\partial c}{\partial z} = W_s\frac{\partial c}{\partial z} + \frac{\partial}{\partial x}\left(\varepsilon\frac{\partial c}{\partial x}\right) + \frac{\partial}{\partial y}\left(\varepsilon\frac{\partial c}{\partial y}\right) + \frac{\partial}{\partial z}(\varepsilon\frac{\partial c}{\partial z})$$

(21.30)

where vertical axis coordinate is represented by '$z$', particles settling velocity by '$w_S$', and suspended load concentration by '$c$'. Similarly, flow velocity components in the $x$, $y$, and $z$-directions are represented by '$u$', '$v$', and '$w$', respectively, and the coefficient of eddy viscosity is represented by '$\varepsilon$'. Ignoring the limited diffusion outside of the vertical diffusion, Eq. (21.30) becomes Eq. (21.31):

$$\frac{\partial c}{\partial t} + u\frac{\partial c}{\partial s} + v\frac{\partial c}{\partial y} + w\frac{\partial c}{\partial n} = w_s\frac{\partial c}{\partial z} + \frac{\partial}{\partial x}\left(\varepsilon\frac{\partial c}{\partial z}\right)$$

(21.31)

The bed load ($s_{bl}$) is given by Eq. (21.32), and suspended load ($s_{sl}$) is given by Eq. (21.33):

$$s_{bl} = k_b \cdot s_{tl}$$

(21.32)

$$s_{sl} = k_s \cdot s_{tl}$$

(21.33)

Most of the bed load transport formulas are based on calibration coefficients $k_b$ and $k_s$. Engelund and Hansen had established the relationship $k_b + k_s = 1$, which is also used in Mike 21C (Dang and Tran 2016).

where '$ks$' is the coefficient of suspended load and '$k_b$' is the coefficient of bed load. Total volume of sediment transport '$S_{tl}$' is given by Eq. (21.34):

$$S_{tl} = 0.05\frac{C^2}{g}\theta^{5/2}\sqrt{(\rho d_s - 1)g d_{50}^3}$$

(21.34)

where the relative density of sediment is represented by $\rho d_s$, and median size of sediment particles by $d_{50}$. Shields parameter ($\theta$) is calculated by Eq. (21.35):

$$\theta = \frac{\tau}{\rho g (\rho d_s - 1) d_{50}} \tag{21.35}$$

### 21.6.3  Morphological Module

Before solving sediment transport equation in morphological module, hydrodynamic solution needs to be obtained first. Next, river bed and hydrodynamic model are applied (DHI 2017). The equation for calculating bank erosion rate ($E_b$) in m/s is Eq. (21.36):

$$E_b = -\alpha \frac{\partial z}{\partial t} + \frac{\beta S}{n} + \gamma \tag{21.36}$$

The additional sediment from bank erosion source ($\Delta S$) can be calculated using Eq. (21.37):

$$\Delta S = E_b (h + h_b) \tag{21.37}$$

where '$h_b$' is the height of bank above the water level, '$S$' is sediment transport near bank, '$z$' represents local bed level, and $\alpha$, $\beta$, and $\gamma$ are parameters defined in model.

Variation of bed level is calculated based on sediment continuity equation in Cartesian coordinate system (DHI 2017) using Eq. (21.38):

$$(1 - p) \frac{\partial z}{\partial t} + \frac{\partial S_m}{\partial x} + \frac{\partial S_y}{\partial y} = \Delta s_e \tag{21.38}$$

In Eq. (21.38), total volume of the sediment transported along x and y directions are represented by '$S_x$' and '$S_y$', respectively. Porosity of the bed is represented by '$p$', and supply of excess sediment from bed erosion by '$\Delta S_e$'.

### 21.6.4  Limitations of MIKE 21C

Both subcritical and supercritical flows can be modeled using MIKE 21C with reasonable accuracy but a supercritical upstream boundary cannot be modeled which defaults to critical depth. Further, complex flow patterns within the hydraulic jump cannot be modeled using MIKE 21C. Determining the capability of a depth-averaged

2-D model to simulate hydrodynamics and sediment transport through river-training structures is still unsolved.

Unlike HEC-RAS and CCHE2D, MIKE 21C is a not available in public domain. It is a copyrighted model and need to purchase a model license from DHI.

## 21.7    Discussion and Conclusions

HEC-RAS 1-D model needs less data and less computational efforts. It is the most stable numerical scheme for solving the problems of alluvial river. It is suitable for basin scale (long reach) and longer duration analysis. Spatially and depth-wise variation of Manning's roughness could not be accounted in 1-D model. But it cannot differentiate the curved channel and assumes whole river reach as a straight channel. HEC-RAS 2-D models can produce detail flood mapping considering spatially varied roughness coefficient. 2-D models are usually limited to shorter time duration and smaller spatial extent than 1-D models.

Similarly, CCHE2D is a 2-D model; it can handle supercritical, subcritical, and their transition. From CCHE2D models, the morphological changes can be computed considering the effects of bed slope and secondary flow in curved channel. But CCHE2D is a highly descriptive model, and it needs more data. It has less accuracy for sharp changing locations.

MIKE 21C is a 2-D model; it has the ability of using both rectilinear and curvilinear computational grid and can account bank erosion. It has the capability to model both supercritical and subcritical but accuracy on supercritical and hydraulic jumps is less. Further, MIKE 21C is not available in public domain. In case of long-term and long-reach sediment analysis, HEC-RAS gives the good results.

Owing to the speedy developments in mathematical models, computer simulation has become an effective tool for studying flow and sediment transport in alluvial river. The review of available models shows that numerous models are existing with various features. Saint Venant equations are used by all the models but have different sediment predictors, energy slope relations, and aggradation/degradation equations. Several complexities have been associated with natural river, due to its size, discharge variation, sediment properties, sediment concentration, and other geographical, meteorological, social factors. No model can reflect all these factors. Therefore, the models cannot have universal applicability. Hence, the choice of the model for particular river depends upon the characteristic of river, type of problem, ability of model to simulate the problem effectively, availability of data for model calibration, and validation and overall available time and budget for solving the problems. Number of models are available with different features. Furthermore, the selection of the type of the model depends upon the scale and objective of the study. Typically, 1-D sediment models are applied on a regional or river scale, whereas 2-D models are used at a local scale.

**Acknowledgements** First author is thankful for the financial support received from the ITEC, Ministry of External Affairs, Government of India during the period of study. The authors are also thankful to the Department of Water Resources Development and Management, IIT Roorkee for providing the required facilities.

# References

Ackers P, White WR (1973) Sediment transport: new approach and analysis. J Hydraul Div 99(hy11):2040–2060

Ahmadi MM, Ayoubzadeh SA, Montazeri NM, Samani JM (2009) A 2D numerical depth-averaged model for unsteady flow in open channel bends. J Agri Sci Technol (JAST) 11(4):457–468

Arora KR (2002) Irrigation water power and water resources. Standard Publishers Distributors

Brunner GW (1995) HEC-RAS River Analysis System. Hydraulic reference manual. Version 1.0, US Army Crops of Engineer, Hydrologic Engineering Center, Davis CA

Brunner GW (2016) HEC-RAS River analysis system. Hydraulic reference manual version 5.0, US Army Crops of Engineers, Institutes of Water Resources, Hydrologic Engineering Center, Davis, CA

Cao Z, and Carling PA (2002). Mathematical modelling of alluvial rivers: reality and myth. Part 1: General review. In proceedings of the institution of civil engineers-water and maritime engineering, vol 154, no 3. Thomas Telford Ltd, pp 207–219

Chang HH (1984) Modeling of River Channel changes. J Hydraul Eng 110(2):157–172

Church M (2006) Bed material transport and the morphology of alluvial river channels. Annu Rev Earth Planet Sci 34:325–354

Copeland RR (1993) Numerical modeling of hydraulic sorting and armoring in alluvial rivers. University of Iowa

Dang TA, Tran TH (2016) Application of the Mike21C model to simulate flow in the lower Mekong river basin. SpringerPlus 5(1):1982

Delft-Hydraulics (1999) Delft3D users' manual. Delft Hydraulics, The Netherlands

DHI (1993) MIKE 21 short description. Danish Hydraulic Institute, Hørsholm, Denmark

DHI (2017) Mike 21C curvilinear model for river morphology. User Guide and Reference Manual, Danish Hydraulic Institute

Dyhouse G, Hatchett J, Benn J (2003) Floodplan modeling using HEC-RAS. Haestad Press

Einstein HA (1950) The bed-load function for sediment transportation in open channel flows (No.1026). US Government Printing Office

Engelund F, Hansen E (1967) A monograph on sediment transport in alluvial streams. TEKNISK-FORLAG Skelbrekgade 4 Copenhagen V Denmark

Galappatti G, Vreugdenhil CB (1985) A depth-integrated model for suspended sediment transport. J Hydraul Res 23(4):359–377

Galappatti R (1983) A depth integrated model for suspended transport. Report 83-07, communications on hydraulics, Department of Civil engineering. Delft University of Technology

Grade RJ, Ranga Raju KG (2000) Mechanics of sediment transportation and alluvial stream problems, 3rd edn. New Age International Publishers

Hamrick JM (1992) Three-dimensional environmental fluid dynamics computer code: theoretical and computational aspects. Virginia Institute of Marine Science, College of William

Hervouet JM, Bates P (2000) The telemac modelling system special issue. Hydrol Process 14(13):2207–2208

Huybrechts N, Villaret C, Hervouet JM (2010) Comparison between 2D and 3D modelling of sediment transport: application to the dune evolution. River Flow, 887–893

Jia Y, Wang SS (1999) Numerical model for channel flow and morphological change studies. J Hydraul Eng 125(9):924–933

Meyer-Peter E, Müller R (1948) Formulas for bed-load transport. In: IAHSR 2nd meeting, Stockholm, Appendix-2. IAHR

Molinas A, Yang TC (1986) Computer program user's manual for GSTARS (generalized stream tube model for alluvial river simulation). US Department of Interior, Bureau of Reclamation, Engineering and Research center

Olsen NR, Skoglund M (1994) Three-dimensional numerical modeling of water and sediment flow in a sand trap. J Hydraul Res 32(6):833–844

Papanicolaou AT, Elhakeem M, Krallis G, Prakash S, Edinger J (2008) Sediment transport modeling review—current and future developments. J Hydraul Eng 134(1):1–14

Rubey WW (1933) Settling velocity of gravel, sand, and silt particles. Am J Sci, XXV 148:325–338

Thomas WA, McAnally Jr WH (1985). User's manual for the generalized computer program system open-channel flow and sedimentation TABS-2. main text and Appendices A through (No. WES/IR/HL-85-1). Army Engineer Waterways Experiment Station Vicksburg MS Hydraulics Lab

Thomas WA, Prashum AL (1977) Mathematical model of scour and deposition. J Hydraul Div 103 (ASCE 13132 Proceeding)

Toffaleti FB (1968) A procedure for computation of the total river sand discharge and detailed distribution, bed to surface (No. TR-5). Committee on Channel Stabilization (Army)

Toombes L, Chanson H (2011) Numerical limitations of hydraulic models. In: Proceedings of the 34th world congress of the international association for hydro-environment research and engineering: 33rd hydrology and water resources symposium and 10th conference on hydraulics in water engineering. Engineers Australia, 2322 pp.

Trush WJ, Scott MM, Luna BL (2000) Attributes of an alluvial river and their relation to water policy and management. Proc Natl Acad Sci 97(22):11858–11863

USACE (2010) HEC-RAS River Analysis System. User's Manual. Version 4. Hydrulic Engineering Center. Davis, CA: U.S. Army Corps of Engineers (USACE)

Van Rijan LC (1993) Principles of sediment transport in rivers, estuaries and coastal seas, vol 1006. Aqua publications, Amsterdam

Vanoni VA (1975) Sedimentation engineering, ASCE manuals and reports on engineering practice, vol 54. Am Soc Civ Eng, Reston, VA

Vieira DA, Wu W (2002) One-dimensional channel network model CCHE1D Version 3.0: User's manual. National Center for Computational Hydroscience and Engineering, University of Mississippi, Oxford, MS

Walstra DJ, Van Rijn LC, Aarninkhof SG (1998) Sand transport at the middle and lower shoreface of the Dutch coast: simulations of SUTRENCH-model and proposal for large-scale laboratory tests. Z2378

Yang CT (1973) Incipient motion and sediment transport. J Hydraul Div 99(10):1679–1704

Yang CT (1996) Sediment Transport. Theory and Practice McGraw-hill, New York

Ye J, McCorquodale JA (1997) Depth-averaged hydrodynamic model in curvilinear collocated grid. J Hydraul Eng 123(5):380–388

Zhang Y (2005) CCHE2D-GUI–graphical user interface for the CCHE2D model user's manual– version 2.2. National Center for Computational Hydroscience and Engineering, Mississippi, US

# Chapter 22
# Design and Sensitivity Analysis of High Head Regulating Radial Gate Using Microsoft Excel Spread Sheet

**Roshan Kumar Deo, Thanga Raj Chelliah, S. K. Shukla, and Mayank Jain**

## 22.1 Introduction

Gates are closure devices in which a leaf is driven across the fluid way to regulate discharge or flow of water. Gates have multiple uses and locations in dams, canals, waterways, etc. These structures restrict the flow of water on a need basis. It could be manually operated, powered or automatic. Different gates are being used under the following category of head over the Sill as: Low Head (H < 15 m), Medium Head (15 < H < 30 m), High Head (H > 30 m).

In any case, operating device/structure is required to open and close the gate, often known as hoist. Hoists come in many forms such as rope drum, hydraulic, screw, etc. Gates and hoists based on different principles and mechanisms are being used for particular discharge of water through spillways, sluices, intakes, regulators, ducts, tunnels, etc. Right selection of gates and their hoisting arrangement ensures safety of the structure and control over the structure effectively. The most commonly used gates nowadays are fixed wheel gates, radial gates and slide gates. The entire work

R. K. Deo (✉) · T. R. Chelliah · S. K. Shukla
Department of Water Resources Development and Management, Indian Institute of Technology Roorkee, Roorkee 247 667, Uttarakhand, India
e-mail: roshandeo6@gmail.com

T. R. Chelliah
e-mail: thangfwt@gmail.com

S. K. Shukla
e-mail: skshukla.tehri@gmail.com

M. Jain
Tehri Hydro Development Corporation Limited, Rishikesh, Uttarakhand, India
e-mail: mayank230667@gmail.com

© The Editor(s) (if applicable) and The Author(s), under exclusive license
to Springer Nature Switzerland AG 2021
A. Pandey et al. (eds.), *Hydrological Extremes*, Water Science
and Technology Library 97, https://doi.org/10.1007/978-3-030-59148-9_22

consisting design, fabrication and erection of particular gates requires specific technical knowledge, skill, time and selection of gate depends upon type of application, head and location of gate for the particular purpose.

*Jeremiah Burnham Tainter* invented Tainter gates (Radial gates) in 1886. He was an employee of Knapp, Stout, and Co., a large lumber company, Wisconsin, USA. A task was given to Mr Tainter to make a floodgate that could be easily opened and closed for applications in river lumber transport.[Sluice Gates, Roller Gates, And Tainter Gates: Tainter Gates—History & Origin; Kavi Pool, Daniel Polverari, Nolan Platt]. (docslide, n.d.).

Radial gate consists of the leaf (or skin) in the form of a circular arc with the Centre of curvature at the hinge or trunnion. The curved part of the gate faces the upper level of water and the tip pointing towards the lower level of pool. The straight sides, the trunnion arms, extend back from the ends of the cylinder sections and meet at a trunnion hub, which actually is a pivot point about which the gate rotates. The design of the Radial gate is started with calculating every pressure force acting through the centre of the imaginary circle, which the gate is a section of, so that all resulting pressure force acts at the hinge or trunnion of the gate. When a radial gate is closed, water load bears on convex (upstream) side and when the gate is rotated, the rush of water passing under the gate helps to open the gate. The closure of the gate requires less effort than other gates because of the rounded face, long radial arms and trunnion bearings. Radial gates are usually operated from above with a gearbox, chain, or electric motor assembly placed on a bridge situated above the piers. (IT Kharagpur).

Design of radial gate requires the determination of size of different components required for operation effectively and design begins with the known value of input parameters like clear width of opening, full reservoir level (FRL), Sill level, Centre line of Trunnion and radius to inside skin plate.

Microsoft Excel is a spreadsheet program of the Microsoft Office suite of applications in which, the arrangement of data into columns and rows creates intersections called cells. Spreadsheets present tables of values (i.e. Data in the form of numbers, strings or formulas) arranged in rows and columns that can be manipulated mathematically using both basic and complex arithmetic operations and functions. Spreadsheets give quicker results and thus save much time than manual calculation and provide both accuracy of calculations and flexibility in presentation. Microsoft Excel is very useful for analyzing data, finding information, preparing charts. The appearance of every part of the spreadsheet remains under user total control, i.e. the user can specify the font style and size used for numbers and text, define the border width and cell size of tables, add images, and colour everything with a brilliant palette. The result or output from the excel spreadsheet can be exported in a variety of file formats applicable for printed documents, group presentations, websites and reports.

Sensitivity analysis is used as an approach for identification impact of different independent variables on a dependent variable under particular assumptions. This technique is used within particular boundaries that depend on one or more input variables. Sensitivity analysis is the study determination of uncertainty in the output of a mathematical model or system (numerical or otherwise) can be apportioned to different sources of uncertainty in its inputs. The process of recomputing results/outcomes under different possible alternatives to determine the impact of a variable input in a sensitivity analysis can be useful for many purposes, including:

- Testing the robustness of the performance of a model/system in the presence of uncertainty.
- Increased understanding of the relationships between input and output variables in a system/model.
- Uncertainty reduction, through the identification of model inputs responsible for the significant uncertainty in the output.
- Searching for errors in the model (by encountering unexpected relationships between inputs and outputs).
- Model simplification—fixing model inputs that have no effect on the output, or identifying and removing redundant parts of the model structure.
- Enhancing communication from modellers to decision-makers (e.g. by making recommendations more credible, understandable, compelling or persuasive) in optimization and calibration stage by focusing on the sensitive parameters of the model.

There are a large number of approaches for performing a sensitivity analysis like Local methods, Scatter plots, Regression analysis, Variance-based methods, Variogram-based Methods, Screening. One of the simplest and most common approaches is that of changing one-factor-at-a-time (OFAT or OAT), which involves moving one input variable, keeping others at their baseline (nominal) values and returning the variable to its nominal value, then similarly repeating for each of the other inputs.

This appears a logical approach as any change observed in the output will unambiguously be due to the single variable changed. Furthermore, by changing one variable at a time, one can keep all other variables fixed to their central or baseline values. This increases the comparability of the results when several input factors are changed simultaneously. Even the modeller can immediately find the responsible input factor for the failure in case of model failure under OAT analysis.

In this paper; 'Microsoft Excel' spreadsheet has been developed for the design of radial with particular input parameter (variables); that may be a useful tool to design similar gate in lesser time if input parameters (variables) are changed. Sensitive analysis has been done using one-factor-at-a-time (OFAT or OAT) to determine the impact on output and presented graphically in terms of weight of gate versus input parameters.

**Advantages and Disadvantages of Radial Gate**

*Advantages*:

- Gate bottom lip provides good hydraulic discharge profile.
- The lifting point of gate is farther away from the point of rotation, i.e. the trunnion, the hoist capacity is reduced compared with the fixed wheel gate.

- The height of the pier for locating the hoist is reduced compared with the fixed wheel gate.
- In case of large openings, by adopting inclined arms, weight of the gate is reduced and hoist capacity required is small.
- There are not many machined components as compared with the wheeled gates, thereby fabrication as well as erection is simple compared with fixed wheel gate.
- For large size gates, there is an option of using hydraulic hoist or rope drum hoist.
- No groove is required in the pier.
- For very large gates, by keeping the Centre of curvature of the skin plate slightly above the Centre line of the trunnion, it is possible to reduce the hoist capacity required to lift the gate.
- Use of spherical plain bearings in the trunnion allows easy alignments of the trunnion.
- In cold regions, it is easier to install heaters on the radial gate body in absence of slots that can accumulate ice.

*Disadvantages*:

- The length of pier is increased Gate bottom.
- There is concentrated load on the pier (only at two points).

## 22.2  Brief Description and Design of Components of Radial Gate Skin Plate (Gate Leaf)

The skin plate of a radial gate consists of plates that have been bent into the form of an arc. Load due to Thrust of water is taken by the convex face of gate and transfer to other components of gate. The radius of curvature of gate is generally H to 1.25H, where H is the vertical distance between the sill and the top of gate. A 1.5 mm corrosion allowance is provided while deciding the thickness. During raising or lowering of gate, the upstream face of the gate rubs against the top seal.

The skin plate is supported on a number of vertical stiffeners. The thickness of the skin plate could be uniform throughout the height of the gate, in case of small height of gate. The minimum thickness of skin plate used is 10 mm. The thickness of the skin plate could vary, with a thicker plate at the bottom and thinner at the top, if the height of the gate is large. Normally, the maximum bending moment in the skin plate is considered as $wL^2/10$, where w is the uniformly distributed load and L is the maximum span. By proportionate the spans, i.e. adjusting the values of a11, a21, and a31; the maximum bending moment can be reduced to $wL^2/12$. If the stiffeners are vertical, the skin plate and the stiffeners would be welded on the shop floor. Depending upon the width of the opening, the assembly of the stiffeners and skin plate is made of standard widths such as 2.0 or 2.5 m. (Technical Specification For Gates & Hoists In Spillways, Under Sluices, Power Intakes, Draft Tubes, And Penstock Liners.)

## Stiffener

The skin plate is supported by stiffeners equally spaced which may be either horizontal or vertical or both. The spacing of the stiffeners may vary from 300 to 500 mm centre to centre distance.

The stiffeners are spaced either vertically or horizontally. If the stiffeners are spaced horizontally, they are supported on horizontal girders. The number of girders, i.e. supports could be 2–4. The spacing of the horizontal girders could be adjusted such that, the bending moment in the stiffeners over the supports would be almost equal.

Considering the stiffener as a straight beam with the loading as in figure below, the bending moment at the supports can be calculated using 'moment distribution method'.

Reaction on sill (assuming centre of gravity, cg is at 0.85 R from trunnion),

$W_s = 0.85 \times W_g/L$, Where L is length of sill beam.

Bending moment in the stiffener at bottom girder would be equal to $W_g \times$ distance of bottom girder from sill $\times$ spacing of stiffener.

## Girder

A number of Horizontal girders are kept to take up the water thrust from the skin plate. Girders are generally plate girders, with webs, web stiffeners and flanges. Welds should be checked. Drain holes are provided to prevent water collection and rusting. They are required to be kept clean. The number of horizontal girders could be 2–4 depending upon height of the gate. In the case of vertical stiffeners, the number of horizontal girders may be 2–4 but in case of horizontal stiffeners with vertical diaphragms, the number of girders could be reduced to a minimum of two numbers.

The load on the girder will be different for different girders. If the arms are parallel, the girders are subjected to bending and shear. If the arms are inclined, in addition to bending and shear, the girders are subjected to axial load, which is the horizontal component of the axial load in the arms and hence connection between the arms and the horizontal girder is to be a rigid connection. The arms and the horizontal girders shall be analyzed as a frame. In case of inclined arms, the centre of gravity line of the arms shall meet the centre of gravity line of girder at 1/5 the span from end, in which case the bending moment at the support will be almost equal to the bending moment at the mid-span. The girder is to be checked for the stresses at the following locations.

- at the support on the cantilever side
- at the support on the span side
- at mid-span.

The bending moment at the cantilever, at span and at mid span are to be arrived by analyzing the girder and the arms as a frame (Sahu and Ajmera, 2017).

### Radial Arm

Starting from the trunnion hub; Radial arms are connected to the vertical end supports of skin plate in the case of small size gates to horizontal girders.

### *Parallel arms*

Parallel arms are provided for small gates where water thrust is not much. These are straight and parallel to each other and are connected to vertical end supports of skin plate or the horizontal girder, at either end.

### *Inclined arms*

In the case of larger gates inclined arms are provided. The arm assembly is connected to the horizontal girder at above 1/5 of span from the end, resulting in substantial saving in weight and size of the horizontal girders. In case of inclined arms, a lateral force is developed at the trunnion. This lateral force can be absorbed by providing a trunnion tie in tension. Alternately, the lateral force could be transferred to the pier through a thrust block embedded in concrete on each side of the opening. The arms are welded to the trunnion hub.

The Arms are critical members as the failure can take place normally in compression members. The l/r ratio is to be kept such that the actual compressive stress in the arm is less than the permissible value for that l/r ratio. The arms are to be designed for the following forces:

- The axial force caused by the reaction from the girders
- The bending moment at the girder and column intersection due to the frame action
- The bending moment near the trunnion caused by the friction developed in the trunnion.

The combined compressive stress and bending stress is to be checked by the unity formula. The weight of the gate is mostly carried by the wire ropes of the hoisting system and the balance weight will be transferred to the trunnion.

### Trunnion Assembly

Horizontal Trunnion Assembly consists of trunnion hub connected to the arms, trunnion bracket mounted on anchor/yoke girders, and trunnion pin acting as a pivot or hinge. In case of inclined arms, side thrust due to inclined arms could be tackled by suitable anchorage or by providing, the tie girder between two trunnions of a gate. As far as possible, the trunnion shall be located at H/3 from sill level, in which case the resultant hydraulic force would be horizontal. This would eliminate the force on the vertical anchors, when gate is in closed position.

Trunnion hub is a heavy steel casting. The castings of trunnion assembly need to be checked for soundness. Blow holes and cracks should be avoided. Generally, phosphor-bronze bearing metal bushing is fitted to trunnion hub. Trunnion pins are made up of cast steel or forged carbon steel with hard chrome or nickel plating.

In case of inclined arms, the radial load is taken by the trunnion bush/hub and the lateral load is to be taken by providing a trunnion tie. Alternately, the lateral load can be transferred to the concrete pier by providing thrust transfer plate. The trunnion is subjected to vertical forces, uplift as well as downward forces. Anchor bolts are to be provided to take care of the uplift forces.

### Trunnion Pin

The material of the trunnion pin may be cast steel to IS 1030, 30 Cr 13, 20 Cr 12 to IS 1570. The pin can be made hollow if cast steel is used. If the pin is of cast steel, the outer surface is hard chromium plated. The pin is to be locked against rotation.

The pin is to be checked for bending moment considering the span as the Centre to Centre distance of the arms of the trunnion bracket, which are the supporting members for the trunnion pin. The load is to be uniformly distributed a length equal to the length of the bushing.

### Trunnion hub

The material of the hub is normally cast steel to IS 1030. The inside diameter is to be equal to the outside diameter of the bushing/spherical plane bearing. The tolerance between the hub and the bushings is to be 0.3 times the diameter of the pin. The thickness is kept normally uniform. The thickness can be reduced on downstream side to reduce the weight of the hub.

### Trunnion Bush

The material of the bush is aluminium bush to IS 305 or self-lubricating bronze bush. The bush is subjected to compressive stress. The actual compressive stress is calculated by dividing the maximum radial load by the projected area of the bush. The projected area will be equal to the product of the length of the bush and the inside diameter. The length of the bush is normally kept between 0.8 and 1.2 times the diameters of the bush for better distribution of the load. The section of the bush is calculated by the formula $t = (0.08 \times dp + 3)$ where dp is diameter of the pin in mm. (Technical Specification For Gates & Hoists In Spillways, Under Sluices, Power Intakes, Draft Tubes, And Penstock Liners.).

## 22.3   Methodology

### 22.3.1   Design

Design of Radial gate, i.e. size of various components and their respective location in the assembly can be done by developing Microsoft excel spreadsheet. In these two sheets are linked, i.e. *design sheet* and *report sheet*. In design sheet, users have access to size different components and check whether the proposed/assumed size is safe and within the allowable permissible limits/stresses or not. The design begins with the given input parameters and then determination of permissible stresses succeeding determination of sizes of each component. The best design is the determination of

the minimum size which is safe and within permissible stresses for the material
employed for the component design. Furthermore, cells have been distinguished
with colour: *Green*—Feed/Input cell (where given input variables/parameters and/or
proposed inputs are put.); *Yellow*—output cells (where output are seen as a result
of inputs fed in the input cell, i.e. green cells.); *Red*—Check/decision cell (where
message 'OK/change the input value or dimension of proposed section' will be
displayed.). The developed Microsoft Excel spreadsheet will be very applicable for
user in designing radial gate in lesser time and with very less computational analysis.
It is also an advantage for user that design report is prepared simultaneously once
user go through design sheet and user need not to do any effort in the report sheet.
The undermentioned chart is shown below to describe the methodology in detail
(Figs. 22.1, 22.2, 22.3, 22.4, 22.5, 22.6, 20.7, 20.8 and 20.9).

| Step 1 : | | (Technical information) | |
|---|---|---|---|
| | Input | Green : | Clear width of opening(W), Clear height of opening(H), Radius factor, FRL, MWL,Sill Level, Top seal level, Trunnion level, Top gate Level |
| | | | Accessibility : Accessible Inaccessible water contact condition : dry / wet |
| | Output | Yellow : | Radius to outside of skin plate |

| Step 2 : | | (Materail specification) | |
|---|---|---|---|
| | Intake | Green : | Skin plate :Stainless steel /structural steel Trunnion pin : cast steel / forged steel |
| | (Note : All the parts are taken as per indicated IS code or reference in the intake sheet.) | | |

| Step 3: | | (Permissible stresses) | |
|---|---|---|---|
| Input | Green: | | Proposed thickness of Skin plate |
| Output | Yellow: | | Yield point (YP) & Ultimate tensile strength (UTS) in MPa |
| Input | Green: | | Adopted value: either YP & UTS obtained or as per manufacturer recommendation |
| Output | Yellow: | | permissible stresses factors for FRL & MWL |

| Step 4: | | (Skin plate) | |
|---|---|---|---|
| Input | Green: | | Proposed location of girder L1, L2, L3 |
| Output | Yellow: | | d1,d2,d3,d4,d5,$\beta$1,$\beta$2,$\theta$1,$\theta$2,$\theta$3,$\theta$4,$\alpha$ Total curve length |
| Input | Green: | | corrosion allowance not considered /corrosion allowance |
| Output | Yellow: | | proposed stiffener location a11, a12, a13 actual stress & allowable stress Total curve length |
| Check | Red: | | OK / change the thickness of skin plate Total curve length |

| Step 5: | | (Stiffener) | |
|---|---|---|---|
| Input | Green : | Assumed wt. of gate | |
| Output | Yellow: | Reaction, SF, BM for all 3 girders under % increase in BM at MWL over FRL | |
| Input | Green: | Proposed section dimension for stiffener | |
| Output | Yellow: | Section properties e.g. area, Ixx, Y1, Z1 etc. Bending stresses, shear stresses, combined stresses & permissible stresses | |
| Check | Red : | OK / change the dimension of the section proposed | |

| Step 6: | | (Girder) | |
|---|---|---|---|
| Output | Yellow: | Re, SF, BM at support & centre (appears after fixing of stiffener design) | |
| Input | Green: | Proposed section dimension for girder | |
| Output | Yellow: | Section properties e.g. area, Ixx, Y1, Z1 etc. Bending stresses, shear stresses at support and centre & permissible stresses | |
| Check | Red: | OK / change the dimension of the section proposed | |

| Step 7: | | (Arm) | |
|---|---|---|---|
| Input | Green: | Arm type (Inclined /Parallel) | |
| Output | Yellow: | Length, load on arm, Clockwise fixed end moment at support | |
| Input | Green: | Proposed section dimension for arm | |
| Output | Yellow: | Section properties e.g. area, Ixx, Y1, Z1 etc. Stiffener factors and distribution factors Bending stresses, compressive stresses | |
| Check | Red: | OK / change the dimension of the section proposed | |

| Step 8: | | (trunnion pin and Trunnion bush) | |
|---|---|---|---|
| Input | Green: | length of bush, length of hub, dia, of pin | |
| Output | Yellow: | Bush ID, OD & thickness Bearing, shear and direct stresses | |
| Check | Red: | OK / change the dimension of the section proposed | |

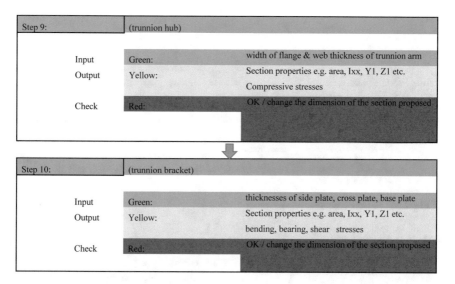

| Step 9: | (trunnion hub) | |
|---|---|---|
| Input | Green: | width of flange & web thickness of trunnion arm |
| Output | Yellow: | Section properties e.g. area, Ixx, Y1, Z1 etc. |
| | | Compressive stresses |
| Check | Red: | OK / change the dimension of the section proposed |

| Step 10: | (trunnion bracket) | |
|---|---|---|
| Input | Green: | thicknesses of side plate, cross plate, base plate |
| Output | Yellow: | Section properties e.g. area, Ixx, Y1, Z1 etc. |
| | | bending, bearing, shear   stresses |
| Check | Red: | OK / change the dimension of the section proposed |

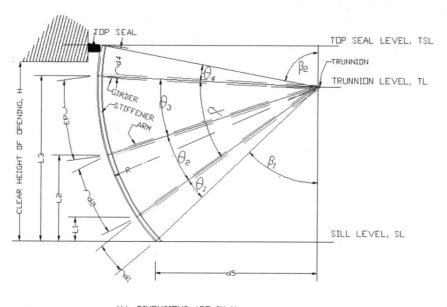

ALL DIMENSIONS ARE IN M

**Fig. 22.1** Location of different components of radial gate in closed position

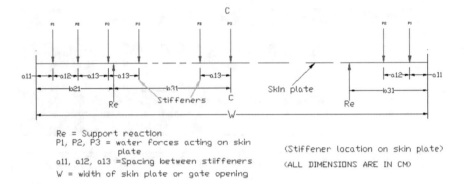

Re = Support reaction
P1, P2, P3 = water forces acting on skin
        plate
a11, a12, a13 =Spacing between stiiffeners
W = width of skin plate or gate opening

(Stiffener location on skin plate)

(ALL DIMENSIONS ARE IN CM)

**Fig. 22.2** Stiffener location on skin plate

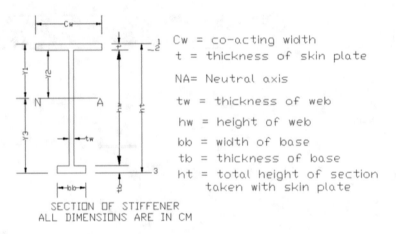

Cw = co-acting width
t = thickness of skin plate
NA= Neutral axis
tw = thickness of web
hw = height of web
bb = width of base
tb = thickness of base
ht = total height of section
        taken with skin plate

SECTION OF STIFFENER
ALL DIMENSIONS ARE IN CM

**Fig. 22.3** Section of stiffener

### 22.3.2 Sensitivity Analysis

In order to determine the impact of input variables, the design as per the above
methodology using Microsoft Excel Spreadsheet has been done. The input variables
(parameters) are varied as following and the corresponding output in terms of total
weight of gate is determined. The graphs are plotted with varied input parameters

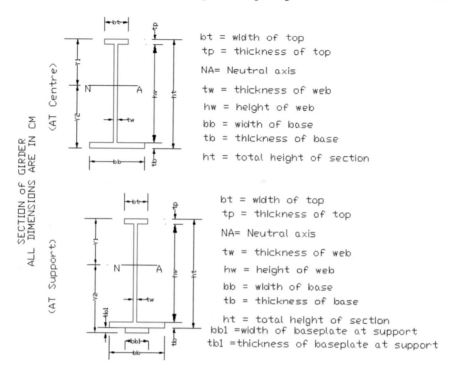

**Fig. 22.4**  Section of girder

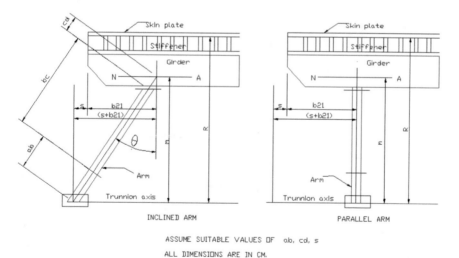

**Fig. 22.5**  Location of skin plate, stiffener, girder, trunnion

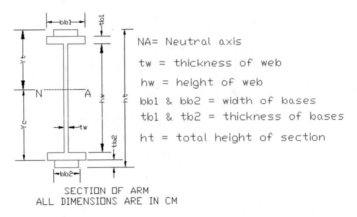

NA= Neutral axis

tw = thickness of web

hw = height of web

bb1 & bb2 = width of bases

tb1 & tb2 = thickness of bases

ht = total height of section

SECTION OF ARM
ALL DIMENSIONS ARE IN CM

**Fig. 22.6** Section of arm

**Fig. 22.7** Section of
trunnion arm

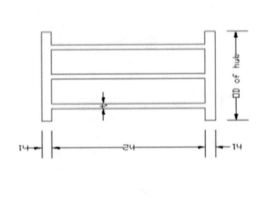

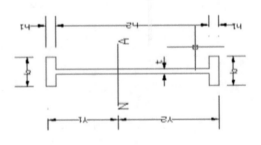

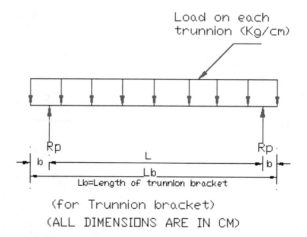

**Fig. 22.8** Showing load and reaction on trunnion

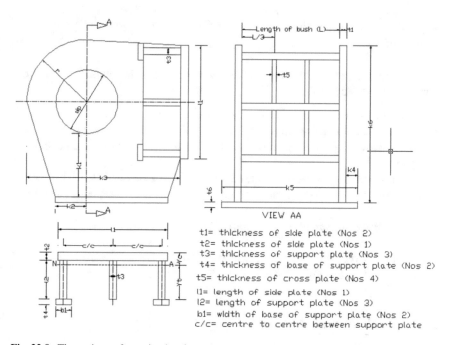

t1= thickness of side plate (Nos 2)
t2= thickness of side plate (Nos 1)
t3= thickness of support plate (Nos 3)
t4= thickness of base of support plate (Nos 2)
t5= thickness of cross plate (Nos 4)

l1= length of side plate (Nos 1)
l2= length of support plate (Nos 3)
b1= width of base of support plate (Nos 2)
c/c= centre to centre between support plate

**Fig. 22.9** Three views of trunnion bracket

versus total weight of gate. This one-factor-at-a-time (OFAT or OAT) approach of sensitive analysis involves moving one input variable, keeping others at their baseline (nominal) values and returning the variable to its nominal value, then similarly repeating for each of the other inputs. The input variables taken for the design are:

| Clear width of opening, W | | 4.5 | M |
|---|---|---|---|
| Clear height of opening, H | | 6 | M |
| Radius factor,RF | | 1.5 | |
| Radius to outside of skin plate, R | | 9 | |
| Full reservoir level, FRL | EL | 830 | M |
| Max water level, | | | |
| MWL | EL | 835 | M |
| Design Head, H at | FRL | 130 | M |
| | MWL | 135 | M |
| Sill level, SL | EL | 700 | M |
| Trunnion level, | | | |
| TL | EL | 707.9 | M |
| Top seal level, | | | |
| TSL | EL | 706.65 | M |
| Top gate level | EL | 706.76 | M |
| No. of horizontal | | | |
| /Arm | | 3 | |
| Accessibility | | Inaccessible | |
| Water Contact condition | | Dry | |
| Density of water | | 1 | gm/cc |

The result (output) comes in terms of weight of gate is as in Table 22.1. The input variables are varied as follows:

- Radius factor (RF) output varied by ±1 (one) unit, i.e. 1.4 and 1.6
- Keeping area of opening constant, i.e. (4.5 × 6 = 27 m²) in order to have same discharge, width (W) of opening varied by −1 (one), −2 (two) unit, i.e. 2.5 and 3.5 m
- Keeping area of opening constant, i.e. (4.5 × 6 = 27 m²) in order to have same discharge, height (H) of opening varied by −1 (one), −2 (two) unit, i.e. 2.5 and 3.5 m
- Head over the sill varied by ±10 (ten), ±20 (twenty) unit, i.e. 110, 120, 140 and 150 m
- A graph between varying input parameters (RF, W, Height, Head) in x-axis and total weight of gate (WT) have been plotted as shown in Table 22.2.

## 22.4 Analysis and Discussion

- It is not possible to reduce Radius factor (RF) further below 1.4 keeping all other input variables constant. Geometrically, the minimum possible Radius factor (RF) is 1.4 for the given conditions. The radius factor (RF), thus, was increased from its minimum possible value to 1.6 and the variation in the output (in terms of total weight of gate in Kg) is presented in graph (Radius factor vs. weight). It is seen that the weight of gate goes on decreasing with increasing RF from this minimum possible value up to 1.5 and then increases again.

**Table 22.1** Weight Calculation of Radial gate designed at given input condition

| S. N. | Components | Material | Area (A) m² | Length (L) m | Width (W) m | thickness (t) m | Volume (V) m³ | Quantity (Qt) Nos. | Density m³/Kg | Weight Kg |
|---|---|---|---|---|---|---|---|---|---|---|
| 1 | Skin plate | Stainless steel | | 8.387 | 4.5 | 0.032 | 1.208 | | 8000 | 9661.56 |
| 2 | Stiffener | Structural steel | 0.0158 | 8.387 | | | 0.133 | 13 | 7850 | 13522.74 |
| 3 | Girder | Structural steel | | | | | | | | |
| | Bottom girder A | | 0.066 | | 4.5 | | 0.3 | | 7850 | 2331.45 |
| | Middle girder B | | 0.066 | | 4.5 | | 0.3 | | 7850 | 2331.45 |
| | Top girder C | | 0.066 | | 4.5 | | 0.3 | | 7850 | 2331.45 |
| 4 | Arm | Structural steel | | | | | | | | |
| | Bottom Arm | | 0.0984 | 5.30 | | | 0.522 | 2 | 7850 | 8187.86 |
| | Middle Arm | | 0.0984 | 5.30 | | | 0.522 | 2 | 7850 | 8187.86 |
| | Top Arm | | 0.0984 | 5.30 | | | 0.522 | 2 | 7850 | 8187.86 |
| 5 | Trunnion pin | Cast steel | 0.2827 | 1.05 | | | 0.30 | 2 | 7700 | 2281.62 |
| 6 | Trunnion bush | Self lubricating bronze | 0.1021 | 1.05 | | | 0.11 | 2 | 6500 | 695.52 |
| 7 | Trunnion hub | | | | | | | | | |
| | Hub | Cast steel | 0.4976 | | 0.9 | | 0.448 | 2 | 7700 | 3448.56 |
| | Trunnion Arm | Structural steel | 0.1096 | 1.60 | | | 0.175 | 6 | 7850 | 8259.46 |
| 8 | Trunnion bracket | Structural steel | | | | | | | | |
| | Side plate1 | | | 1.38 | 1.38 | 0.10 | 0.192 | 2 | 7850 | 1503.39 |
| | Side plate2 | | | 1.46 | 1.49 | 0.12 | 0.261 | 4 | 7850 | 8196.91 |
| | Support plate | | | 0.40 | 1.05 | 0.10 | 0.042 | 6 | 7850 | 1974.43 |
| | Cross plate | | | 0.42 | 0.40 | 0.03 | 0.005 | 8 | 7850 | 340.83 |
| | Base of support | | | 0.12 | 1.05 | 0.05 | 0.006 | 4 | 7850 | 197.44 |
| | Base plate | | | 1.44 | 1.38 | 0.05 | 0.100 | 2 | 7850 | 784.29 |
| | | | | | | | | | Total | 82425 |

**Table 22.2** Sensitivity analysis

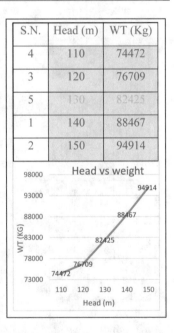

| S.N. | RF | WT (Kg) |
|---|---|---|
| 1 | 1.4 | 86940 |
| 3 | 1.5 | 82425 |
| 2 | 1.6 | 84453 |

Radius factor vs weight

| S.N. | H (m) | W (m) | WT (Kg) |
|---|---|---|---|
| 2 | 10.8 | 2.5 | 168360 |
| 1 | 7.71 | 3.5 | 114449 |
| 3 | 6 | 4.5 | 82425 |

Width vs weight

| S.N. | W (m) | H (m) | WT (Kg) |
|---|---|---|---|
| 4 | 4.5 | 6 | 82425 |
| 1 | 3.86 | 7 | 95509 |
| 2 | 3.38 | 8 | 120792 |
| 3 | 3 | 9 | 148775 |

Height vs weight

| S.N. | Head (m) | WT (Kg) |
|---|---|---|
| 4 | 110 | 74472 |
| 3 | 120 | 76709 |
| 5 | 130 | 82425 |
| 1 | 140 | 88467 |
| 2 | 150 | 94914 |

Head vs weight

- Keeping area of opening (A) constant in order to have fixed discharge, if we go on changing width of opening (W), there will be corresponding change in height of opening (H). Therefore, increasing width of opening (W) by 1 (one) unit, i.e. up to 5.5 m, the corresponding height of opening (H) needs to be 4.9 m to have constant area of opening (A), i.e. $6 \times 4.5 = 27$ m². And further we go on increasing width

(W), there will be corresponding decrease in height of opening (H). But it is not geometrically possible to reduce height of opening (H) below 5.26 m keeping all other variables constant. Therefore, width of opening (W) was reduced from its given value, i.e. 4.5 m up to ±2 (two) units, i.e. up to 2.5 m and the variation in the output (in terms of total weight of gate in Kg) is presented in graph (width vs. weight).

- As it is not geometrically possible to reduce the height of opening (H) below 5.26 m keeping all other variables constant. In order to have significant impact in outcome, height of opening (H) was increased from its given value, i.e. 6 m by 3 (two) units, i.e. up to 9 m and the variation in the output (in terms of total weight of gate in Kg) is presented in graph (height vs. weight).
- Common problems encountered as mentioned in various Case studies regarding failure of radial gates can be carefully incorporated in the design.
- Only cost-effective and efficient design is not adequate; proper inspection, fabrication, erection/installation and operation and maintenance are also necessary for long run and successful operation as well as function of radial gate.

## 22.5  Conclusion

Based on the above analysis, it is observed that:

- The weight of gate will be minimum, i.e. least size of components gate required in design of radial gate at little higher value than minimum possible Radius factor (RF) which, in fact, differ based on other given input variable. It is seen that the weight of gate goes on decreasing with increasing RF from this minimum possible value up to 1.5 and then increases again.
- Larger the width (W) of gate for particular discharge and keeping other input parameters (variables) constant, the smaller the size of the components in design of radial gate and hence more economical design.
- Smaller the height of opening(H) particular discharge and keeping other input parameters (variables) constant, the smaller the size of the components in design of radial gate and hence more economical design.
- Lower the head over sill, i.e. lower water pressure, smaller size of components is required in the design of radial gate and vice versa.
- There is no ideal maintenance-free design of radial gate. Design should be such that there would be minimum maintenance requirement including long life, minimum travel/overturn of gate, simpler hoisting arrangement, convenient for inspection, operation and maintenance, less cost.
- The design of radial gate done using Microsoft excel spreadsheet that reduces many iterations and helps in getting better results and saves time and with less effort. Thus, the aim of the objectives was fulfilled successfully.

# References

docslide (n.d.) https://docslide.us/documents/design-of-radial-gate.html

(n.d.) https://www.ivt.ntnu.no/ept/fag/tep4195/innhold/Forelesninger/forelesninger%202006/5%
20-%20Hydro%20Power%20Plants.pdf

(n.d.) https://www.bgstructuralengineering.com/BGSMA/ContBeams/BGSMA_CB_0201.htm

(n.d.) https://www.ijasre.net/print_articale.php?did=2796

(n.d.) Technical specification for gates & hoists in spillways, under sluices, power intakes, draft tubes, and penstock liners. Koteshwar Hydro Electric Project

Bureau of Indian Standards, Manak Bhawan 9, Bahadur Shah Zafar Marg, New Delhi-2. (IS:808-1989, IS:226-1975). Dimensions of structural steel (column, Channel, Angle). Bureau of Indian Standards

Bureau of Indian Standards, Manak Bhawan 9, Bahadur Shah Zafar Marg, New Delhi-2. (IS:1030-1982). Cast steel-specification

Bureau of Indian Standards, Manak Bhawan 9, Bahadur Shah Zafar Marg, New Delhi-2. (IS:2062-2011). Hot rolled medium and high tensile structure steel-specification

Bureau of Indian Standards, Manak Bhawan 9, Bahadur Shah Zafar Marg, New Delhi-2 . (IS:4622-1992;). Recommendations for structural design of fixed wheel gates

Bureau of Indian Standards, Manak Bhawan 9, Bahadur Shah Zafar Marg, New Delhi-2;. (IS:4623-2000; ). Recommendations for structural design of radial gates, Third revision

Davis, Calvin Victor and Sorensens, Kenneth (1969) Hand book of applied hydraulics

Department of Mechanical Engineering, Institute of Technology, Hydrabad (n.d.) Design and analysis of spillway Radial gate. Int J Adv Sci Res Eng (IJASRE).

Continuous Beam Analysis (2007, 2008). In: Quimby B (ed) A beginner's guide to structural mechanics/analysis

Kharagpur IT (n.d.) Hydraulic structures for flow diversion and storage, Version 2 CE. Module 4

Lewin J (n.d.) Hydraulic gates and valves: In free surface flow and submerged outlets, 2nd edn. D engg., FICE, FIMechE, FCIWEM, Thomas Telford Ltd

Madanaiah P, Consultant, Centre for good Governance (n.d.) Maintenance of hydraulic gates: important parametrs

Mehta DK (n.d.) Various types of gates, their important components and their functions. https://www.slideshare.net/IEIGSC/various-types-of-gates-their-important-components

Naidu NK (n.d.) Operation and maintenance of Radial gates

Nigam PS (1987) Handbook of hydroelectric engineering. Nemchand and Bros, Roorkee

Novak P, Moffat AIB, Nallury C, Narayana R (1996) Hydraulic structures. E & FN SPON, London, Glasgow, Weinheim, New York, Tokyo, Melbourne, Madras

Punmia BC (1992) Strength of material and theory of structures. Laxmi Publications, 7/21, Aansari Road, Daryaganj, New Delhi-2

Rajput RK (2003) Strength of materials (Mechanics of Solids), 3rd edn. Chand S & Company Ltd

Sahu RA (n.d.) Design of Radial gate. Int J Adv Sci Res Eng (IJASRE)

Stutsman RD, Ahl T, Brogdon C, Dunbar S (1993) ASCE manuals and reports on engineering practice No. 79

Underwood JH (1989) Applications and limitations of finite element analysis to armament components. Technical Report ARCCB-TR-89018, July

Varshney RS (1986) Hydropower structures. Nemchand and Bros

# Part III
# Intelligent Irrigation Water Management

# Chapter 23
# Model to Generate Crop Combinations for Tribal Farmers in Palghar, Maharashtra, India

**Aniket Deo, Amit Arora, and Upendra Bhandarkar**

## 23.1  Introduction

Tribal farmers in Palghar district, Maharashtra practice rain-fed rice cultivation which is their staple food and after Kharif season, almost 90% farmers generally migrate to nearby cities for jobs or choose to become construction laborers in local areas but work availability is a serious concern and most farmers end up with no source of income for the rest of the year. Promoting horticulture cultivation (vegetables) during Rabi season can be an impressive source of livelihood for tribals, however, unavailability of water resources is a key challenge for this concept. Constructing water structures can elucidate this issue but water resource management will play a key role in defining the sustainability of this model.

Many researchers across various disciplines have addressed the issue of sustainable and optimal use of agricultural resources by developing mathematical models for decision-making in agricultural systems (Gal et al. 2011) (Singh 2012). It is understood from the literature and suggested by Hussain et al. (2007) that water resource management can be optimized by implementing an appropriate cropping pattern.

Singh (2012) developed two management models using linear programming (LP) and chance constrained linear programming (CCLP) to maximize the net annual return from obtained optimal cropping pattern in a canal command area in India.

A. Deo (✉) · A. Arora
Centre for Technology Alternatives for Rural Areas (CTARA), IIT Bombay, Mumbai 76, India
e-mail: aniketdeo1992@gmail.com

A. Arora
e-mail: amitarora3@gmail.com

U. Bhandarkar
Department of Mechanical Engineering, IIT Bombay, Mumbai 76, India
e-mail: bhandarkar@iitb.ac.in

© The Editor(s) (if applicable) and The Author(s), under exclusive license to Springer Nature Switzerland AG 2021
A. Pandey et al. (eds.), *Hydrological Extremes*, Water Science and Technology Library 97, https://doi.org/10.1007/978-3-030-59148-9_23

Optimal planting schedule was developed by Darby-Dowman et al. (2000) using two-stage stochastic programming wherein they considered profit maximization as the objective function. Similarly, studies were done by Jabelli et al. (2016), Benli and Kodal (2003) optimized cropping pattern by considering an objective function of either maximizing the profits or income. Raju and Kumar (1999) used multi-objective linear programming on a case study of Sri Ram Sagar Project, Andra Pradesh, India with three conflicting objectives of net benefits, agriculture production and labor employment to select the best compromise irrigation plan. Another multiobjective model was prepared by Groot et al. (2012) who designed farming systems to evaluate farm performance indicators and its consequences of adjustment in farm management by setting three objectives of maximizing profit, organic matter balance and minimize the labor requirement and soil nitrogen loss. Xevi and Khan (2005) used multiobjective linear programming to determine the optimum crop area concerning profit, variable cost and groundwater pumping requirement as objectives. A multi-objective linear fuzzy programming model was prepared by Zeng et al. (2010) for crop area planning, which focused on optimal cropping pattern in water stress region of China. They considered three objectives of maximizing net return, minimizing evapotranspiration and yield maximally satisfying a certain goal.

This research focuses on generating a set of cropping pattern alternatives with respect to a farm pond (as water reservoir), which can be recommended to tribal farmers and adopted according to his/her requirements of subsistence, income generation and resource allocation. Maximum studies (in this field) are done after analyzing the existing cropping pattern and calculating an optimal cropping pattern for a scenario through mathematical models. However, for our purpose of research, it would be impractical to suggest one optimal cropping pattern because of the hybrid behaviors of farmer. Therefore this research focuses on generating cropping patterns combinations that are applicable for this region. Among a list of vegetables that can be cultivated, alternatives of suitable crop combinations and their cropped areas are generated using a linear programming model which satisfies the objective of profit maximization desirable to farmers subjected to constraints of water availability, land availability and economics. This model is able to predict the economic output of various crop combinations, which in else case would not have been possible to predict without actual field experimentations. This model can be further applied to various other locations by changing the model coefficients applicable to that location. The model is iterated for three cases of cropping patterns being two crop, three crop and four crop per season. Similarly, two scenarios of irrigation methods being furrow and drip are developed for each case, i.e., a total of six cases are developed to understand the change in crop selection and cropped areas when different methods are applied.

## 23.2   Methodology

Participatory study including surveys, focused group discussions, group meetings and farm measurements were conducted to understand the current agriculture system and farmer's perspective of agriculture as a source of livelihood. This process was necessary to develop a realistic model with systemic approach. Farmer's opinion was taken and a list of vegetables was noted which can be cultivated and sold in the local market. The list of 15 vegetables comprises of Cluster beans, Carrot, Cucumber, Eggplant, Onion, Capsicum, Radish Spinach, Tomato, Okra, Bitter gourd, Chilly, Pumpkin, Cow pea, Coriander leaves. Literature study and field-level investigation gave important tribal farm properties which were used to derive farmer's objective and constraints. Important properties that are associated with a tribal agriculture system are productivity, profitability, input cost, crop duration, water sufficiency, diversity of crops and market flexibility (Alexandratos 1995).

### 23.2.1   Objective Function

**Maximizing Net Benefit**: Productivity and profitability are two different aspects that are often taken under the same virtue (Alexandratos 1995). They may or may not be directly proportional to an agriculture system because high yield does not necessarily fetch high profits because of the fluctuating market rates and input costs. Hence it is important to consider profit or net benefit, i.e., the difference in income generated and input costs. Equation 23.1 represents the objective function.

$$MaxZ_A = \sum_{k=1}^{n} P_K X_K \qquad (23.1)$$

where,

$P_k$   is the profit per m$^2$ for the kth crop.
$X_k$   is the decision variable and cropped area of kth crop.
$n$     is the number of crops in the cropping pattern.

### 23.2.2   Constraints

**Land Constraint**: Tribals have limited land holdings, therefore, the total agricultural area considered in this model has a bound. Equation 23.2 represents the land constraint.

$$\sum_{k=1}^{n} X_K \leq L \tag{23.2}$$

An average land holding in this region (as calculated from primary data) is 1.4 acres. The land is fully utilized during the Kharif season for rice cultivation whereas partially utilized during Rabi season due to resource constraints. The maximum land available for Rabi cultivation (L) is considered 1 acre (4046 m$^2$) in this model.

**Water Constraint**: The crop water requirement $W_k$ (m) cannot exceed the water available in the water source. Equation 23.3 represents the water constraint.

$$\sum_{k=1}^{n} W_K X_K \leq \emptyset X_P D_P \tag{23.3}$$

where,

$X_p$   is the farm pond area (m$^2$).
$D_p$   is the depth of farm pond (m).
$\emptyset$   is the evaporation loss percentage.

Irrigation unavailability and inefficiency is a major constraint limiting most farmers during Rabi crop cultivation. Under the Government of Maharashtra scheme ("*Magel tella shet tal*"), a farmer can avail a farm pond of 30 × 30 × 3 m dimension with 70% subsidy on excavation. The current research uses this farm pond model as water source for irrigation which is recharged by rainwater (average rainfall in this region is 2600 mm). The total water available for crops is the volume of water minus evaporation losses in pond. The evaporation losses ($\emptyset$) calculated from evapotranspiration values account to almost 50% of the water in pond. The water discharged from the water source will exceed the crop water requirement because of the losses in irrigation system, hence surplus water considering efficiency of irrigation system is considered.

**Budget Constraint**: Considering low purchasing power of the tribals, the cost of cultivation $C_k$ (INR) is a limiting factor for agriculture. Equation 23.4 represents the budget constraint.

$$\sum_{k=1}^{n} C_K X_K \leq B \tag{23.4}$$

where,
   B is the total cost of cultivation (INR) in the cropping pattern.

Farmer's survey and group meeting conducted to assess their livelihood profile concluded with shared understanding of the people that a farmer can invest an amount of INR.10000/- for a season of second crop. This is the maximum budget considered for this model.

**Minimum Productivity Constraint**: Currently, a few farmers take second crop majorly for subsistence while surplus is sold in the market. The concept of multi cropping diversifies the farmer's portfolio when he/she approaches the market. It is learned from the field level investigation that the costumer footfall increases when they see multiple vegetable at the farmer's stall. Another advantage of multicropping is continuous income due to variable crop intervals. Therefore, considering multiple cropping the cropped area of each crop should at least produce a minimum quantity.

$$p_K X_K \geq S \qquad (23.5)$$

where,

$p_k$   is the productivity per $m^2$ of kth crop.
S     is the minimum quantity of vegetable in kg per season and is considered 200 for the current model.

## 23.3   Model Simulation

A linear programming method is chosen and the calculation is done using MATLAB software (Optimization Toolbox).

The model is simulated for 2 crop, 3 crop and 4 crop pattern among the list of 15 vegetables where $^{15}C_2$, $^{15}C_3$ and $^{15}C_4$ combinations are generated, respectively. The crop patterns are iterated with two cases with different irrigation method being furrow and drip. A total of six cases are developed, each crop pattern being iterated for two irrigation method (Table 23.1).

### 23.3.1   Model Assumptions

- Farm pond is recharged by rainwater only.
- The sowing date of crops is assumed to be 15 November and no rainfall is considered thereafter.
- Wholesale market rates are considered for all crops.
- Noncompanion crop combinations are omitted during simulation of the model.

**Table 23.1** Cases in model

| Crop pattern | Irrigation method |
|---|---|
| Two cropping pattern | Drip |
| Three cropping pattern | Furrow |
| Four cropping pattern | |

- The overall efficiency of drip irrigation is 90% while that of furrow irrigation is 40%.
- Productivity of crops in furrow irrigation and sprinkler irrigation is same.
- Cost of cultivation of 30 × 30 × 3 m³ farm pond is INR.130000/. Cost of low cost drip irrigation system is INR.25000/- per acre. The infrastructure cost includes the cost of construction of farm pond (and cost of drip system in case of drip irrigation system).

### 23.3.2 Model Parameters

**Crop Water Requirement**: The evapotranspiration value for this region is taken from a study done in Dahanu weather station. The net water requirement is calculated by considering sowing date and duration of crop. Crop water requirement is calculated using Eq. 23.6 (Allen et al. 1998)

$$ET_O \times K_C = Crop\ water\ requirement \qquad (23.6)$$

where Eto is reference evapotranspiration, Kc is crop coefficient at a particular growth stage (Table 23.2).

**Cost of cultivation**: A standard procedure available in literature was used to calculate cost of cultivation of each crop (Alexandratos 1995). Costs associated with seeds, land preparation, fertilizer, pesticides, labor and transportation are included in calculating cost of cultivation. Farmer's survey was conducted to understand the local

**Table 23.2** Crop water requirement and crop duration (days) of selected vegetables

| Crop | Duration | Water requirement (mm/season) |
|---|---|---|
| Cluster beans | 90 | 264.37 |
| Carrot | 100 | 308.63 |
| Cucumber | 105 | 297.72 |
| Eggplant | 130 | 447.75 |
| Onion/dry | 150 | 602.74 |
| Capsicum | 165 | 668.43 |
| Radish | 35 | 107.13 |
| Spinach | 60 | 151.05 |
| Tomato | 135 | 470.82 |
| Okra | 100 | 302.62 |
| Bitter gourd | 105 | 310.83 |
| Chilly | 125 | 424.07 |
| Pumpkin | 105 | 312.42 |
| Cow pea | 100 | 274.88 |
| Coriander leaves | 110 | 417.49 |

**Table 23.3** Cost of cultivation of selected vegetables according to local context

| Crop | Cost of cultivation (INR/acre) |
|---|---|
| Cluster beans | 19,178.04 |
| Carrot | 19,946.78 |
| Cucumber | 22,981.28 |
| Eggplant | 19,623.1 |
| Onion/dry | 22,253 |
| Capsicum | 20,594.14 |
| Radish | 16,952.74 |
| Spinach | 18,652.74 |
| Tomato | 21,039.2 |
| Okra | 18,935.28 |
| Bitter gourd | 19,906.32 |
| Chilly | 16,790.9 |
| Pumpkin | 17,438.26 |
| Cow pea | 19,056.66 |
| Coriander leaves | 16,629.06 |

farming practices and expenditures. Triangulation of data was done by consulting farmers and local agricultural officers. Table 23.3 present below is specific for this region.

**Net benefit**: APMC (Agriculture Produce Market Committee) data were used to analyze last 3 year's market rate and an average value was considered for this study. The yield multiplied by market rates gave the gross returns for cultivation. The difference in gross return and cost of cultivation gave net benefit for each crop. Table 23.4 presents the profit per acre of each crop.

## 23.4  Results and Discussion

The model iterated all possible combinations of available crops and gave cropped area and profit as output. Using the model results, cost of infrastructure (pond and drip system/channels) and payback period is calculated. The results show wide range of profits which can be attained from different combination of crops, hence it becomes important from a farmer's perspective to select the best combination which would fetch the maximum profit. The figure (Figs. 23.1, 23.2, 23.3, 23.4, 23.5, 23.6) shows evidence that selection of a wrong combination may reduce the profit.

The horizontal axis on the graphs denotes the crop combination index which is the representation of a particular combination in $^{15}C_n$ values, n being 2, 3, and 4 for two, three and four cropping pattern, respectively. The grids on Y-axis of graph can be considered to differentiate crop combinations into four categories based on

**Table 23.4** Net returns
expected from selected
vegetables

| Crop | Profit (INR/acre) |
| --- | --- |
| Cluster beans | 49,100.3 |
| Carrot | 98,029.23 |
| Cucumber | 33,418.04 |
| Eggplant | 140,511.3 |
| Onion/dry | 106,593.9 |
| Capsicum | 122,597 |
| Radish | 53,348.87 |
| Spinach | 1264.858 |
| Tomato | 116,284.6 |
| Okra | 34,622.62 |
| Bitter gourd | 52,925.59 |
| Chilly | 86,565.26 |
| Pumpkin | 80,978.22 |
| Cow pea | 32,817.73 |
| Coriander leaves | 13,061.62 |

**Fig. 23.1** Net returns for
two crop combinations in
drip irrigation method

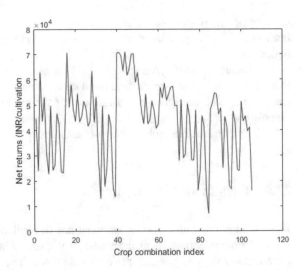

profit values. The first category being the most recommended crop combinations
falls within the upper most grid on Y-axis. Similarly, the categories: Moderately
recommended, least recommended, not recommended falls within second, third, and
fourth range from top, respectively.

Figures 23.1, 23.2, 23.3, 23.4, 23.5, 23.6 show the variation in profits when two
different methods of irrigation is used. The difference in values is apparent because
the irrigation water requirement for both the methods is different hence the available

**Fig. 23.2** Net returns for
three crop combinations in
drip irrigation method

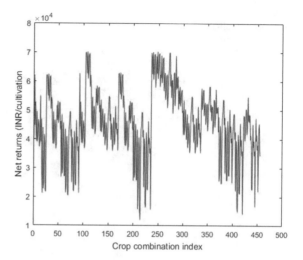

**Fig. 23.3** Net returns for
four crop combinations in
drip irrigation method

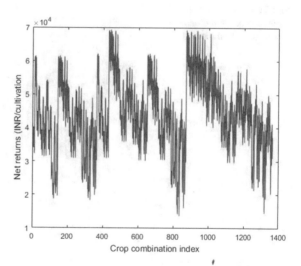

water for crops reduces in case of furrow irrigation due to low overall efficiency.
Lesser water availability is leading to less cropped area and consecutively low profits.

Tables 23.5, 23.6, 23.7, 23.8, 23.9, 23.10 show the top five examples of crop combinations in most recommended and least recommended category for two crop, three crop and four crop pattern in drip and furrow irrigation methods. The crop combination changes in most recommended category when cultivated with two different irrigation systems because water availability plays a major role in selection of crops. It is observed that two crop pattern fetches the maximum profit and lowest payback period among the three cropping patterns. This is because the crop with maximum net benefit is allotted a higher area making higher profits whereas in case of three

**Fig. 23.4** Net returns for
two crop combinations in
furrow irrigation method

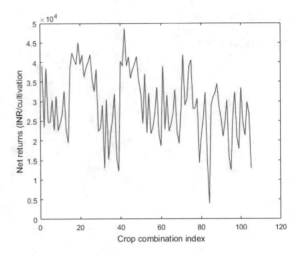

**Fig. 23.5** Net returns for
three crop combinations in
furrow irrigation method

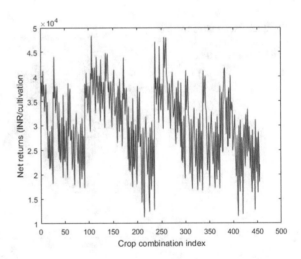

and four crop patterns there is a constraint of having minimum 200 kg production
for other crops as well within the same water availability. The payback period as
calculated by the ratio of cost of infrastructure and profit indications that cultivation
through furrow irrigation will require higher payback period time because of low
profits. Similarly, the payback increases as number of crops increases because the
infrastructure cost remains almost the same while profit reduces.

**Fig. 23.6** Net returns for four crop combinations in furrow irrigation method

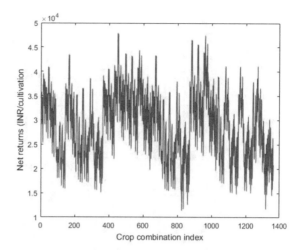

**Table 23.5** Two crop combination examples in drip irrigation

| Crop Comb Index | Crop 1 | Crop 2 | Crop 1 area (m²) | Crop 2 area(m²) | Profit (INR) | Infrastructure cost (INR) | Payback period (years) |
|---|---|---|---|---|---|---|---|
| *Most recommended* | | | | | | | |
| 44 | Eggplant | Tomato | 1991.09 | 66.62 | 71,062.12 | 47,050.77 | 0.66 |
| 41 | Eggplant | Capsicum | 1922.58 | 133.25 | 70,805.68 | 47,047.87 | 0.66 |
| 16 | Carrot | Eggplant | 79.95 | 1981.19 | 70,740.63 | 47,055.15 | 0.67 |
| 40 | Eggplant | Onion | 1971.79 | 79.95 | 70,583.42 | 47,042.44 | 0.67 |
| 48 | Eggplant | Pumpkin | 1963.82 | 111.04 | 70,422.89 | 47,073.30 | 0.67 |
| *Not recommended* | | | | | | | |
| 32 | Cucumber | Spinach | 1557.40 | 249.84 | 12,941.48 | 46,713.73 | 3.61 |
| 39 | Cucumber | Coriander leaves | 1518.85 | 333.12 | 13,620.37 | 46,774.54 | 3.43 |
| 83 | Spinach | Cow pea | 249.84 | 1880.37 | 15,330.07 | 47,146.47 | 3.08 |
| 105 | Cow pea | Coriander leaves | 1833.82 | 333.12 | 15,949.83 | 47,194.55 | 2.96 |
| 79 | Spinach | Okra | 249.84 | 1890.18 | 16,252.80 | 47,159.08 | 2.90 |

**Table 23.6** Three crop combination examples in drip irrigation

| Crop Comb Index | Crop 1 | Crop 2 | Crop 3 | Crop 1 area (m²) | Crop 2 area (m²) | Crop 3 area (m²) | Profit (INR) | Infrastructure cost (INR) | Payback period (years) |
|---|---|---|---|---|---|---|---|---|---|
| *Most recommended* | | | | | | | | | |
| 248 | Eggplant | Capsicum | Tomato | 1851.12 | 133.25 | 66.62 | 70,238.73 | 47,041.45 | 0.67 |
| 108 | Carrot | Eggplant | Tomato | 79.95 | 1909.73 | 66.62 | 70,173.68 | 47,048.49 | 0.67 |
| 239 | Eggplant | Onion | Tomato | 1900.33 | 79.95 | 66.62 | 70,016.46 | 47,036.02 | 0.67 |
| 105 | Carrot | Eggplant | Capsicum | 79.95 | 1841.22 | 133.25 | 69,917.24 | 47,045.99 | 0.67 |
| 273 | Eggplant | Tomato | Pumpkin | 1892.36 | 66.62 | 111.04 | 69,855.94 | 47,066.68 | 0.67 |
| *Not recommended* | | | | | | | | | |
| 214 | Cucumber | Spinach | Coriander leaves | 1316.20 | 249.84 | 333.12 | 12,024.68 | 46,838.76 | 3.90 |
| 420 | Spinach | Cow pea | Coriander leaves | 249.84 | 1589.15 | 333.12 | 14,043.33 | 47,201.72 | 3.36 |
| 410 | Spinach | Okra | Coriander leaves | 249.84 | 1597.43 | 333.12 | 14,823.15 | 47,212.47 | 3.19 |
| 213 | Cucumber | Spinach | Cow pea | 249.84 | 249.84 | 1578.72 | 14,946.89 | 47,078.27 | 3.15 |
| 235 | Cucumber | Cow pea | Coriander leaves | 249.84 | 1532.17 | 333.12 | 15,566.65 | 47,126.80 | 3.03 |

**Table 23.7** Four crop combination examples in drip irrigation

| Crop Comb Index | Crop1 | Crop2 | Crop 3 | Crop 4 | Crop1 area (m²) | Crop2 area (m²) | Crop3 area (m²) | Crop4 area (m²) | Profit (INR) | Infra-structure cost (INR) | Payback period (years) |
|---|---|---|---|---|---|---|---|---|---|---|---|
| *Most recommended* | | | | | | | | | | | |
| 443 | Carrot | Eggplant | Capsicum | Tomato | 79.95 | 1769.76 | 133.25 | 66.62 | 69,350.29 | 47,039.57 | 0.68 |
| 873 | Eggplant | Onion | Capsicum | Tomato | 1760.36 | 79.95 | 133.25 | 66.62 | 69,193.07 | 47,027.09 | 0.68 |
| 434 | Carrot | Eggplant | Onion | Tomato | 79.95 | 1818.97 | 79.95 | 66.62 | 69,128.02 | 47,034.15 | 0.68 |
| 934 | Eggplant | Capsicum | Tomato | Pumpkin | 1752.39 | 133.25 | 66.62 | 111.04 | 69,032.54 | 47,057.78 | 0.68 |
| 468 | Carrot | Eggplant | Tomato | Pumpkin | 79.95 | 1811.00 | 66.62 | 111.04 | 68,967.50 | 47,064.81 | 0.68 |
| *Not recommended* | | | | | | | | | | | |
| 835 | Cucumber | Spinach | Cow pea | Coriander leaves | 249.84 | 249.84 | 1287.49 | 333.12 | 13,660.15 | 47,133.60 | 3.45 |
| 825 | Cucumber | Spinach | Okra | Coriander leaves | 249.84 | 249.84 | 1294.21 | 333.12 | 14,291.95 | 47,142.45 | 3.30 |
| 1320 | Spinach | Okra | Cow pea | Coriander leaves | 249.84 | 1095.15 | 499.68 | 333.12 | 14,577.95 | 47,209.13 | 3.24 |
| 824 | Cucumber | Spinach | Okra | Cow pea | 249.84 | 249.84 | 1084.66 | 499.68 | 15,476.39 | 47,085.72 | 3.04 |
| 860 | Cucumber | Okra | Cow pea | Coriander leaves | 249.84 | 1037.88 | 499.68 | 333.12 | 16,073.32 | 47,133.83 | 2.93 |

**Table 23.8** Two crop combination examples in furrow irrigation

| Crop Comb. Index | Crop 1 | Crop 2 | Crop 1 area (m²) | Crop 2 area (m²) | Profit (INR) | Infrastructure cost (INR) | Payback period (years) |
|---|---|---|---|---|---|---|---|
| *Most recommended* | | | | | | | |
| 42 | Eggplant | Radish | 878.46 | 1368.99 | 48,558.68 | 42,789.30 | 0.88 |
| 19 | Carrot | Radish | 1558.31 | 551.19 | 45,023.53 | 42,721.70 | 0.95 |
| 16 | Carrot | Eggplant | 1633.66 | 79.95 | 42,357.82 | 42,527.71 | 1.00 |
| 25 | Carrot | Pumpkin | 1637.24 | 111.04 | 41,890.47 | 42,544.70 | 1.02 |
| 71 | Radish | Tomato | 1340.62 | 841.87 | 41,872.79 | 42,757.47 | 1.02 |
| *Not recommended* | | | | | | | |
| 39 | Cucumber | Coriander leaves | 1346.64 | 333.12 | 12,198.04 | 42,511.13 | 3.49 |
| 95 | Okra | Coriander leaves | 1324.81 | 333.12 | 12,412.16 | 42,500.43 | 3.42 |
| 105 | Cow pea | Coriander leaves | 1458.50 | 333.12 | 12,905.54 | 42,565.94 | 3.30 |
| 32 | Cucumber | Spinach | 1557.40 | 249.84 | 12,941.48 | 42,573.59 | 3.29 |
| 79 | Spinach | Okra | 249.84 | 1659.67 | 14,280.31 | 42,623.71 | 2.98 |

**Table 23.9** Three crop combination examples in furrow irrigation

| Crop Comb. Index | Crop 1 | Crop 2 | Crop 3 | Crop 1 area ($m^2$) | Crop 2 area ($m^2$) | Crop 3 area ($m^2$) | Profit (INR) | Infrastructure cost (INR) | Payback period (years) |
|---|---|---|---|---|---|---|---|---|---|
| *Most recommended* | | | | | | | | | |
| 106 | Carrot | Eggplant | Radish | 79.95 | 833.40 | 1327.03 | 48,377.30 | 42,785.83 | 0.88 |
| 256 | Eggplant | Radish | Tomato | 808.95 | 1366.75 | 66.62 | 48,029.57 | 42,786.78 | 0.89 |
| 260 | Eggplant | Radish | Pumpkin | 809.11 | 1335.04 | 111.04 | 47,924.72 | 42,793.09 | 0.89 |
| 237 | Eggplant | Onion | Radish | 764.39 | 79.95 | 1395.95 | 47,058.78 | 42,785.79 | 0.91 |
| 259 | Eggplant | Radish | Chilly | 682.17 | 1398.19 | 199.87 | 46,403.12 | 42,805.36 | 0.92 |
| *Not recommended* | | | | | | | | | |
| 214 | Cucumber | Spinach | Coriander leaves | 1219.88 | 249.84 | 333.12 | 11,229.14 | 42,571.44 | 3.79 |
| 410 | Spinach | Okra | Coriander leaves | 249.84 | 1200.10 | 333.12 | 11,423.10 | 42,561.75 | 3.73 |
| 420 | Spinach | Cow pea | Coriander leaves | 249.84 | 1321.21 | 333.12 | 11,870.04 | 42,621.09 | 3.59 |
| 225 | Cucumber | Okra | Coriander leaves | 249.84 | 1079.02 | 333.12 | 12,372.43 | 42,502.42 | 3.44 |
| 445 | Okra | Cow pea | Coriander leaves | 399.74 | 1018.42 | 333.12 | 12,756.67 | 42,546.18 | 3.34 |

Table 23.10 Four crop combination examples in furrow irrigation

| Crop Comb Index | Crop 1 | Crop 2 | Crop 3 | Crop 4 | Crop 1 area (m²) | Crop 2 area (m²) | Crop 3 area (m²) | Crop 4 area (m²) | Profit (INR) | Infrastructure cost (INR) | Payback period (years) |
|---|---|---|---|---|---|---|---|---|---|---|---|
| *Most recommended* | | | | | | | | | | | |
| 451 | Carrot | Eggplant | Radish | Tomato | 79.95 | 763.88 | 1324.79 | 66.62 | 47,848.20 | 42,783.31 | 0.89 |
| 455 | Carrot | Eggplant | Radish | Pumpkin | 79.95 | 764.04 | 1293.08 | 111.04 | 47,743.35 | 42,789.62 | 0.90 |
| 962 | Eggplant | Radish | Tomato | Pumpkin | 739.59 | 1332.79 | 66.62 | 111.04 | 47,395.61 | 42,790.57 | 0.90 |
| 881 | Eggplant | Onion | Radish | Tomato | 694.87 | 79.95 | 1393.70 | 66.62 | 46,529.67 | 42,783.27 | 0.92 |
| 885 | Eggplant | Onion | Radish | Pumpkin | 695.03 | 79.95 | 1361.99 | 111.04 | 46,424.82 | 42,789.57 | 0.92 |
| *Not recommended* | | | | | | | | | | | |
| 825 | Cucumber | Spinach | Okra | Coriander leaves | 249.84 | 249.84 | 954.31 | 333.12 | 11,383.38 | 42,563.73 | 3.74 |
| 1320 | Spinach | Okra | Cow pea | Coriander leaves | 249.84 | 399.74 | 881.12 | 333.12 | 11,721.17 | 42,601.32 | 3.63 |
| 835 | Cucumber | Spinach | Cow pea | Coriander leaves | 249.84 | 249.84 | 1050.62 | 333.12 | 11,738.78 | 42,610.92 | 3.63 |
| 860 | Cucumber | Okra | Cow pea | Coriander leaves | 249.84 | 399.74 | 747.82 | 333.12 | 12,625.41 | 42,536.01 | 3.37 |
| 824 | Cucumber | Spinach | Okra | Cow pea | 249.84 | 249.84 | 960.00 | 499.68 | 14,409.62 | 42,648.13 | 2.96 |

## 23.5  Conclusion

Choice of an appropriate cropping pattern in horticultural cultivation plays a major role in the economics of the farming system. Decision-making in cropping pattern selection for tribal farmers in Palghar district, Maharashtra is addressed by developing a mathematical model for creating crop combinations, which satisfies objectives and constraints applicable in this region and categorizing these combinations into four categories being most recommended, moderately recommended, least recommended and not recommended on the basis of profits for one season cultivation. Understanding different motivations of tribals for farming, scenarios of two, three and four cropping pattern with two different irrigation methods are developed to provide choice of cropping pattern alternatives. The iteration of crops in cropping pattern provide wide variety of crop combinations which can be chosen by the farmers, however the categorization of model results into four categories helps in suggesting farmers regarding the crop combinations. The results of the scenarios suggest that crop selection, cropped area, and profit changes when cropping pattern or irrigation methods are changed. Drip irrigation turns out to be more profitable than furrow irrigation in all the cropping pattern while two crop pattern gives most profit. This model can be further improved by reducing the assumptions and introducing uncertainty attribute in the system.

## References

Alexandratos N (ed) (1995) World agriculture: towards 2010: an FAO study. Food & Agriculture Org.

Allen RG, Pereira LS, Raes D, Smith M (1998) Crop evapotranspiration-guidelines for computing crop water requirements-FAO irrigation and drainage paper 56. FAO Rome 300(9):D05109

Benli B, Kodal S (2003) A non-linear model for farm optimization with adequate and limited water supplies: application to the South-east Anatolian Project (GAP) region. Agric Water Manag 62(3):187–203

Darby-Dowman K, Barker S, Audsley E, Parsons D (2000) A two-stage stochastic programming with recourse model for determining robust planting plans in horticulture. J Oper Res Soc 83–89

Groot JCJ, GJM Oomen, WAH Rossing (2012) Multi-objective optimization and design of farming systems. Agric Syst 110:63–77

Hussain SS, Mudasser M (2007) Prospects for wheat production under changing climate in mountain areas of Pakistan–an econometric analysis. Agric Syst 94(2):494–501

Jebelli J, Paterson B, Abdelwahab A (2016) A linear programming model to optimize cropping pattern in small-scale irrigation schemes: an application to Mekabo Scheme in Tigray, Ethiopia. Int J Environ Agric Res 2(8):24–34

Le Gal P-Y, Patrick D, Faure G, Novak S (2011) How does research address the design of innovative agricultural production systems at the farm level? A review. Agric Syst 104(9):714–728

Raju KS, Nagesh Kumar D (1999) Multicriterion decision making in irrigation planning. Agric Syst 62(2):117–129
Singh A (2012) An overview of the optimization modelling applications. J Hydrol 466:167–182
Xevi E, Khan S (2005) A multi-objective optimisation approach to water management. J Environ Manag 77(4):269–277
Zeng X, Kang S, Fusheng Li Lu, Zhang, and Ping Guo, (2010) Fuzzy multi-objective linear programming applying to crop area planning. Agric Water Manag 98(1):134–142

# Chapter 24
# Development of Mathematical Model for Estimation of Wetting Front Under Drip Irrigation

**Harsh Vardhan Singh and Vishwendra Singh**

## 24.1 Introduction

Water scarcity already affects every continent. Around 1.2 billion people, or almost one-fifth of the world's population, live in areas of physical scarcity, and 500 million people are approaching this situation (Singh et al. 2006; Swain et al. 2015). Another 1.6 billion people around the globe face economic water shortage, where countries lack the necessary infrastructure to take water from rivers and aquifers (Watkins 2006; Verma et al. 2016). Approximately, one-third of the world's population live with water stress, i.e. in that area where the withdrawal of freshwater exceeds 10% of the renewable storage (Elliott et al. 2014; Swain et al. 2017a). If the same consumption patterns continue, two out of three people on earth will live under water-stressed conditions by the year 2025 (Karlberg and de Vries 2004; Swain et al. 2018). The conditions will be detrimental for countries like India with increasing population, higher food demand, increasing temperatures and changing precipitation patterns (Swain et al. 2017b). The economy of India being dependent on agriculture, it is necessary to focus on proper planning and management of water availability and water use, so as to increase the agricultural yield. As reported by Directorate of Economics & Statistics, Department of Agriculture & Cooperation, Ministry of Agriculture, Government of India, 48% of total geographical area of the country is covered by agricultural lands, of which, only one-third area is irrigated (Swain et al. 2017a). Irrigation is crucial for

H. V. Singh (✉)
Department of Water Resources Development and Management, Indian Institute of Technology Roorkee, Roorkee, India
e-mail: singhharsh127@gmail.com

V. Singh
Jain Irrigation Pvt. Ltd., Jalgaon, India
e-mail: vishwendrasingh3009@gmail.com

A. Pandey et al. (eds.), *Hydrological Extremes*, Water Science and Technology Library 97, https://doi.org/10.1007/978-3-030-59148-9_24

347

increasing productivity on the existing cultivable lands. Projected per-hectare irrigation consumption along with lack of available water for agricultural production, anthropogenic consumption, industrial projects and ecological use has intensified its significance (Elliott et al. 2014). There are many methods of practising irrigation, of which, drip irrigation is considered to have the highest application efficiency (Elmaloglou and Diamantopoulos 2007). Drip irrigation is the strongest way of irrigation to conserve the water in the water-scarce areas (Michael 1978). Per drop more crops can be obtained from this method of irrigation and thus, the government agencies of many countries are working on introducing farmers with this method of irrigation (Singh et al. 2006).

The geometry of the wetted soil volume under trickle irrigation takes a spherical or ellipsoidal-like shape when water is applied from a point source, whereas that takes a cylindrical-like shape when water is applied from a line source. Water flowing out from a buried emitter moves vertically and laterally into the soil and wets a soil volume, whose shape can be assumed similar to a truncated ellipsoid (Zur 1996). A table was presented by Keller and Bliesner (1990) to show the spacing between emitters and lines under an emitter with a discharge of 4 L/h, for a number of soils, rooting depths and degrees of soil profile homogeneity. Amir and Dag (1993) published field data on the width and depth dimensions of wetted soil volumes under moving emitters and reported a good agreement with the empirical model presented by Schwartzman and Zur (1986), which related the wetted depth and the wetted width to soil hydraulic conductivity, emitter discharge and total volume of water in soil. Elmaloglou and Malamos (2006) suggested a simple alternative empirical method that can be applied to numerical results of a cylindrical flow model to estimate surface radius and depth of the wetted soil volume under a surface point source.

The objective of the present study is to determine the wetting front dimensions of drip emitters and establish the relationship between emitter discharge and wetting front movement in vertical and horizontal directions with three different discharge rates of 2, 4 and 8 L per hour (lph).

## 24.2   Materials and Methods

### 24.2.1   Description of Experimental Setup

To conduct the experiment for determination of wetting front dimension and to establish a relationship of emitter discharge with wetting front movement, a lysimeter was erected using a transparent perplex sheet of 4 mm thickness. The dimension of the lysimeter was 76.2 cm length, 76.2 cm width and 76.2 cm height, respectively. The reinforcement was provided to the lysimeter using 4 Clan wide and 5 mm thick iron strips. The soil of irrigation research station SHIATS was brought to the water resource engineering laboratory, air-dried and sieved (using 2 mm sieve) to maintain homogeneity. All foreign materials like grass and root trashes were removed.

## 24.2.2   Filling of Soil in Lysimeter

The soil brought from the irrigation research station SHIATS was filled in the lysimeter maintaining the same bulk density (1.3 gm/cm$^3$). To maintain the bulk density, soil was filled in lysimeter in layers of 10 cm each, calculating the required mass of the soil for the column volume up to 10 cm depth.

## 24.2.3   Mass Calculation

Density = mass/volume.
   For 10 cm filling lysimeter volume will be;

Volume = l * b * h
Volume = 76.2 * 76.2 * 10
Volume (cm$^3$) = 58,064.4

Mass of soil required to maintain bulk density will be;

Mass = density * volume
Mass = 1.3 * 58,064.4
Mass = 75,483.72 gm = 75.4834 kg.

## 24.2.4   Installation of Drip Irrigation System in Lysimeter

After filling the soil in lysimeter and maintaining the bulk density, one lateral line of 12 mm diameter was installed near one of the walls of the lysimeter. Drippers of 2, 4 and 8 lph were connected one by one. The lateral line was connected with the water source and the drippers were checked for the rated discharge before starting the experiment. Once the checking is done, the system is run and readings were taken at every 15 min for different discharges through emitters. The pressure within the lateral was set at 1 bar.

## 24.2.5   Measurement of Wetting Front

The horizontal and vertical dimensions of the wetting front movement through 2, 4 and 8 lph emitter discharge at regular interval were measured. For measurement of the vertical dimension, graph paper strips were placed at all four faces (wall) of the lysimeter. On each wall, several graph paper trips were placed at 6 cm distance each, for precise measurement of wetting front dimension and shape. For measurement of horizontal wetting front, a thin strip of the graph paper was placed on the soil surface and the movement of the wetting front by visual observation was marked by Jolt graph strip at regular time interval. This experiment was conducted and readings were taken at a gap of 15 min for two hours, i.e. 15, 3, 45, 60, 75, 90, 105 and 120 min, to facilitate the various volumes of water discharge through selected drippers.

### 24.2.6  Measurement of the Moisture Content at Various Points of Wetting Front

Following the measurement of wetting front, soil sample was collected at various depths and at various distances in vertical and horizontal direction, respectively. The collected samples were analyzed for moisture content on dry basis.

### 24.2.7  Analysis of Data Collected from Lysimeter

Horizontal and vertical dimensions of wetting front in different coordinates under 2, 4 and 8 lph dripper discharge measured from the lysimeter were used for graphical representation to establish the relationship between dripper discharge and wetting front dimension and also to find the shape of wetting front for a particular dripper discharge in a given soil condition.

### 24.2.8  Mathematical Modelling

The mathematical model was developed for the horizontal (X) and vertical (Y) movements of wetting front for the inputs of time and varying dripper discharges separately as below:

$$X = f(V_t, t) = C_x v_t^{ax} t^{bx} \tag{24.1}$$

$$Y = f(V_t, t) = C_y v_t^{ay} t^{by} \tag{24.2}$$

where, x is the dimension of wetting front in horizontal direction (cm), y is the dimension of wetting front in vertical direction (cm), $v_t$ is the volume discharged through dripper (litres) after time t (min), t is the water application time (min) and Cx, Cy, a and b are empirical constant.

## 24.3   Results and Discussion

### 24.3.1   Development of Equation to Estimate Horizontal Wetting Front Movement for 2 lph Dripper Discharge

(a)  **Horizontal wetting front versus volume discharged**

The horizontal distance of wetting front measured at surface around the dripper at various time interval for 2 lph dripper discharge was utilized to develop the mathematical equation for estimating the horizontal movement. The volume of water discharged through dripper and development of wetting zone was plotted as shown in Fig. 24.4. The graphical representation exhibits strong power relation between two parameters ($R^2 = 0.98$) and given by Eq. (24.3). This equation can be used to develop dimensions of wetting zone in horizontal direction with various dripper discharges in medium loam soil.

$$X = 11.37 \, v_t^{0.4159} \tag{24.3}$$

where, X = Dimension of wetting front in horizontal direction (cm).
  $v_t$ = Volume discharged through dripper (litres) after time t (minutes).
  The relation between two parameters is given by the mathematical Eq. (24.4). The graphical representation between wetting front and duration of water application shows strong power relation ($R^2 = 0.98$).

$$X = 2.763 \, t^{0.4159} \tag{24.4}$$

where, X = horizontal movement of wetting front (cm).
  t = water application time (minutes).

(b)  **Horizontal wetting front versus volume discharged and duration of water application**

Combining Eqs. (24.3) and (24.4) gives the dimension of wetting front in horizontal direction due to the volume of water discharged by emitter in a given time duration. These equations are represented both in linear and power form in Eqs. (24.5) and (24.6), respectively.

$$X = 5.6049 \, v_t^{0.208} t^{0.208} \tag{24.5}$$

$$X = 5.685 \, v_t^{0.416} + 1.3815 \, t^{0.416} \tag{24.6}$$

where, X = horizontal movement of wetting front (cm),

$v_t$ = volume discharged through dripper (litre) in time t (minutes) and
t = water application time (minutes).

### 24.3.2 Development of Equation to Estimate the Vertical Wetting Front Movement

#### (a) Vertical wetting front versus volume discharged

The vertical distance of wetting front measured below emitter point at various time interval for 2 lph dripper discharge was utilized to develop the mathematical equation for estimating the vertical movement. The volume of water discharged through dripper and development of wetting zone was plotted as shown in Fig. 24.8.

The graphical representation exhibits strong power relation between two parameters ($R^2 = 0.94$) and given by the following Eq. (24.7). This equation can be used to develop dimensions of wetting zone in vertical direction with various dripper discharges in medium loam soil.

$$Y = 4.0598 \, v_t^{0.6338} \tag{24.7}$$

where, Y = Dimension of wetting front in vertical direction (cm).
$v_t$ = Volume discharged through dripper (litres) after time t (minutes).

#### (b) Vertical wetting front versus duration of water application

The mathematical equation showing the relation between two parameters, i.e. vertical wetting front and duration of water application exhibits strong power relationship ($R^2 = 0.99$).

$$Y = 0.4703 \, t^{0.6338} \tag{24.8}$$

where, Y = dimension of wetting front in vertical direction (cm),

#### (c) Vertical wetting front versus volume discharge and duration of water application

Combining Eqs. (24.7) and (24.8) gives the dimension of wetting front in vertical direction influenced by volume of water discharged by emitter in given time duration. This can be represented by two equations, i.e. linear and power equation and both the equations show strong correlation between the wetting front and dripper discharge and time.

$$Y = 1.3817 \, v_t^{0.3169} t^{0.3169} \tag{24.9}$$

$$Y = 2.0299 \, v_t^{0.6338} + 0.23515 \, t^{0.6338} \tag{24.10}$$

where, Y = Dimension of wetting front in vertical direction (cm),
 V = Volume discharged through dripper (litre) after time t (min), and
 t = Duration of water application (min).

The observed data for vertical and horizontal wetting front were analyzed and the results so obtained are discussed below.

### (d)  Wetting front in horizontal and vertical direction

The dimensions of wetting front measured in horizontal and vertical direction through 2, 4 and 8 lph emitter discharge after 15, 30, 45, 60, 75 90, 105, 120 min duration are given in Table 24.1. The tabular data clearly reveal that maximum total depth of wetting front after 120 min by 8 lph emitter was 25.2 cm, whereas that by 4 lph and 2 lph emitter were 21.1 cm and 19.6 cm, respectively, for the same time duration of water application in medium loam soil. This depth of wetting front was attained after 16 L of volume discharged through 8 lph, 8 L of volume discharged through 4 lph emitter and 4 L of volume discharged through 2 lph emitter.

The horizontal distance of wetting front for 8 lph dripper discharge was measured 4.7 cm, 10.5 cm, 13 cm, 16 cm, 18 cm, 19.3 cm, 20.5 cm, and 21.6 cm earn after 15 min, 30 min, 45 min, 60 min, 75 min, 90 min, 105 min and 120 min, respectively. For the 4 lph dripper discharge, horizontal wetting front was measured 3.6 cm, 4.9 cm, 8 cm, 11 cm, 12.3 cm, 13.7 cm, 15.2 cm and 16.8 cm after the same duration of water application. Similarly, for 2 lph dripper, observations of horizontal wetting front movement with 15 min interval were 3.2 cm, 3.8 cm, 4.5 cm, 52, cm, 6.8 cm, 8.4 cm, 10 cm, 11.6 cm. Graphical representation of wetting front formation and development with increase of irrigation duration are given in Figs. 24.1, 24.2. 24.3 and 24.4.

### (e)  Validation of developed equations

Figure 24.1 clearly shows the dimension of wetting front development in vertical direction for 2 lph dripper discharge. It can be observed that the vertical distance was increasing very rapidly in the beginning but slowly at later stages. A similar trend was observed for 4 lph dripper discharge, as shown in Fig. 24.2. This is due to the reason that, initially the moisture content of soil was very less and rate of infiltration was high. As the moisture content of the soil has increased resulting in reduced rate of infiltration and consequentially, the rate of increase of wetting front also reduced. The significant reduction of wetting front depth was observed after 120 min with 4 L of water applied. Figure 24.3 represents the dimension of wetting front in horizontal direction for 2 lph dripper discharge. The result shows a higher rate of wetting front development in the beginning, which slows down later with increase in volume of water.

Similar trend was observed for 8 lph dripper discharge as shown in Fig. 24.4. The maximum lateral movement of wetting front from 2, 4 and 8 lph emitter after 15,

**Table 24.1** Dimension of wetting front in horizontal and vertical direction from 2, 4 and 8 lph emitters at different application durations

| Dripper discharge (lph) | Duration of water application (min) | Volume of water discharge through dripper (litre) | Wetting front in horizontal distance (cm) | Wetting front in vertical distance (cm) |
|---|---|---|---|---|
| 2 | 15 | 0.5 | 8.2 | 3.2 |
| | 30 | 1 | 11.6 | 3.8 |
| | 45 | 1.5 | 13.8 | 4.5 |
| | 60 | 2 | 16.0 | 5.2 |
| | 75 | 2.5 | 16.9 | 6.8 |
| | 90 | 3 | 17.8 | 8.4 |
| | 105 | 3.5 | 18.7 | 10.0 |
| | 120 | 4 | 19.6 | 11.6 |
| 4 | 15 | 1 | 10.5 | 3.6 |
| | 30 | 2 | 14.5 | 4.9 |
| | 45 | 3 | 16.2 | 8.0 |
| | 60 | 4 | 17.8 | 11.0 |
| | 75 | 5 | 18.9 | 12.3 |
| | 90 | 6 | 20.0 | 13.7 |
| | 105 | 7 | 20.6 | 15.2 |
| | 120 | 8 | 21.1 | 16.8 |
| 8 | 15 | 2 | 13.2 | 4.7 |
| | 30 | 4 | 16.2 | 10.5 |
| | 45 | 6 | 18.2 | 13 |
| | 60 | 8 | 19.2 | 16 |
| | 75 | 10 | 20.2 | 18 |
| | 90 | 12 | 21.7 | 19.3 |
| | 105 | 14 | 23.2 | 20.5 |
| | 120 | 16 | 25.2 | 21.6 |

30, 45, 60, 75, 90, 105 and 120 min' application duration is in shape of parabola. An average distance in horizontal direction (wetted radius) is 10–15 and 20–25 cm for 2 lph and 4 lph, respectively. This wetted radius shall be the deciding factor for lateral spacing in drip irrigation horticultural crops in rows.

The graphical representation of the wetting front under 2, 4 and 8 lph emitter discharge for different time durations can be matched with the shape of the root zone of different crops generally grown in agro climatic zone of Allahabad region. The matching of wetting front and the shape of the root zone for various crops will lead to optimal utilization of nutrient and water uptake, and hence, will minimize the losses. Moreover, it will also help to sustain the natural resources, reduction in weed growth and enhancement of crop yield.

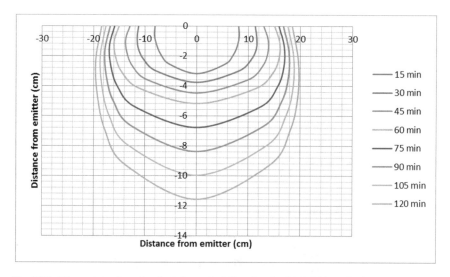

**Fig. 24.1**  Dimension of wetting front in vertical direction for 2 lph dripper discharge

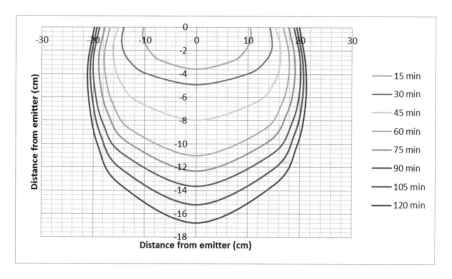

**Fig. 24.2**  Dimension of wetting front in vertical direction for 4 lph dripper discharge

(f)  **Moisture content at various nodes of wetting front**

Figure 24.5 represents the percent moisture content at various nodes of wetting front after 16 L of water discharged. From the figure, it is clear that moisture content at depth 5, 10 and 15 cm below the dripper was 40%, 44% and 60%, respectively. This indicates that the moisture level at different points is distributed quite evenly in vertical and horizontal direction. This distribution of moisture in wetting zone

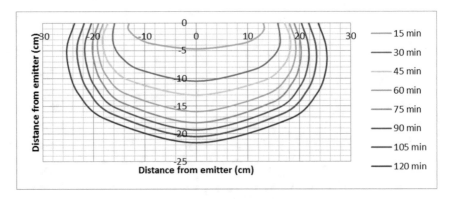

**Fig. 24.3** Dimension of wetting front in vertical direction for 8 lph dripper discharge

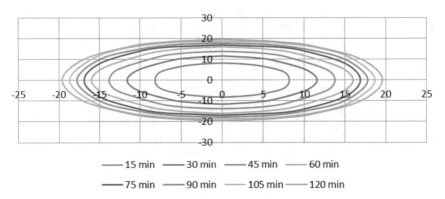

**Fig. 24.4** Dimension of wetting front in horizontal direction for 2 lph dripper discharge

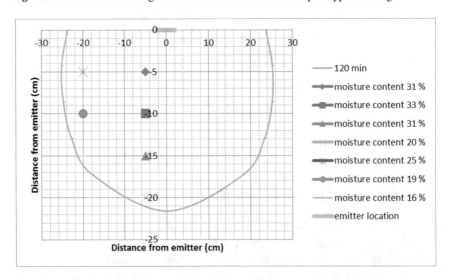

**Fig. 24.5** Moisture content (%) at various nodes from the emitter after 16 L discharge

provides more opportunity for optimum utilization of water and nutrient uptake. Laterally at 15 cm left from the dripper and at a depth of 5 cm from the surface, moisture content was 64%. Similarly, at 10 cm left from the dripper and at a depth of 5 cm, moisture content was measured to be 55%. The highest moisture content of 71% was measured at 10 cm left from the dripper at the depth of 10 cm. These results show that, in the wetting zone, the moisture content after the application of 4 L of water and 120 min' duration, increased from 44 to 71%. This variation may be due to the error in simulating the field condition in the laboratory, but the measured moisture content is within the available moisture content in medium loam soil.

### (g)   Horizontal wetting front versus duration of water application

The relationship between the wetted zone and volume discharged through dripper is developed as shown in Fig. 24.6. The plot showing a relationship between the wetted zone and duration of water application is also developed and represented in Fig. 24.7. It is evident from both the plots (Figs. 24.5 and 24.6) that the wetted front possessed an excellent match with both the parameters.

To check the efficiency of the developed mathematical equation for estimation of dimension of wetting front in horizontal direction for dripper discharge of 2 lph, a comparison study was performed between observed and calculated dimension of wetting front in horizontal direction given in Table 24.2. The graphical representations of these comparisons are shown in Fig. 24.7.

Value of $R^2$ for both the Eqs. (24.5) and (24.6) showed exquisite agreement ($R^2 = 0.99$). The comparison of both the equation shows that these can be effectively used for estimating dimension of wetting front in horizontal direction by 2 lph drippers in medium loam soils.

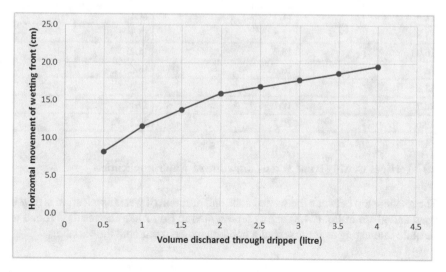

**Fig. 24.6** Relation between horizontal movement of wetting front (cm) and volume discharged through dripper (litre) for 2 lph dripper discharge

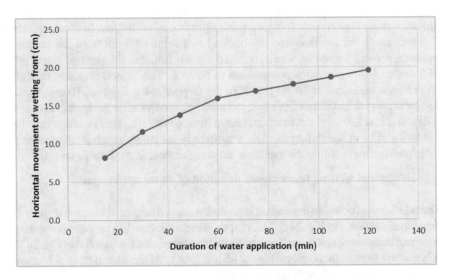

**Fig. 24.7** Relation between horizontal movement of wetting front (cm) and duration of water application (min) for 2 lph dripper discharge

**Table 24.2** Comparison of observed and calculated data of horizontal movement of wetting front for 2 lph dripper discharge by developed Eqs. (24.5) and (24.6)

| Observed dimension of wetting front in horizontal direction (cm) | Calculated dimension of wetting front in horizontal direction estimated by Eq. (24.5) (cm) | Calculated dimension of wetting front in horizontal direction estimated by Eq. (24.6) (cm) |
|---|---|---|
| 8.2 | 8.5 | 8.5 |
| 11.6 | 11.4 | 11.4 |
| 13.8 | 13.5 | 13.5 |
| 16.0 | 15.2 | 15.2 |
| 16.9 | 16.6 | 16.6 |
| 17.8 | 18.0 | 18.0 |
| 18.7 | 19.1 | 19.1 |
| 19.6 | 20.2 | 20.2 |

## (h)  Vertical wetting front versus duration of water application

The relationship between the wetted zone and duration of water application is developed and given in Eq. (24.8) and is represented in Fig. 24.8. The result is found to be quite interesting as it possessed a good match between both the variables.

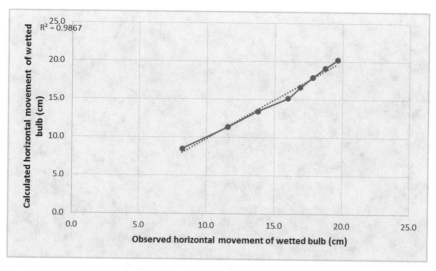

**Fig. 24.8** Comparison of observed and calculated data of horizontal movement of wetting front by developed Eqs. (24.5) and (24.6)

## (i)  Validation of developed equations

To validate the developed mathematical models, observed wetting front in vertical direction for dripper discharge of 2 lph was used and checked for agreement with modelled results. The vertical wetting front was calculated for various discharges in different duration of water application and compared with the results of the lysimetric study. The comparison of the results between calculated and observed values is given in Table 24.3 and co-relation between them is shown in Fig. 24.9. It can be noticed that the results possessed an excellent correlation with that of the observed data.

**Table 24.3** Comparison of observed and calculated data of vertical movement of wetting front for 2 lph dripper discharge by developed Eqs. (24.9) and (24.10)

| Observed dimension of wetting front in vertical direction (cm) | Calculated dimension of wetting front in vertical direction estimated by Eq. (24.9) (cm) | Calculated dimension of wetting front in vertical direction estimated by Eq. (24.10) (cm) |
| --- | --- | --- |
| 3.2 | 2.6 | 2.6 |
| 3.8 | 4.1 | 4.1 |
| 4.5 | 5.2 | 5.2 |
| 5.2 | 6.3 | 6.3 |
| 6.8 | 7.3 | 7.3 |
| 8.4 | 8.1 | 8.1 |
| 10.0 | 9.0 | 9.0 |
| 11.6 | 9.8 | 9.8 |

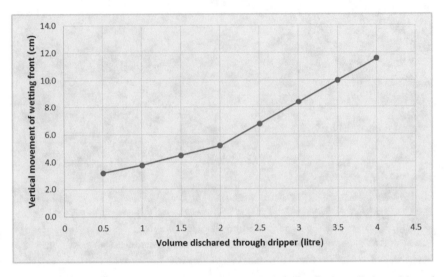

**Fig. 24.9** Relation between vertical movement of wetting front (cm) and volume discharged through dripper (litre) for 2 lph dripper discharge

From the results, it is clear that both Eqs. (24.9) and (24.10) can be used effectively for estimating the shape of vertical wetting front in medium loam soil which can be matched with root zone of the crop for optimum utilization of water and nutrients.

## 24.4 Conclusion

Lysimetric experiment was conducted to measure the wetting front in horizontal and vertical direction for 2, 4 and 8 lph dripper discharge at various durations. The measured results were used to develop the mathematical models to establish the relationship between the wetting front and volume of water discharged in given time duration in medium loam soils. The correlation between the calculated values through the developed models and measured results from lysimeter shows a strong power relation. The following conclusions were drawn from the study.

1. In case of 2 lph drippers, the horizontal distance of wetting front after 15 cm, 30 min, 45 min, 60 min, 75 min, 90 min, 105 min and 120 min were 8.2 cm, 11.6 cm, 13.8 cm, 16 cm, 16.9 cm, 17.8 cm, 18.7 cm, and 19.6 cm, respectively. The shape of wetting front is shown in Fig. 24.4. The distance is almost equal in all the quadrants and symmetrical in shape.
2. In case of 4 lph drippers, the horizontal distance of wetting front after 15 min, 30 min, 45 min, 60 min, 75 min, 90 min, 105 min and 120 min were 10.5 cm, 14.5 cm, 16.2 cm, 17.8 cm, 18.9 cm, 20 cm, 20.6 cm and 21.1 cm, respectively.

Regarding the shape of the wetting front, distance is almost equals in all the quadrants and symmetric in shape.

3.  In case of 8 lph drippers, the horizontal distance of wetting front after 15 min, 30 min, 45 min, 60 min, 75 min, 90 min, 105 min and 120 min were 13.2 cm, 16.2 cm, 18.2 cm, 19.2 cm, 20.2 cm, 21.7 cm, 23.2 cm and 25.2 cm, respectively. The distance is almost equals in all the quadrants and symmetric in shape.

4.  In case of 2 lph drippers, the vertical distance of wetting front after 15 min, 30 min, 45 min, 60 min, 75 min, 90 min, 105 min and 120 min were 3.2 cm, 3.8 cm, 4.5 cm, 5.2 cm, 6.8 cm, 8.4 cm, 10 cm and 11.6 cm, respectively. The shape of wetting front is shown in Fig. 24.1. The distance is almost equals in third and fourth quadrants and symmetric in shape.

5.  In case of 4 lph drippers, the vertical distance of wetting front after 15 min, 30 min, 45 min, 60 min, 75 min, 90 min, 105 min and 120 min were 3.6 cm, 4.9 cm, 8 cm, 11 cm, 12.3 cm, 13.7 cm, 15.2 cm and 16.8 cm, respectively. The shape of wetting front is shown in Fig. 24.2. The distance is almost equals in third and fourth quadrants and symmetric in shape.

6.  In case of 8 lph drippers, the vertical distance of wetting front after 15 min, 30 min, 45 min, 60 min, 75 min, 90 min, 105 min and 120 min were 4.7 cm, 10.5 cm, 13 cm, 16 cm, 18 cm, 19.3 cm, 20.5 cm and 21.6 cm, respectively. The shape of wetting front is shown in Fig. 24.3. The distance is almost equals in third and fourth quadrants and symmetric in shape.

The result obtained through this study can be used to predict the horizontal and vertical movement of wetting front for the sound designing of drip irrigation system and can also be used in determining irrigation periods for various crops to match their varying root zone temporally. The matching of the root zone for various crops will lead to optimal utilization of nutrient and water uptake and thus will minimize the losses.

# References

Amir I, Dag J (1993) Lateral and longitudinal wetting patterns of very low energy moving emitters. Irrig Sci 13(4):183–187

Elliott J, Deryng D, Müller C, Frieler K, Konzmann M, Gerten D, Eisner S et al (2014) Constraints and potentials of future irrigation water availability on agricultural production under climate change. Proc Natl Acad Sci 111(9):3239–3244

Elmaloglou ST, Malamos N (2006) A methodology for determining the surface and vertical components of the wetting front under a surface point source, with root water uptake and evaporation. Irrig Drain J Int Comm Irrig Drain 55(1):99–111

Elmaloglou S, Diamantopoulos E (2007) Wetting front advance patterns and water losses by deep percolation under the root zone as influenced by pulsed drip irrigation. Agric Water Manag 90(1–2):160–163

Karlberg L, de Vries FWP (2004) Exploring potentials and constraints of low-cost drip irrigation with saline water in sub-Saharan Africa. Phys Chem Earth Parts A/B/C 29(15–18):1035–1042

Keller J, Bliesner RD (1990) Sprinkle and trickle irrigation. Avi Book, New York

Michael AM (1978) Irrigation: theory and practice. Vikas Publishing House

Schwartzman M, Zur B (1986) Emitter spacing and geometry of wetted soil volume. J Irrig Drain Eng 112(3):242–253

Singh DK, Rajput TBS, Sikarwar HS, Sahoo RN, Ahmad T (2006) Simulation of soil wetting pattern with subsurface drip irrigation from line source. Agric Water Manag 83(1–2):130–134

Swain S, Patel P, Nandi S (2017a) Application of SPI, EDI and PNPI using MSWEP precipitation data over Marathwada, India. In: Geoscience and remote sensing symposium (IGARSS) 2017, pp 5505–5507

Swain S, Patel P, Nandi S (2017b) A multiple linear regression model for precipitation forecasting over Cuttack district, Odisha, India. In: 2nd International conference for Convergence in Technology (I2CT) 2017. IEEE, pp 355–357

Swain S, Verma M, Verma MK (2015) Statistical trend analysis of monthly rainfall for Raipur district, Chhattisgarh. Int J Adv Eng Res Stud IV(II):87–89

Swain S, Verma MK, Verma MK (2018) Streamflow estimation using SWAT model over Seonath River Basin, Chhattisgarh, India. In: Hydrologic modelling, water science and technology library, vol 81. Springer Singapore, pp 659–665

Verma M, Verma MK, Swain S (2016) Statistical analysis of precipitation over Seonath river basin, Chhattisgarh, India. Int J Appl Eng Res 11(4):2417–2423

Watkins K (2006) Human development report 2006—beyond scarcity: power, poverty and the global water crisis. UNDP human development reports (2006)

Zur B (1996) Wetted soil volume as a design objective in trickle irrigation. Irrig Sci 16(3):101–105

# Chapter 25
# Challenges of Food Security in Tanzania: Need for Precise Irrigation

**Mitthan Lal Kansal and Deogratius Nyamsha**

## 25.1  Introduction

Food security is a standout among the most pressing challenges in developing nations, especially those from Sub-Saharan Africa, Southern and Eastern Asia (Schindler et al. 2014). In developed countries, the issue is relieved by giving focused on food security interventions, including food aid as immediate food relief, food stamps, or indirectly through subsidized food production (Mwaniki 2006). Increasing productivity through irrigated agriculture system in a country is of vital important for food supplies, wages creation, public investment and expenditure. Nowadays, microirrigation systems are adopted due to their economical water use and enhanced yield compared with surface irrigation systems. These endeavors have fundamentally diminished food insecurity and hunger in these countries.

Cooperative for Assistance and Relief Everywhere (CARE) enables the understanding of hunger complexity and under-nutrition based on availability, access of nutritious food, food reliability source, capacity of farmers to produce and crops market (CARE 2013). World Development Indicator report (WDI 2016) highlights that more than 25 years ago, the share of the world's population suffering from hunger has fallen. This has made undernourished population to become almost half globally from 19 to 11%. Hence, the ongoing efforts to reduce hunger by 2030 will not be successful if the current trend is going to continue.

M. L. Kansal (✉) · D. Nyamsha
Department of Water Resources Development & Management, Indian Institute of Technology Roorkee, Roorkee, India
e-mail: mlk@wr.iitr.ac.in

D. Nyamsha
e-mail: dnyamsha@gmail.com

D. Nyamsha
Singida District Council, Singida, Tanzania

A. Pandey et al. (eds.), *Hydrological Extremes*, Water Science and Technology Library 97, https://doi.org/10.1007/978-3-030-59148-9_25

363

In Tanzania, food production at the national level in average years certifies the demand but fluctuates between years of excess in good season and deficit in bad weather. At the households level, the government and other agencies have made efforts to support agricultural productivity to attain food security (URT 2011a). When food availability is considered satisfactory, still access is a challenge to households producing less than 30% of the requirements. Various indicators are used to assess food security status, including dietary diversity and food consumption, dietary supply adequacy, consumer price index, per capita GDP, the prevalence of undernourishment, prevalence of stunting and wasting in children under-five and mortality rate and can be quantified through Global Hunger Index (GHI), food security index and household surveys.

Keeping this in view, the present study aims at assessing food security challenges in Tanzania and emphases the need for adopting precise irrigation for improving food security. Precise irrigation in agriculture involves water application in an efficient manner to the farm. It includes modern systems for monitoring and to control soil water shortage by developing suitable environment for plants.

## 25.2   Description of the Study Area

United Republic of Tanzania (URT) lies between latitude $01°$ and $11°$ S and longitude $29°$ and $41°$ E including mainland and offshore Islands like Zanzibar, Mafia and Pemba. The country has a total area of 94.73 million hectares of which about 5.7% is covered by three large lakes; Victoria, Tanganyika and Eyasi. The cultivable area is about 44 million hectares, or 46.5% of the total area. The population is estimated at over 52 million (2017 estimate) people with annual growth rate of 3.1% (NBS 2018). Figure 25.1 is a map showing the location of Tanzania.

The annual average rainfall ranges from 500 to 3000 mm in most area of the country with central part receiving annual average rainfall from 500 to 800 mm. The mean temperature ranges from 15 to 30 °C. Socio-economic activities include agriculture, mining, tourism and industries. The agricultural sector contributes to about 25.88% of the national economy and source of employment to almost 75% of the people (Chongela 2015).

## 25.3   Status and Challenges of Food Security in Tanzania

Food security exists when all people at all times have physical, social and economic access to sufficient, safe and nutritious food, which meets dietary needs and food preferences for an active and healthy life (FAO et al. 2013; WSFS 2009). Likewise, insecure food exists when people do not have adequate food access both physically, socially and economically. Basically, food security is described as a condition that is related to individuals on food consumption. In Tanzania, more efforts are done to

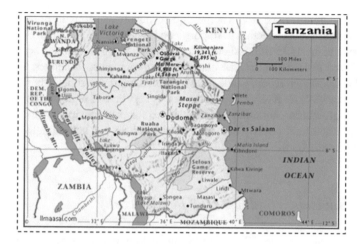

**Fig. 25.1** Location map of Tanzania (www.ilmaasai.com)

reduce food insecurity including initiatives on irrigation and promotion of drought tolerant crop varieties in drought prone zones.

Food security is a complex term to measure because it concern with production and distribution of food, contrary to hunger. Definitely, food security means food availability for the population whereas famine or hunger is the consequence of undernourishment to an individual or households. There are factors that influence food security such as social, economic, political and physical phenomenon. The main determinants of food security are availability, access, utilization and stability (WSFS 2009) as outlined in Fig. 25.2.

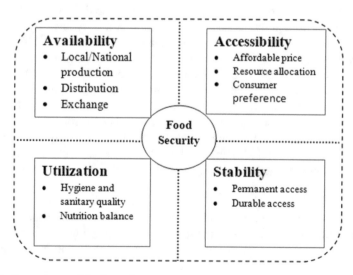

**Fig. 25.2** Determinants of food security

Food availability concern with both sufficient quantity and quality of food supplied through agricultural production or import. Food access to an individual or household is a measure to meets nutritious diet at a certain level. The available and accessed food when utilized should meet all the nutritional requirements. Moreover, food secure population must have access to adequate food all the times without experiencing any shocks. Tanzania's food security is portrayed by regular and regional food shortages despite its growth in economic and agriculture for past previous years and hence the rate of household poverty and food shortages have not substantially decreased (Schindler et al. 2014).

## 25.3.1 Indicators and Status of Food Security

The dimensions and interlinked variable indicators associated with food security make it difficult to measure using single indicators. At higher level, food security means balance between food demand and supply; however, household level can be measured directly by the use of household surveys on dietary intake. Food security at an individual level is easily measured by use of anthropometric information. Dietary diversity and food frequency are the most valuable indicators of energy intake or nutrient adequacy (Weismann et al. 2006). To assess the condition of food security in Tanzania indicators used are dietary diversity, food consumption, adequacy of dietary energy supply, consumer price index, GDP, prevalence of undernourishment, mortality rate, and prevalence of stunting and wasting in children under the age of five.

**Dietary energy consumption**

The number of food groups consumed in a given specific time without taking into consideration consumption frequency is referred to as dietary diversity (Weismann et al. 2008). The most nutrient consumed in Tanzania is carbohydrates, obtained from maize, cassava and rice as well as green vegetables, beans and peas (Mazengo et al. 1997; Kalinjuma et al. 2013). Carbohydrate comprises 75%, protein 10.5% and fat is only 12.5% of the energy intake. In rural and urban areas, Ugali prepared from maize or cassava flour is the main dish consumed after rice. The main difference in food consumption of urban and rural communities is that, in urban areas individuals usually eat three meals a day while in rural areas people normally take two meals a day. As highlighted by Kalinjuma et al. (2013), more than 50% of household in Singida, Dodoma, Iringa and Njombe consume maize.

**Prevalence of undernourished population**

Undernourished population is an indication for chronic food deprivation. The GHI 2016 figure even suggests a deterioration of the situation in Tanzania having 32.1%. This indicates that about 16.35 million people were consuming inadequate food compared with 14.5 million people in 2008.

**Dietary energy adequacy**

Average dietary adequacy enables to determine if undernourished population is due to inadequate food supply or bad distribution in a country. In developing countries like Tanzania, a vast majority of dietary energy requirement falls between 2100 and 2400 kcal/capita/day. Kakwani and Son (2015) indicated the use of average energy requirements of 2,100 cal/capita per day adopted from USDA where 67 developing countries used it including Tanzania. Estimate of adequate dietary energy supply is given by Eq. 25.1.

$$ADESA = \frac{DES}{MDER} \times 100\% \tag{25.1}$$

where; ADESA, average dietary energy supply adequacy (%); DES, dietary energy supply (kcal/capita/day) and MDER, minimum dietary energy requirement (kcal/capita/day).

ADESA derived from major food supply in Tanzania increased from 103% in 2012 to 105% in 2016 indicating an improvement in dietary supply. In fact, there is 5% more available calories, which indicates adequate caloric supply if food distribution was made for the entire population in Tanzania. MDER increased from 2163 to 2205 kcal in 2012 and 2016, respectively, yet far higher than 2100 kcal (8,800 kJ), the minimum requirement as recommended by USDA. Table 25.1 shows food availability in Tanzania for 2016/17.

The value of MDER made undernourishment to increase from 16.1% in 2012 to 16.80% in 2016. Hence, the current undernourished situation is attributed mainly due to unequal distribution of food among the population not due to insufficient supply of food.

**GDP per capita**

Gross domestic product per capita is an indicator that signifies the proportion of people leaving in extremely poverty measured in terms of daily income obtained by dividing total GDP to the total population including effects of inflation. In 2016, GDP per capita was found as 867 US$ about 7% of the world's GDP average (NBS-Tanzania 2017a). The Tanzania's GDP increased from 651.9 US$ in the year 2006 to 867 US$ in the year 2016 and it has been increasing for previous years. Therefore, purchasing power for Tanzania is still in the rank of lower income level countries.

**Table 25.1** Food availability in 2016/17 (URT 2017)

| Description | Food commodity (MT) | | |
|---|---|---|---|
| | Cereals | Noncereals | Total |
| Production | 9,457,108 | 6,715,733 | 16,172,841 |
| Requirement | 8,355,767 | 4,803,560 | 13,159,326 |
| Deficit/Surplus (-/+) | 1,101,341 | 1,912,173 | 3,013,515 |

## Consumer Price Index (CPI)

The value of CPI decreased to 109.10 in June from 109.26 in May 2017 with an average value of 85.48 from 2009 to 2017 (NBS-Tanzania 2017b). During December 2016, the CPI was 105.4. According to Tanzania National Bureau of Statistics (NBS), inflation rate was at level of 5.2% in 2016, down from 5.6% in 2015 and this has affected most of the consumer in purchasing goods and services while price of food commodity currently is still high. Generally, CPI compares food price over a range of time in a country determining the ability of purchasing food after increase in price. Both CPI and GDP per capita measure the ability to access or buy food.

## Prevalence of stunting in Children under-five

WHO (2010) described stunting as impaired growth condition and development that children experience from poor nutrition and repetitive infection. The proportions of children with low height for age indicate effects of under nutrition and infectious diseases even before births. This study observed that stunting in children under-five was 34.7% in 2016 indicating a 'high prevalence' situation for health. In fact, this result shows an improvement from 'very high prevalence' situation of 49.7% and 43.0% in the year 1992 and 2008, respectively.

## Prevalence of under-five wasting

Children suffering from wasting depict indication for acute under nutrition caused by insufficient food intake or incidence to infectious diseases. The study illustrated a prevalence of wasting in children under-five of 3.8% in 2016 in 'acceptable' range being higher than 2.7% in 2008. The thresholds for stunting and wasting are given in Table 25.2.

## Under-five mortality rate

This determines the number of children under 5 years of age who die before or by the age of 5 per 1000 live births in a year. This indicator reveals progress toward ensuring children right to life and getting nutritional diet. In 2016, the Tanzania average value was 4.9%, being improvement from an average of 16.3% in 1992.

**Table 25.2** Threshold for stunting and wasting of under-five children (WHO 2010)

| Stunting of under-five | | Wasting of under-five | |
|---|---|---|---|
| Threshold (%) | Prevalence | Threshold (%) | Prevalence |
| <20 | Low | <5 | Acceptable |
| 20–29 | Medium | 5–9 | Poor |
| 30–39 | High | 10–14 | Serious |
| ≥40 | Very high | ≥15 | Critical |

## 25.3.2  Global Hunger Index (GHI)

Hunger can be defined as a condition in which an individual, household member fails to eat sufficient food requirements. The factors that contribute to a high GHI include low-income level and poverty, violent conflicts or war, lack of power and defectively fail to delivered physical condition and nutrition programs (WHO 2010). GHI was firstly set up by International Food Policy Research Institute (IFPRI) to assess hunger globally, monitor the progress of the MDGs and interpret its trends (Wiesmann et al. 2015). Globally, countries are ranked depending on 100 point scale GHI score-based severity scale as indicated in Table 25.3. The value of GHI of 0 indicates no hunger while 100 bad situations, but in practice, it is very difficult to reach 0 and 100 values.

In 2016 GHI report, GHI is calculated using four component indicators in three steps (Grebmer et al. 2016). The first step is to identify indicator values, i.e. proportion of undernourishment (PUN), prevalence of children under-five with wasting (CWA) and stunting (CST) and mortality rate of under-five children (CMR) in percentage. Secondly, to find standardized score; and lastly, aggregating standardized scores to get index. The weights assigned to proportion of undernourishment and children mortality rate is 1/3, and prevalence of children with stunting and wasting is 1/6 on the aggregated scores.

$$GHI = \frac{1}{3}SPUN + \frac{1}{6}SCWA + \frac{1}{6}SCST + \frac{1}{3}SCMR \qquad (25.2)$$

where; GHI, Global hunger index; SPUN, standardized PUN value; SCWA; standardized CWA value; SCST; standardized CST value; SCMR; standardized CMR value. The threshold values for various indicators considered for 2016 GHI are shown in Table 25.4.

**Table 25.3**  GHI severity scale

| S/N | Severity level | Value |
|-----|----------------|-------|
| 1 | Low | ≤9.9 |
| 2 | Moderate | 10.0–19.9 |
| 3 | Serious | 20.0–34.9 |
| 4 | Alarming | 35.0–49.9 |
| 5 | Extremely alarming | ≥50.0 |

**Table 25.4**  GHI threshold for 2016 (Grebmer et al. 2016)

| Indicators | Observed maximum | Threshold |
|------------|------------------|-----------|
| Undernourishment | 76.5 | 80 |
| Wasting | 26.0 | 30 |
| Stunting | 68.2 | 70 |
| Mortality rate | 32.6 | 35 |

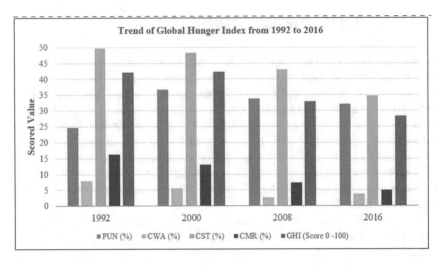

**Fig. 25.3** Indicators of global hunger index in Tanzania

For 2016, the standardized values of indicators are assessed as follows:

SPUN = 1/80(PUN * 100) = (1/80) * 32.1 * 100 = 40.125
SCWA = 1/30(CWA * 100) = (1/30) * 3.8 * 100 = 12.667
SCST = 1/70(CST * 100) = (1/70) * 34.7 * 100 = 49.571
SCMR = 1/35(CMR * 100) = (1/35) * 4.9 * 100 = 14
Therefore, GHI = **28.4,** i.e. "**serious**."

The GHI score for Tanzania in 2008 was 32.9, and hence, one can say that there is improvement by a score of 4.5 which brought Tanzania at 96 out of 118 countries. It is worth noting that the GHI score was 42.4 and 42.1 in years 2000 and 1992, respectively. The GHI indicators and its variation in Tanzania are shown in Fig. 25.3.

The variation in GHI shows that the trend is improving, but still lot more is required. The world over GHI severity of 2016 is shown in Fig. 25.4.

### 25.3.3 Challenges of Food Security

The strength of Tanzania's economy depends on emerging sectors like mining, tourism and fisheries; however, primarily is highly dependent on agriculture. It is likely that factors like climate change, etc. will result in damaging infrastructure, crop losses and will result in increase in food prices if appropriate technologies are not used. Further, in Tanzania, despite of huge land for agriculture (44 Mha), cultivation is done on a very small land (around 5.1 Mha) which is about 5% of the total surface area. In addition, undeveloped agriculture is dominated by overdependence

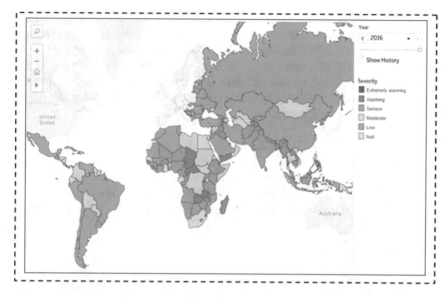

**Fig. 25.4**  World over GHI severity (Grebmer et al. 2016)

on subsistence farming, limited income opportunities, unutilized irrigation potential and limited market access due to poor infrastructure. Further, due to population pressure and low outputs from agriculture, farmers are moving to peripheral land to increase cultivation level. The country is not a food-deficit country at present but there are several issues that may affect the ability of the country to feed the people in future. These are discussed as follows:

**Population growth and food demand**

The population projections of Tanzania show that it will have an annual growth of about 3.1% as shown in Table 25.5. The growing population will continue to add pressure on land while other competing demand for land will also hinder the country's food security (Mliga 2012). Hence efforts are needed to increase availability and diversify food production for the population.

**Undeveloped agriculture**

Agriculture sector is predominantly characterized by over reliance on primary agriculture and affected by low soil fertility, postharvest losses, minimal use of farm inputs, environmental degradation and inadequate food storage. There is decline in

**Table 25.5**  Population projection of Tanzania 2016–2020 (NBS 2018)

| Year | 2016 | 2017 | 2018 | 2019 | 2020 |
|---|---|---|---|---|---|
| Population (million) | 50.94 | 52.55 | 54.20 | 55.89 | 57.64 |
| Growth rate (%) | – | 3.1 | 3.1 | 3.1 | 3.1 |

farm input investment such as fertilizers, seeds and technology adoption. This reveals the constraints faced by farmers to access and implement technologies for increasing productivity and access to financing (WFP 2012). However, the government has put efforts on use of modern agricultural machine still about 70% of the farmers use hand hoe, 20% use oxen plough and only 10% used tractor. Further, most of the irrigated lands are served by traditional methods of surface irrigation with unlined canals leading to low efficiency.

## Limited income opportunities

In Tanzania, smallholder farming families represent the majority of the population and an even larger majority of the poor and food insecure population. The agricultural outputs and incomes obtained are not sufficient and stable enough over time to improve livelihood, ensure food security and make farmers invest in business to increase household incomes.

## Unutilized irrigation potential

The level of irrigation development in Tanzania is still very low resulting into marginal use of available potential for irrigation due to various reasons including inadequate financial capital and human resource. Tanzania has abundant land for irrigation about 29.4 million hectares that could be irrigated, but only 0.6 million hectares of this land is currently being utilized for irrigation.

## Barriers to market access

Barrier to market access is a problem that farmers have to overcome. This is due to poor infrastructure and barrier in penetrating the market as a result of limited resource base, lack of information and insufficient policies. Tanzania has moderate agricultural market linkages compared with other domestic sectors with poorly developed marketing system for both domestic and international markets (Kiratu et al. 2011). Poor infrastructure literally limits the markets to which farmers can profitably take their produce by increasing the cost of transportation.

Approximately 70% or rural people are food insecure. The solutions to improving food security rely on increasing availability, access, utilization and stability of food at all levels. Strategies for ensuring food security in Tanzania have to include widespread of knowledge transfer by capacity building. Focus should be on training and research development and access to capital and infrastructure development. Food insecurity cannot be reduced without transforming agriculture and lives of rural people meanwhile sourcing possibilities of creating off-farm opportunities (Mwaniki 2006). Therefore, efforts should be undertaken to reduce the number of farmers who are not motivated and who look up farming activity just because they have no more options. Furthermore, obstacles related to trade for both internal and external market need to ensure that food surplus reaches to areas with food deficit.

There is a need to modernize irrigation and accelerate implementation of the national irrigation plan, which envisages increase in irrigated land to 1 million by 2020. This has to be achieved only if the government continues financing irrigation, supporting farmer-led irrigation and fair allocation of water. Given low efficiency of surface irrigation due to water losses during conveyance, distribution and application adopting precise irrigation such as drip irrigation selectively would help to save water and energy meanwhile promoting food security.

## 25.4  The Need for Precise Irrigation

Precise irrigation includes both accurate crop water demand assessment and correct application of water at the required time (Smith et al. 2010). It involves accurate and precise application of water to meet specific demand to individual plants and minimize the adverse impacts (Misra et al. 2005; Raine et al. 2007). In FIGARO irrigation, precision irrigation technology is defined as irrigating with preciseness or as targeting of water according to the requirement for crop production (Tsakmakis et al. 2013). Precise irrigation worldwide is not widespread. Most of the world approximately 82% of the people do not irrigate at all only rely on rainwater for agricultural production (Kansal 2017). The remaining 18% use surface irrigation, microirrigation, sprinkler, and center pivot and water spray gun. The various methods of precision irrigation and their advantage and disadvantages have been detailed.

### 25.4.1  Advantages of Precise Irrigation

(a) The system is able to apply a define volume of water to match crop water requirements as it reduces surface runoff and deep percolation;
(b) High application efficiencies generally in the range of 80–95% compared with 50–90% of surface irrigation;
(c) Less labor requirement compared to surface irrigation, however it depends on system and degree of precision;
(d) Fertilizer and other plant chemicals can be applied during system operation;
(e) The system works on a range of topography however there must be some land forming;
(f) Other advantages include increased crop yield, improved crop quality, reduce energy cost and can be applied for all crops.

### 25.4.2  Disadvantages of Precise Irrigation

(a) The system involves high capital investment; however the operating expenses are reduced;
(b) Require some form of energy source to operate such as electric and diesel;
(c) Water has to be filtered to prevent clogging with sediments and poor water quality can affect the durability of the facilities;

(d) Technical capacity is required to operate and maintain the system as compared to surface irrigation.

### 25.4.3 Methods of Precise Irrigation

Conventional methods of irrigation for over 40 years have contributed to the insight of selecting appropriate methods that could be used to increase the application efficiency of water (Reinders 2011). The methods vary depending on their construction, functioning, costs and performance. In general, three types of irrigation systems have been identified: flooding such as border, basin and furrow; moving systems including center pivot, linear and traveling gun; and stationary systems like sprinkler either permanent or portable, micro-systems and drip irrigation systems. Among the irrigation systems, the main types of irrigation application systems commonly used as precise methods are sprinkler, center pivot and linear move and drip irrigation systems.

**Drip irrigation system**

Drip irrigation is defined as application of water through the use of point or line source emitter on or below the soil surface at low pressure about 20–200 kPa and low emitter discharge of 1–30 L/hour (Dasberg and Or 1999). It consists of system of pipes and emitters that deliver water drop by drop to the root zone of the plant or crops. The emitters are placed so as to produce a wetting pattern along the crop row or at every plant. Drip irrigation has been considered as the most efficient method of irrigation compared with the rest.

**Sprinkler irrigation**

Technology such as sprinkler, microsprinkler and drip irrigation has the capability of applying the desired quantity of water with high precision and uniformity. Sprinkler technology applies water to plant under pressure through small orifices or nozzles called sprinklers. The system comprises of network of pipes, sprinklers and pump for developing the required pressure. Sprinkler system when properly designed can achieve high values of water application and water distribution efficiencies (Michael 2007; Howell 2003). It is more efficient compared with surface irrigation methods, however is more costly to install and operate. In sprinkler systems, water is sprayed into the air, losing considerable amounts to evaporation. Sprinkler systems are classified as rotating and perforated systems. In rotating head sprinkler, nozzles of small size are attached on risers, pipe at fixed interval on the lateral laid to the ground surface while perforated systems holes are drilled along pipe length to allow water spray under pressure and are designed for relatively low pressure.

**Center pivot irrigation**

Center pivot is another type of overhead sprinkler system that consists of a series of pipe segments joined together mounted on trowel having sprinklers positioned along

**Table 25.6** Efficiencies of different irrigation methods (Howell 2003)

| S/N | Irrigation method | Lower | Mean | Upper |
|---|---|---|---|---|
| 1 | Surface irrigation | 25 | 40 | 55 |
| 2 | Subsurface drip | 75 | 90 | 95 |
| 3 | Microirrigation | 70 | 85 | 95 |
| 4 | Sprinkler irrigation | 60 | 75 | 85 |
| 5 | Center pivot | 75 | 80 | 90 |
| 6 | Lateral move | 60 | 70 | 80 |
| 7 | Automated irrigation | 75 | 90 | 95 |

its entire length. The system has a machine that carries water in circular patterns and covers the entire field (Ahmet 2009). The field area to be served is dependent on the machine's length which allows irrigation within short-time spell. Design features of center pivot system include underground mainline or well in center, pivot point, laterals, towers and trusses. Advantages associated with center pivot in agriculture include its ability to irrigate long-distance field given that surface topography is flat, however, the system has proved to have high initial cost.

**Linear move irrigation**

Fields that are not appropriate to be used due to its topography and shape by center pivot, lateral move irrigation systems can be adopted. The method can be designed to move in a straight line as described by its name (Hill and Robert 2000; Agri-Africa 2016). Linear move composes of a series of trusses that suspend the irrigation system and move laterally in the direction of the rows, motor and pumps are mounted on a cart-like equipment to supply channel that travels with the machine at a length between 500 m and 1000 m. Linear move irrigates rectangular fields and requires additional management compared with center pivot. There are four types of low-pressure system; LEPA—low energy precision application, LPIC—low pressure in-canopy, LESA—low elevation spray application and MESA—medium elevation spray application. The types operate at low pressure ranging from 0.5 to 2.5 bars and use fixed sprinklers or drop tubes to apply water (TWDB 2004). Table 25.6 shows the efficiencies of different irrigation methods.

## 25.4.4 Precise Irrigation in Tanzania

Agricultural in the country is mainly rain fed and affected by vagaries of weather resulting from climate change. This invariably has subsequently subjected crop production to be low hence the need for an effective means of increasing production and productivity. Therefore, further scope to promote precise irrigation, example sprinkler and drip systems as they reduce farming costs and energy requirements while improving yield is of high imperative (WBCSD 2014). Currently, precise irrigation in Tanzania covers less than 0.1% of irrigated land while there is plenty of scope for irrigation.

With abundance of water resources and high potential for irrigation in Tanzania, expansion of area under irrigation is inevitable for both cash and food crop production (URT 2011b; FAO Aquastat 2014). Almost 85% of the area suitable for agriculture is used by smallholders' farmers cultivating between 0.2 and 2 ha. Efforts now have to be done to reduce poverty in the country especially rural areas by increasing the growth and development of agricultural sector for the livelihood of the majority of the people. Figure 25.5 shows the demonstration plot of drip irrigation in Tanzania.

When assessing water for irrigation, it is important to identify both underground and surface water. The exercise has to be accomplished at the time of evaluating potential areas, which can suit the need for precise irrigation. Table 25.7 shows water withdrawal per sector in Tanzania.

Due to its high initial cost, yields obtained from precision irrigation will not be sufficient to cover all costs associated at the moment a farmer starts farming. In spatially varied application systems, the risk for economic benefits increases once

**Fig. 25.5** Drip irrigation in Tanzania (*Source* iAGRI 2016)

**Table 25.7** Water withdrawal in per sector in Tanzania (FAO Aquastat 2014)

| Description | Amount (m$^3$/year) |
|---|---|
| Total water withdrawal | $5.18 \times 10^9$ |
| Agricultural water withdrawal | $4.63 \times 10^9$ |
| Irrigation water withdrawal | $4.43 \times 10^9$ |
| Irrigation water requirements | $0.97 \times 10^9$ |
| Total water withdrawal per capita | 144.80 |

**Fig. 25.6** Onion irrigated with drip system

farmers are producing under capacity (iAGRI 2016). Therefore, it remains whether the costs of the system are reduced or simple designs are adopted. Figure 25.6 shows an irrigated onion crop served with drip irrigation system.

### 25.4.5 Economic Issues of Precise Irrigation

Social aspects of precise irrigation include improved life, provision of labor, ability to access and acquire food and contribution toward efforts to food security. In Tanzania, precise irrigation is not yet widely used on a country scale and covers less than 0.1% of irrigated land. The system cost is determined by crop type, row spacing and the total field area irrigated and it can range from US$1000 to 3000 per ha. A good example is a smallholder horticulture development project implemented by Fintrac from 2009 to 2015 as part of the US Government's feed the Future initiative. More than 61,000 rural families across Tanzania benefitted from the project through technology transfer, good agricultural practices training, and business skills development (USAID 2015). About 6,400 farmers had adopted drip irrigation on their farms.

Smallholder farmer in Tanzania investing between US$1,000 and 1,500 in a 0.4 ha drip irrigation system and rotating onion, tomato, and cabbage throughout a year, selling into domestic markets is expected to earn nearly US$6,000 in net income (USAID 2016). When managed well, a farmer is likely to recover the cost of drip irrigation within 1 year or few years. Precision irrigation can be used to reduce unnecessary use of farming inputs like fertilizers and pesticides, labor and water, thereby minimizing the losses of nutrients and plant chemicals to the environment. Table 25.8 shows the annual returns from 0.4 ha of drip irrigation.

**Table 25.8** Annual returns for Tanzania Smallholder 0.4 ha drip set (USAID 2016)

| Cropping | Revenue (US$) | Operation cost (US$) | Net income (US$) |
|----------|---------------|----------------------|------------------|
| Onion    | 4,583         | 1,528                | 3,054            |
| Tomatoes | 5,357         | 3,514                | 1,843            |
| Cabbage  | 2,500         | 1,541                | 959              |
| Total    | 12,440        | 6,583                | 5,856            |

## 25.5  Conclusions

Food security is used to describe a phenomenon that is related to individuals on food consumption and is one of the key agricultural development indicators. As highlighted in this study, there is a short fall in domestically produced food in Tanzania, one of the reasons being increased population growth rate of 3.1% while the growth in food production has been less than 1%. The demand for food to feed the population is greater because of low yield from agriculture.

The major challenges highlighted in this study include undeveloped agricultural technologies, unutilized irrigation potential, limited income opportunities, increased population growth and poor market access. To meet the aforementioned challenges, adapting modern technologies in agriculture is highly emphasized. For example, currently, the use of precise irrigation in Tanzania covers less than 0.1% of irrigated land and there is plenty of scope for precise irrigation. On-farm adaptive studies through participatory programs and farmers' field schools will help in capacity building to adapt scientific technology.

Furthermore, to reduce transitory household food insecurity improved production mechanism, water use efficiency, increasing irrigated land, requirement and creation of proper storages, off-farm opportunities and market access are emphasized. In fact, investments are needed to local institutions and small scale credit schemes to help afford the initial cost required for establishment of precise irrigation. NGOs or private sectors must be encouraged to invest in food security initiatives. Socioeconomic aspects of drip irrigation, sprinkler and center pivot system suggest that when precise irrigation technology is adopted it can improve the water application efficiencies up to a range of 80–90% compared to 40–45% for surface irrigation methods (Dukes and Scholberg 2005). Other studies suggest that precision irrigation can save water up to 50% and average from 8 to 20% for a number of years (Evans and Sadler 2008).

**Acknowledgements** The authors are thankful to the Government of the United Republic of Tanzania, Indian Technical and Economic Cooperation (ITEC) Ministry of Internal Welfare Government of India and the Department of Water Resources Development and Management, Indian Institute of Technology Roorkee for providing necessary support and facilities.

# References

AgriAfrica (2016) Whats new in agriculture: linear move machines. http://www.farmingportal.co.za
Ahmet K (2009) Advantages and disadvantages of center pivot irrigation. Agriculture guide. http://www.agricultureguide.org/
CARE International (2013) Exploring the challenges and solutions to food and nutrition security: findings from the CARE learning tour to South Sudan and Tanzania, Feb 17–22
Chongela J (2015) Contribution of agriculture sector to the Tanzanian economy. Am J Res Commun 3(7):57–70
Dasberg S, Or D (1999) Practical applications of drip irrigation. In: Drip irrigation. Springer, Berlin, Heidelberg, pp 125–138
Dukes MD, Scholberg JM (2005) Soil moisture controlled subsurface drip irrigation on sandy soils. Appl Eng Agric 21(1):89–101
Evans RG, Sadler EJ (2008) Methods and technologies to improve efficiency of water use. Water Resour Res 44(7)
FAO Aquastat (2014) AQUASTAT-FAO's information system on water and agriculture. http://www.fao.org/nr/aquasta
FAO, Ifad and WFP (2013) The state of food insecurity in the world 2013. The multiple dimensions of food security. FAO, Rome
Grebmer K, Berstein J, Prasai N, Amin S, Yohannes Y, Towey O, Thompson, Sonntag A, Patterson F, Nabarro D (2016) 2016 global hunger index: getting to zero hunger. Concern Worldwide, Welthungerhilfe, IFPRI and United Nations, Washington, DC/Dublin/Bonn
Hill, Robert W (2000) Wheelmove Sprinkler irrigation operation and management. All archived Publications, Paper 153. http://digitalcommons.usu.edu/extension_histall/153
Howell TA (2003) Irrigation efficiency encyclopedia of water science. Marcel Dekker, New York, pp 467–472
IAGRI (2016) Innovative agricultural research initiatives. From lab to farm collaborative research. Exposure post, May 19th. http://iagri.org/stories/from-lab-to-farm
Ilmaasai (2017) Tanzania Map. http://ilmaasai.com/tanzania-map/accessed
Kakwani N, Son HH (2015) Measuring food insecurity: global estimates. University of New South Wales, Australia. ECINEQ Working paper: 370
Kalinjuma AV, Mafuru L, Nyoni N, Madaha F (2013) Household food and nutrition security baseline survey for Dodoma, Iringa, Njombe and Singida. Tanzania Food and Nutrition Centre and TAHEA Iringa
Kansal ML (2017) Issues and challenges in water productivity for sustainable agricultural growth in India. J Indian Water Resour Soc 37(1) (Special Issue)
Kiratu S, Markel L, Mwakalobo A (2011) Food security: the Tanzanian case. Series on Trade and Food Security
Mazengo MC, Simell O, Lukmanji Z, Shirima R, Karvetti RL (1997) Food consumption in rural and urban Tanzania. Acta Trop 68:313–326
Michael AM (2007) Irrigation theory and practice, 2nd edn. Vikas Publishing House, pp 579–655
Misra R, Raine S, Pezzaniti D, Chaelsworth P, Hancock N (2005) A scoping study on measuring and monitoring tools and technology for precision irrigation. CRC for irrigation futures. Irrigation Matters Series No. 01/05
Mliga A (2012) Food and water security in Tanzania. Future directions international strategic analysis paper, 1–26 July. http://futuredirections.org.au
Mwaniki A (2006) Achieving food security in Africa: challenges and issues. Cornell University, US Plant, Soil and Nutrition Lab
NBS (2017a) National Bureau of Statistics Tanzania. Tanzania GDP per capita 1988–2017. http://tradingeconomics.com/tanzania/gdp-per-capita
NBS (2017b) National Bureau of Statistics Tanzania. Tanzania consumer price index (CPI) 2009–2017. https://tradingeconomics.com

NBS (2018) National Bureau of Statistics. National population projections. The United Republic of Tanzania, 113 pp

Raine SR, Meyer WS, Rassam DW, Hutson JL, Cook FJ (2007) Soil–water and solute movement under precision irrigation: knowledge gaps for managing sustainable root zones. Irrig Sci 26(1):91–100

Reinders FB (2011) Irrigation methods for efficient water application: 40 years of South African research excellence. Water SA 37(5):765–770

Schindler J, Graef F, Jha S, Sieber S (2014) Food security impact assessment in Tanzania. Trans-SEC Project Policy Brief No. 1, 10th May

Smith RJ, Baillie JN, McCarthy AC, Raine SR, Baillie CP (2010) Review of precision irrigation technologies and their application. National Centre for Engineering in Agriculture Publication 1003017/1, USQ, Toowoomba

Tsakmakis I, Kokkos N, Sylaios G (2013) The theoretical boundaries of precision irrigation for cotton cultivation in Northern Greece. Fexible and PrecIseIrriGationPlAtform to Improve FaRm scale water PrOductivity (FIGARO)

TWDB (2004) Texas Water Development Board. Water conservation-best management practice (BMP) guide. Texas Water Development Board. Water conservation implementation task force, Austin, Texas

URT United Republic of Tanzania (2011a) Tanzania agriculture and food security investment plan (TAFSIP), 2011-2012 to 2020-2021. Main Document, 18th October, 2011

URT United Republic of Tanzania (2011b) Tanzania agriculture and food security investment plan (TAFSIP) 2011-12 to 2020-21. Working Paper No.4 on Priority investments: irrigation development, water resources and land use management

URT United Republic of Tanzania (2017) National food security bulletin, vol 02. Ministry of Agriculture, Livestock and Fisheries

USAID (2015) United states agency for international development. Tanzania Agricultural Productivity Program (TAPP): Final Report 2009/2015

USAID (2016) United states agency for international development. Drip irrigation in smallholder markets: a cross partnership study in Tanzania. Feed the Future program

WBCSD (2014) World business council of sustainable development. Co-optimizing solutions: water and energy for food, feed and fibre. World Business Council for a Sustainable Development, Geneva, Switzerland

WDI (2016) World development indicator. World development indicators highlight: featuring the sustainable development goals. International Bank for Reconstruction and Development/World Bank, Washington, DC

Weismann D, Hoddinott J, Aberman N, Ruel M (2006) Review and validation of dietary diversity. Food security frequency and other proxy indicators of household food security. International Food Policy Research Institute. Report submitted to United Nations World Food Programme, Rome, Italy

Weismann D, Bassett L, Benson T, Hoddinott J (2008) Validation of food frequency and dietary diversity as proxy indicators of household food security. International Food Policy Research Institute. Report submitted to World Food Programme, Rome

Weismann D, Biesalski HK, Grebmer K, Bernstein J (2015) Methodological review and revision of the global hunger index. Center for Development Research, University of Bonn. ZEF Working Paper 139

WFP, World Food Programme (2012) Assessing food security at WFP: towards a unified approach-design phase report. UN World Food Programme

WHO, World Health Organization (2010) Nutrition landscape information system (NLIS) country profile indicators: interpretation guide. WHO Document Production Services, Geneva, Switzerland

WSFS (2009) World summit of food security. Draft declaration of the world summit on food security, Rome, 16–18 Nov

# Chapter 26
# Soil Moisture Depletion-Based Irrigation Technology for Summer Finger Millet Under Midland Situation of Chhattisgarh Plains

**M. P. Tripathi, Yatnesh Bisen, Priti Tiwari, Prafull Katre, Karnika Dwivedi, and G. K. Nigam**

## 26.1  Introduction

Finger millet is an essential cereal crop for subsistence agriculture in the areas of India, Eastern Africa, and Srilanka. The major cereal producing country in the world is India. World finger millet production is 4.5 million tons. Due to lack of awareness, a very important cereal grain *ragi* has not come into the international market in areas of its adaptation. It is well known fact that Chhattisgarh state is known for its unique and distinctive culture and heritage all over the world. Finger millet is an important minor millet grown in Chhattisgarh about 8.04 thousand ha area. The common name "Finger Millet" is derived from the finger-like branching of the panicle. It is used in the production of beer, porridge, soup, bread, cake, and puddings. Flour of malted grain is very useful food for baby and is often fed to diabetic patients in India (Bhatnagar 1952). Finger millet is only one of its kind uses as a prophylaxis for dysentery (Lemordant 1967). The finger millet has got high nutritious value (protein 3.3 g, fiber 3.6 g, minerals 2.7 g, iron 3.9 mg, and calcium 344 mg per 100 gm of grains). They do not form a taxonomic group, but rather a functional or agronomic one.

The productivity level of *kharif* finger millet in Chhattisgarh is only about 185 kgha$^{-1}$ due to unavailability of irrigation in upland and midland areas, whereas no information is available in summer season for growing of finger millet. Finger millet is predominantly grown under rain-fed conditions. Irrigation is the prime requirement for finger millet production if it is grown during *rabi* or summer seasons. In

M. P. Tripathi (✉) · P. Tiwari · P. Katre · K. Dwivedi · G. K. Nigam
Department of Soil and Water Engineering, FAE, IGKV, Raipur, Chhattisgarh, India
e-mail: mktripathi64@gmail.com

Y. Bisen
Department of Soil and Water Conservation Engineering, Dr. PDKV, Akola, Maharashtra, India

A. Pandey et al. (eds.), *Hydrological Extremes*, Water Science and Technology Library 97, https://doi.org/10.1007/978-3-030-59148-9_26

case of inadequate supply of water, there is need for improved technologies based on the water management, developed by the research institutes, is to be very successful in efficient consumption of the existing water resources. In Chhattisgarh, farmers are growing summer rice in 186 thousand ha area (Anonymous 2014) in the canal and tube well commands. Rice is water-loving crop and having high water requirement (900–1200 mm). High value crop like *ragi* that requires less water may be the alternative of summer rice in Chhattisgarh. This can be achieved by increasing the cropping intensity and also by increasing the yield per crop per unit area.

In the coming decades increasing the food production with less water the big challenge, predominantly in countries were limited water and land resources (FAO 2002). The scarcity of water and cost of irrigation worldwide are important to highlight developing methods of irrigation that maximize the water use efficiency (WUE) (Hess 1996). Deficit irrigation is the only mode of maximizing WUE for higher production per unit of water applied (English and Raja 1996). Irrigation scheduling is important on ecological benefits point of view, e.g., reduced losses of fertilizers resulting from a decrease in seepage increase in the soil moisture (Mao 1996).

Keeping the importance of finger millet, pattern of rainfall, and available water in view, an experiment during summer by considering different irrigation treatments based on soil moisture depletion were carried out on finger millet in midland situation of Chhattisgarh plains during three consecutive years, i.e., 2014, 2015 and 2016. This study was undertaken at IGKV research farm to determine the water requirement and to assess the suitability of water management practices on finger millet for maximum production.

## 26.2   Materials and Methods

The experiment was conducted under ITRA-sponsored project entitled "Measurement to Management (M2M): Improved water use efficiency and agricultural productivity through experimental sensornetwork" running under the Department of Soil and Water Engineering, SVCAET & RS, FAE, IGKV, Raipur which is situated at an average height of 290 m above mean sea level with an intersection of $21^0 14'$N latitude and $81^0 42'$E longitude. In the study, area about 85% of annual rainfall occurs during June to September. Mainly month of June to September, variation in rainfall pattern has been high from year to year. The average daily maximum and minimum temperature is about 31 °C and 19 °C, respectively. The relative humidity ranges from 72% (morning) to 54% (evening). Metrological data recorded by the Department of Agrometeorology, Indira Gandhi Krishi Vishwavidyalaya, Raipur were used in this study. A total of 95.0 mm, 81.7 mm and 31.1 mm rainfall was received during the crop growing season of 2014, 2015 and 2016, respectively. Five treatments of Soil Moisture Depletion Level (SMDL) including T1 (30% SMDL), T2 (40% SMDL), T3 (50% SMDL), T4 (60% SMDL) and T5 (70%SMDL) and one treatment of moisture conservation as M1 (soil mulch) and one treatment of no moisture conservation as M2 (without mulch) were considered in this study. These treatments were replicated three times under split plot design, by taking into consideration, moisture conservation as main plot and moisture depletion levels under subplot. The size of each plot was taken as $5.0 \times 5.0$ m$^2$.

Twenty-eight days seedling (Plate 26.1) of finger millet (GPU 28) was transplanted at 10 cm plant to plant spacing and 25 cm row to row spacing (Plates 26.1 and 26.2). Recommended dose of fertilizer NPK 50: 40:20 kg ha$^{-1}$ was applied in all the treatments. Only one-third quantity of nitrogen and full quantity of phosphorous and potash were applied as basal dose at the time of transplanting and remaining two-third quantity of nitrogen was again divided into two equal parts and applied at the time of tillering and primordial stages. Soil mulch operation is shown in Plate 26.3.

Depth of water application was calculated accurately along with required quantity of water given to the field for replenishing its field capacity. Daily moisture content was recorded using Time Domain Reflectometry (TDR) as per the treatments. Soil moisture content was measured at three depths (0–15 cm, 15–30 cm and 30–45 cm) to know the available moisture status of soil which will lead to maximize the yield. The texture of the soil at the site was silt clay loam which was having bulk density as 1.43 gm cc$^{-1}$, field capacity as 24.15% and wilting point as 10.16%. Depth of irrigation was calculated by empirical formula (Eq. 26.1) (Acharya et al. 1989).

**Plate 26.1** Growth of nursery

**Plate 26.2** Transplanting operation

**Plate 26.3** Soil mulching operation

$$D = (FC - PWP).SMDL.BD.RD.10 \qquad (26.1)$$

where, $D$ is irrigation depth (mm), $FC$ is field capacity (%), $PWP$ is permanent wilting point, $SMDL$ is soil moisture depletion level, $BD$ is bulk density (gm cc$^{-1}$)and $RD$ is depth of root zone (cm), Water requirement (WR).

Water requirement was computed by:

$$WR = IR + ER + S \qquad (26.2)$$

where, WR = water requirement (cm), IR = irrigation requirement (cm), ER = effective rainfall (cm) and S = soil profile contribution

### Water use efficiency (WUE)

Water use efficiency denotes the production per unit of water applied. Eventually, WUE of finger millet (*ragi*) crop has been expressed as the ratio of grain yield (kg ha$^{-1}$) to the total amount of water used (cm) of crop (Michael 1978).

*Field water use efficiency: It is the ratio of crop yield (Y) to total water used in the field (WR)*

$$\text{Field water use efficiency} = \frac{Y}{WR} \qquad (26.3)$$

where, FWUE = Field water use efficiency (kg ha$^{-1}$ cm$^{-1}$),
Y = Yield (kg ha$^{-1}$), WR = Water requirement (cm).

### Field water balance

Efficient management of water and soil is required to check and control the field water balance, which is vital for such studies. Analysis of water balance was done on the basis of soil moisture recorded in the field and water applied for each plot considering other parameters such as rainfall (effective) and losses (evaporation and runoff). Water storage changes occurred in the field as well as in the plants. Equation of water balance was used as follows:

**Plate 26.4** Threshing
operation

$$P + I - R - D - ET = \Delta S \qquad (26.4)$$

In which, $P$ = Precipitation (mm), $I$ = Irrigation (mm), $R$ = Runoff from the field (mm), $D$ = Downward drainage out of the root zone (mm), $E$ = Evaporation from the field (mm), $T$ = Transpiration by the crop canopy (mm) and $\Delta S$ = Change in the storage (mm).

All the recommended package and practices of *ragi* cultivation were adopted in this experiment. Yield attributing characteristics were noted time to time for analysis. Crop takes 120 days to mature after sowing. Harvesting was done separately for both net plot as well as border plot. After harvesting of crop threshing operation was done by paddle operated millet thresher (Plate 26.4). Replication-wise grain yield data for all the treatments were recorded and tabulated for advance analysis.

## 26.3   Results and Discussion

The research was carried out during the summer season of the years 2014, 2015 and 2016. The water requirement, WUE, grain yield, benefit ratio, and other yield attributing parameters of summer finger millet (*ragi*) were recorded for different treatments during the study years, respectively.

The plant heights are compared with the total amount of water used in irrigation under different treatments are shown in bar and line diagrams for the years 2014, 2015, and 2016 (Fig. 26.1). It shows that the treatment $M_1T_1$ (Application of water at 30% SMDL) achieves average higher plant height compared with other treatments with water consumption of 38.51 cm which is higher than treatments $M_1T_2$, $M_1T_3$, and $M_1T_4$. Treatment $M_1T_5$ obtains lesser plant height compared with other treatments under soil mulch condition and used more water (41.98 cm). Based on the above result, it can be revealed that the water deficit affects plant height and

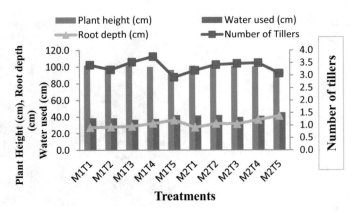

**Fig. 26.1** Interaction effect of plant height, water used number of tillers and root depth under different SMDL with combination of soil mulch and without mulch of finger millet during 2014–16

vegetative growth and these are directly related to production. Interaction between average plant height, number of tillers per plant during 2014–2016 also shows that (Table 26.1 and Fig. 26.1) the treatment $M_1T_4$ (irrigation at 60% SMDL with soil mulching) perform better as compared with other treatments and irrigation water use was also be comparable minimum to other treatments. Chandrasekharappa's (1979) reported that the stress at tillering or flowering reduced the leaf area and number of productive tillers, and delayed flowering.

The statistical analysis of average grain yield of summer finger millet for the period of 2014–2016 shows that there were significant differences between soil mulch and without mulch with combination of different soil moisture depletion levels (Table 26.2). The highest yield was recorded under $M_1T_4$ (irrigation at 60% SMDL) 23.48 q ha$^{-1}$ followed by $M_1T_3$ (irrigation at 50% SMDL) 22.97 q ha$^{-1}$.

**Table 26.1** Yield attributing parameters of summer finger millet (2014–2016)

| Treatments | Plant Height (cm) | Number of tillers | Root depth (cm) | Water used (cm) |
|---|---|---|---|---|
| M1T1 | 100.8 | 4.0 | 26.0 | 38.5 |
| M1T2 | 100.6 | 3.3 | 25.0 | 38.4 |
| M1T3 | 100.8 | 4.3 | 22.7 | 36.6 |
| M1T4 | 099.6 | 4.3 | 28.0 | 37.5 |
| M1T5 | 094.5 | 2.7 | 27.7 | 42.0 |
| M2T1 | 100.4 | 3.7 | 25.3 | 41.6 |
| M2T2 | 100.1 | 3.7 | 31.0 | 42.0 |
| M2T3 | 099.4 | 4.0 | 29.0 | 39.6 |
| M2T4 | 099.2 | 3.3 | 37.0 | 41.1 |
| M2T5 | 090.9 | 3.0 | 36.7 | 45.2 |

**Table 26.2** Two-way table of grain yield (q ha$^{-1}$) for the year 2014–16

| Treatment SMDL | Soil mulch M1 | Without mulch M2 | Mean |
|---|---|---|---|
| 30% | 21.29 | 19.24 | 20.27 |
| 40% | 21.27 | 19.13 | 20.20 |
| 50% | 22.97 | 20.51 | 21.74 |
| 60% | 23.48 | 20.35 | 21.91 |
| 70% | 17.40 | 16.15 | 16.78 |
| Mean | 21.28 | 19.08 | |
| Factors | | SE(m) | C.D. |
| Factor A (Moisture conservation) | | 0.31 | 2.01 |
| Factor B (SMDL) | | 0.50 | 1.51 |
| Factors B at same level of A | | 0.69 | N.S. |
| Factors A at same level of B | | 0.70 | N.S. |

Comparable results have also been the declaration by Rao (1977) in which optimum soil moisture regime and consumptive use of finger millet were determined, the study was undertaken on sandy loam soil during *rabi* season.

The result indicates that the interaction between different treatments shows that the maximum WUE (6.21 kg ha$^{-1}$ mm$^{-1}$) was achieved in $M_1T_4$ (irrigation at 60% SMDL) followed by $M_1T_3$ (irrigation 50% SMDL) 6.10 kg ha$^{-1}$ mm$^{-1}$ whereas the lowest WUE (4.35 kg ha$^{-1}$ mm$^{-1}$) was found in case of $M_1T_5$ (irrigation at 70% SMDL) with soil mulch condition (Table 26.3). In case of no mulch condition, the maximum WUE was found in $M_2T_3$ (irrigation at 50% SMDL) 4.95 kg ha$^{-1}$ mm$^{-1}$ which was followed by 4.89 kg ha$^{-1}$ mm$^{-1}$ in case of $M_2T_1$ (irrigation at 30% SMDL) (Table 26.3 and Fig. 26.2).

Number of irrigation at 60% SMDL with soil mulch was found to be 8 for summer finger millet under midland situation of Chhattisgarh plains. whereas, number of water application 50% SMDL was found to be 10 for finger millet during summer under midland situation of Chhattisgarh plains (Table 26.3).

**Table 26.3** Interaction effect of moisture conservation and soil moisture depletion levels on water use efficiency of finger millet during summer season (2014–2016)

| Treatments Soil Moisture Depletion Level (SMDL) | Water use efficiency (kg ha$^{-1}$ mm$^{-1}$) | | Number of irrigation | |
|---|---|---|---|---|
| | Soil mulch | Without mulch | Soil mulch | Without mulch |
| 30% SMDL | 5.82 | 4.89 | 14 | 15 |
| 40% SMDL | 5.51 | 4.50 | 11 | 13 |
| 50% SMDL | 6.10 | 4.95 | 10 | 11 |
| 60% SMDL | 6.21 | 4.62 | 8 | 9 |
| 70% SMDL | 4.35 | 3.36 | 7 | 8 |

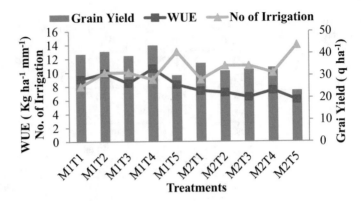

**Fig. 26.2** Interaction effect of grain yield, WUE and no of irrigation under different SMDL with combination of soil mulch and without mulch of finger millet during 2014–16

**Table 26.4** Technoeconomic comparison of summer finger millet and summer rice

| Particulars | Summer rice | Summer finger millet |
|---|---|---|
| Cultivation cost (Rs. ha$^{-1}$) | 48.112 | 35.200 |
| Production (q ha$^{-1}$) | | |
| Grain yield (q ha$^{-1}$) | 42.34 | 26.72 |
| Straw yield (q ha$^{-1}$) | 67.15 | 66.31 |
| Minimum support price (Rs q$^{-1}$) | 1.360 | 1.650 |
| Gross income (Rs. ha$^{-1}$) | 57.582 | 44.088 |
| Net income (Rs. ha$^{-1}$) | 7.970 | 8786 |
| Water requirement (cm) | 113 | 37.5 |
| Benefit cost ratio | 1.19 | 1.25 |

Summer rice experiment also conducted in 2016 for checking the cost-effective viability of summer finger millet (*ragi*) as compared with summer rice. Based on economic analysis, finger millet performs better with less water consumption and gave more benefit as compared with summer rice (Table 26.4).

## 26.4 Conclusions

On the basis of water management study of summer finger millet (*ragi*), it could be concluded that application of 375 mm of irrigation water appropriate for higher production of *ragi*. Irrigation at 60% soil moisture depletion level (eight numbers of irrigations) with soil mulch condition can be recommended for summer *ragi* under midland situation of Chhattisgarh plains. Irrigation at 50% soil moisture depletion level (nine number of irrigations) can also be recommended for *ragi* if water is not the limiting factor during summer under midland situation of Chhattisgarh Plains.

On the basis of water saving, point of view in treatment T4 (60% soil moisture depletion level) is better than all the treatments with less water requirement which is found to be 375 mm and water use efficiency is 6.21 kg ha$^{-1}$ mm$^{-1}$, when this application applies in large area it could lead toward highest productivity of water. Based on pooled data of 3 successive years, it can be concluded that the water requirement of summer *ragi* is 37.5 cm, which can be accomplished by eight numbers of irrigation. It can be concluded that 75.5 cm of water per hectare can be saved if summer rice is replaced by finger millet. The water saving can increase the cultivated area more than three times under summer finger millet as compared with summer rice.

# References

Acharya MS, Gupta AP, Singh J (1989) Irrigation scheduling using soil moisture depletion in wheat. Proc. of National Seminar on irrigation scheduling, pp 2–4, 1–7 March 1989

Anonymous (2014) Agricultural statistics. Department of Agriculture, Government of Chhattisgarh, Raipur

Bhatnagar SS (1952) The wealth of India. Counc Sci Ind Res, New Delhi III:160–166

Chandrasekharappa T (1979) Effect of different duration of soil moisture stress at different growth stages on growth and yield of finger millet (Eleusine coracana) in Visvesvaraya Canal Tract, Mandya. Thesis, University of Agricultural Sciences, Bangalore, India. (1979) Department of Agronomy, University of Agriculture and Science Bangalore Karnataka, India. From Mysore J Agric Sci 14:460

English M, Raja SN (1996) Perspectives on deficit irrigation. Agric Water Mgt 32:1–14

FAO (2002) Deficit irrigation practices. Water Report No. 22. Rome

Hess A (1996) A micro-computer scheduling program for supplementary irrigation. In: Irrigation scheduling: from theory to practice. Proc. ICID/FAO Workshop. Rome

Lemordant D (1967) EthnobotaniqueEthiopienne. A 1 institute Pasture d' Ethiopie

Mao Z (1996) Environmental impact of water-saving irrigation for rice. In: Irrigation scheduling: from theory to practice. Proc. ICID/FAO Workshop. Rome

Michael AM (1978) Irrigation theory and practices. second edition, pp 525–526, Vikash Publication, New Delhi

Rao BR (1977) Studies on the water requirement of Kalyani finger millet (Eleusinecoracana (L.) Gaertn.). S. V. Agricultural College, Tirupati (A. P.). Plate 8: Harvesting of finger millet crop

# Chapter 27
# Challenges of Water Supply Management in Harbour City of Freetown in Western Sierra Leone

**Augustine Amara and Mitthan Lal Kansal**

## 27.1 Introduction

Water is the basic need for the sustainable development of any urban area. Safe and adequate water supply is needed for various institutions, commercial, industries, agriculture, and domestic users. Managing water supply is challenging due to growing and conflicting demand for the purposes of domestic, agricultural, industrial, institutional, and commercial (He et al. 2009; Batten 2016). Further, management of such water supply systems is challenging especially when the urban areas are expanding due to increasing population and there are adverse impacts of climate change (UNICEF/GWP 2015; Howard et al. 2016; Koop and van Leeuwen 2017). Here, managing water supply means the supply of a safe and adequate quantity of water to the end-users in a cost-effective manner with low/no environmental impact. It is desired to inject adequate finances not only to manage urban water supply but also to improve the public health and related costs.

African cities like Freetown, water supply barely meets the domestic requirements. The presence of industrial and commercial centres put more pressure on the available sources in meeting the demand. As the needs for water keep rising, the infrastructures keep deterioration across Africa, particularly in Sierra Leone. In most African cities, structures meant to supply water are in very bad shape and form due to inadequate maintenance and poverty. Poverty in African has limited the delivering of adequate water services to the citizens. The economic, capacity development, water scarcity, institutional performance, financing, policy, political will and strategic innovation,

A. Amara
Ministry of Water Resources, Freetown, Sierra Leone
e-mail: ingamara20131@gmail.com

M. L. Kansal (✉)
Water Resources Development & Management, IIT Roorkee, Roorkee, India
e-mail: mlk@wr.iitr.ac.in

A. Pandey et al. (eds.), *Hydrological Extremes*, Water Science and Technology Library 97, https://doi.org/10.1007/978-3-030-59148-9_27

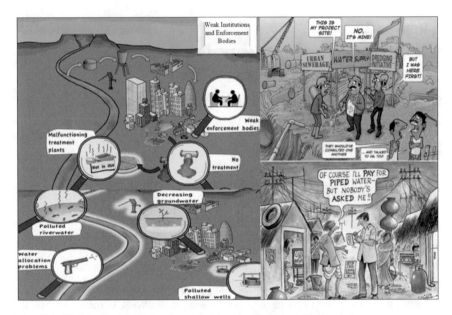

**Fig. 27.1** Weak institutions and enforcement (www.wsp.org/userfiles/image)

stand as a barrier to managing water supply sources in most African cities. African governments are not willing to invest in the provision of water services. There are issues of collaboration and coordination among water authorities and the problem of who works where and when is always an issue as depicted in Fig. 27.1.

The policies and regulatory laws are not harmonised and improvement in water supply services is yet to be achieved. Environmental and trans-boundary water management strategy is not unified and inadequate to cater for better management strategy in Africa. Accessing water in West Africa still remains a challenge and many people go without water for days due to poor management strategies. Poor quality water services in West Africa are directly reflected in the health status of its people. Poor quality water has been the leading cause of death for under-five children in West Africa.

Providing adequate water across Africa could ensure good health and ability to fight various diseases; reduces deaths among children under age five; reduced risk of sexual assault. Water accessibility in Africa can help to improve girl-child school education and hence the literacy rate. Saving precious time in collecting water helps improve economic growth and encourage foreign investment.

Keeping the importance of water supply management in mind, the present study is carried out with the following objectives: (1) assess the water supply and the demand for current and future users, (2) assess the sustainability of existing water supply, (3) assess the deficiency in managing water supply, and (4) suggest the measures to improve water supply management in Freetown.

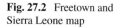

**Fig. 27.2** Freetown and Sierra Leone map

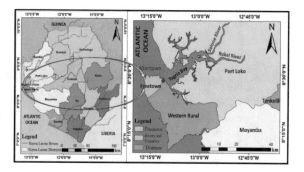

## 27.2  Description of the Case Study

Freetown, located in the west part of Sierra Leone, serves as the capital city. It has developed around a historically strategic deep natural harbour in Africa and partially enclosed by the Atlantic Ocean. It is the major urban, economic, cultural, educational, and political centre. It served as the river's estuary in Sierra Leone with latitude 8° 29′ 2.39″ and longitude 13° 14′ 2.40″ seen in Fig. 27.2.

Freetown has extended in its population after the 11 years of civil unrest (1991–2002). It has grown beyond its limits and developments, not in harmony with the city plan. The population has moved to settle within water catchment areas. In this regards, water sources are exposed to drying up making it difficult for water supply and demand to match. This has built so much pressure on the remaining freshwater sources like Guma Lake, Charlotte, Kongo, and Takayama seen in Plate 27.1. The city has extensions beyond its limit and has evolved into congestion without much expansion in water and sanitation infrastructures.

The climatic condition of the Freetown is tropical with two seasons: wet/raining season from May to October, and a dry season from November to April. The average annual rainfall varies between 2,500 mm and 5,000 mm. The average monthly temperature varies from 24.6 °C to 27.8 °C with a maximum temperature of 32 °C in the month of March.

The average annual relative humidity of Freetown is 80.8% and the average monthly relative humidity ranges from 72% in February to 89% in September as seen in Fig. 27.3. Administratively, Freetown is divided into eight zones (Central I and II, East I, II and III, and West I, II and III) with a current population of approximately 1.1 million.

## 27.3  Present Water Supply System in Freetown

Guma Valley Water Company, supply water to Freetown; covering Allen Town in the East, around the peninsula to Hamilton in the West. It is, however, a wholly autonomous operated water body receiving no financial support from the government,

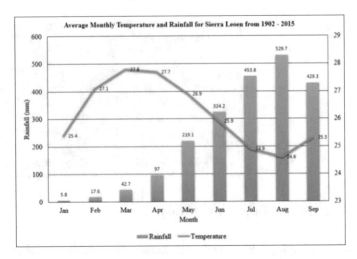

**Fig. 27.3** Average monthly temperature and rainfall for Sierra Leone from 1901 to 2015 (https://sdwebx.worldbank.org)

whether Ministry of Finance or other Ministries. GVWC has to generate its own finance through revenue collections from the sale of water (The Government of Republic of Sierra Leone and Ministry of Energy 2010).

The city gets its water supply from the following sources: (1) Guma Lake; (2) Kongo, Takayama and Sugar Loaf; (3) Charlotte and Bathurst seen in Fig. 27.4.

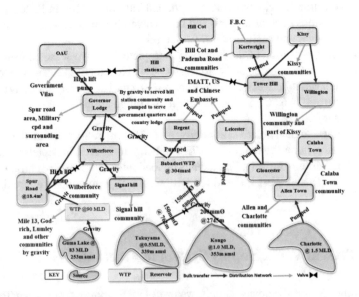

**Fig. 27.4** Schematic diagram of water sources and distribution plan

The Guma Lake located 20 km away from Freetown is a major input source and delivers 83 MLD. The reliable output of Guma Lake is 80 MLD, Babadori reservoir (sources of Kongo, Takayama and Sugar Loaf dry-up mostly in the drying season) is 1.5 MLD and Charlotte and Bathurst sources are 1.5 MLD, respectively. Most of the places situated on a high ground are supplied from the Babadori reservoir (Wilberforce, Hill Station, Gloucester, Leicester Peak and the Fourah Bay College). The Guma Lake treated water is connected with the Babadori treated water at Governor's Lodge. The Charlotte and Bathurst sources supply Allen Town reservoirs and by extension to the Calaba Town communities.

The distribution system servicing water has cross-connections between the distribution network and the bulk transfer system. The pressure will fail to rise in the Bulk Transfer system thereby reducing the pressure and some areas going without a water supply. One main pipe, 800 mm diameter run 200 m from the Guma treatment plant (at 208 m above mean sea level-ASL) works and join two transmission mains (700 and 550 mm) running 16 km comprising of steel and ductile pipes seen in Fig. 27.5.

The pipes carry treated water from the Guma treatment works under gravity to the Spur Road reservoir. From the Spur Road reservoir, the treated water is distributed to the water users. The 550 mm main supply water to communities between the Spur Road reservoir and the Guma treatment works seen in Fig. 27.6. In most of the areas, people are faced with the challenge of low pressure at the certain time of the day why some go for months without water supply seen in Fig. 27.7.

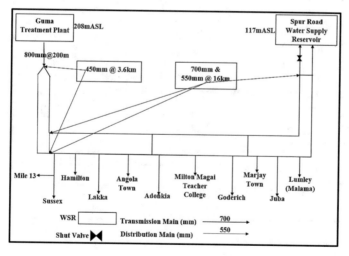

**Fig. 27.5** Supply network from treatment works to Spur Road of Freetown

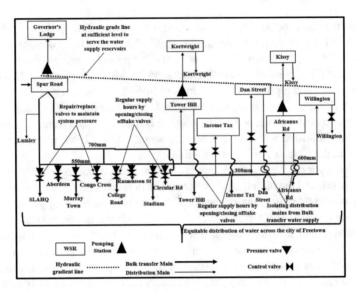

**Fig. 27.6** Hydraulic grade line from Spur Road to Far East of Freetown

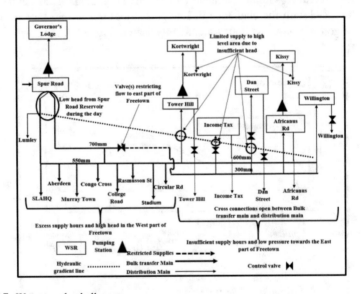

**Fig. 27.7** Water supply challenges

## 27.4 Assessment of Water Supply Scheme in Freetown

GVWC in providing water services faced a lot of challenges. The system was established in the 1960s and since then no tangible improvement on expansion and major repairs. The water demand and supply of the existing water supply system including the population has been estimated for 2008, 2010, 2012, 2014, and 2017 seen in Table 27.1.

The current water supply system including, the sources are under tremendous pressure to meet the desired output. The expected water demand and supply for the current water supply from 2008 to 2017 is seen in Fig. 27.8.

Guma supply hardly caters for the water supply requirement of the city as demand keeps rising due to urbanisation. Water supply system sustainability indicates that meeting the water supply needs of all water users considering better provisions for the future generation. Sustainable water services account for the Social, Economic and Environmental aspect. In this study, the assessment of water supply sustainability in Freetown is carried out using the methodology shown in Fig. 27.9.

The entire process includes the development of sustainability index tool using indicator questions, data collection, and weights generation using SPSS tool. The indicator questions are developed considering institutional, management, financial,

**Table 27.1**  Estimated population, supply and demand

Existing water supply in Freetown

|  | 2008 | 2010 | 2012 | 2014 | 2017 |
|---|---|---|---|---|---|
| Population forecast | 884220 | 945770 | 1011610 | 1082030 | 1196950 |
| Demand (MLD) | 88 | 95 | 101 | 108 | 120 |
| Supply (MLD) | 72 | 63 | 74 | 80 | 86 |
| Consumption | 81 | 67 | 73 | 74 | 72 |

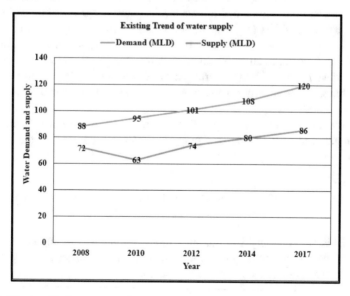

**Fig. 27.8**  The trend in the water supply

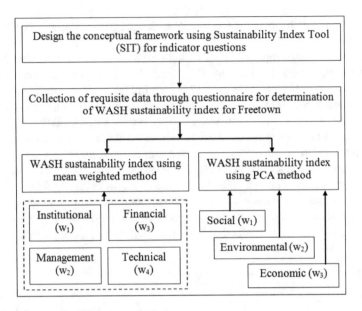

**Fig. 27.9** Water sustainability assessment

and technical aspects. The tool was developed globally to measure the progress in sustainable services by individual countries in water supply interventions.

There are a series of indicator questions developed to determine the progress at the institutional level, managerial, financial, and technical levels. The indicator questions targeted stakeholders at institutional levels, relevant legislator bodies, decision-makers, service providers (e.g. water committee, school or utility) and community members. Indicator questions for each intervention in 2017 relating to institutional, management, financial, and technical as shown in Fig. 27.10 below. The questionnaire was developed on 'Google form platform' with set of 32 indicator questions and distributed electronically, responses were generated through the same electronics means.

The answers from the indicator questions were used to determine indicator scores and aggregated to give the water supply sustainability index using the following equation:

$$\text{WSI} = \sum_{n=1}^{n} (W_i S_i) \tag{27.1}$$

where WSI is Water supply Sustainability Index, $W_i$ is the weighing factor equal to the ratio of the variance of each factor to total cumulative variance coefficients in the equation, and $Si$ is scored value of each indicator.

The overall sustainability index $= W_1 + W_2 + W_3 + W_4$

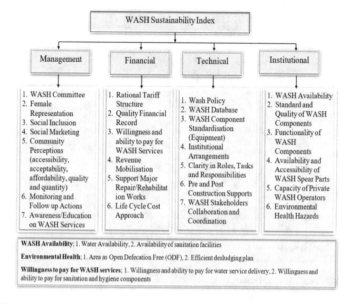

**Fig. 27.10**  Indicator questions

$$= (0.249 * 0.446) + (0.242 * 0.570) + (0.241 * 0.411)$$
$$+ (0.269 * 0.443) = 0.467 = 46.7\% \tag{27.2}$$

Also SI using PCA $= W_1 + W_2 + W_3$
$$= (0.40 * 0.552) + (0.540 * 0.244)$$
$$+ (0.560 * 0.204) = 0.467 = 46.7\% \tag{27.3}$$

Graphically, the components of WSI are shown in Figs. 27.11 and 27.12.

The Guma water is yet to meet the demand as supply management still remains to be a problem in most communities. Many communities in Freetown are yet to have access to water. When the water supply was increased by 95%, supply only satisfies demand in 2020 as seen in Fig. 27.13.

## 27.5    Challenges of Water Supply in Freetown

The existing water supply conditions in Freetown are the result of poor institutional framework, inadequate management style, limited financial investment poor quality design and construction (Technical), making it unsustainable. Freetown has grown rapidly and water supply infrastructures extending to some of the areas are uncoordinated, making it difficult to extend water supply services in the city. Available water

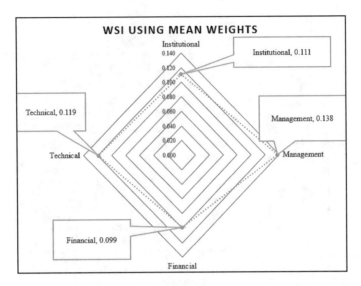

**Fig. 27.11** Water sustainability index

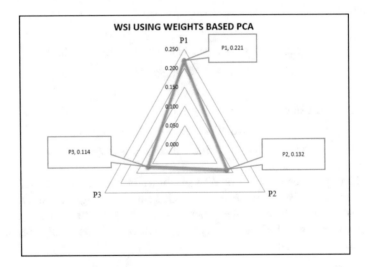

**Fig. 27.12** Water sustainability index using PCA

sources are under great pressure in trying to coup with the increasing population and variation in climatic conditions. Overall, one can categorise the challenges of water supply in Freetown as technical and managerial issues. These challenges are summarised as follows:

(A)   The various technical issues are:

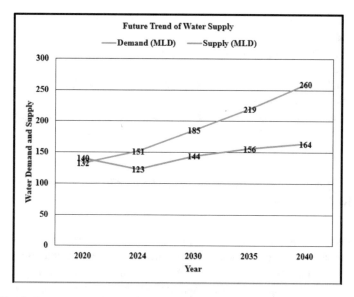

**Fig. 27.13**   The Future trend of water supply to Freetown

**Plate 27.1**   Water supply
sources and treatment works

- Inadequate water conservation strategies (unpermitted water withdrawals and diversions) and catchment protection measures.
- Weak leak detection system.
- Poor water quality monitoring and modelling.
- Poor infrastructures and state of maintenance for water supply structures (rehabilitation and reconstruction of water supply schemes).
- Inability to quantify Current and Future Risks associated with water supply management like water demand, available supply, increasing population and climate change impacts.

(B)   The various managerial issues are:

- The lack of adequate funding and limited investment in water supply management.
- Weak leadership and low capacity in water supply management.
- Weak institutions and water supply management enforcement bodies.
- Lack transparency and low payment mobilisation skills.
- Weak participation of community members and low awareness.
- Weak financial planning and management strategies in water supply management.

## 27.6   Suggestions to Improve Water Supply Management in Freetown

Access to water supply in Freetown poses crucial issues for the people. Inadequate water supplies of poor quality affect human health. Poor water administrative, managerial, institutions, capacity and financing remain a challenge. Appropriate and timely investment in managing the water supply is expected to reduce the spread of diseases and cost. The prevention of water-related illness, death and associated community's health-care costs can help burst a nation's economy and strengthen the human resource capacity. The GVWC can consider the following suggestions and if taken into consideration could be expected to enhance the water supply in Freetown:

(1) Reclaiming stormwater and wastewater for reuse and recycle, more so stormwater running into the ocean. Using appropriate technology to reclaim this water and providing appropriate treatment can to large extend provide water to water users along the coastal line.

(2) Instituting regulations to protect the catchment areas by creating buffers and restrict the cutting down of vegetation around catchment areas, watershed, wetland and the like.

(3) Regulate agricultural, industrial, and human wastes from entering the water body. Wastewater or industrial wastes to be treated before discharged into water bodies.

(4) Institutionalised water quality surveillance and monitoring and reduce leakages in the distribution network system in Freetown.

(5) Develop policies that cater to water supply management strategies, integrated water resources management including environmental protection.

(6) Develop regulations to monitor the amount of water abstracted and diverted from water bodies.

(7) Employ measures to reduce the sediment load in water reservoirs and increased awareness of water supply management.

## 27.7 Conclusions

In Africa, the population is rising rapidly and climate change impacts are visible. Water supply management challenges in the harbour city of Freetown are already massive besides the expectation of further degradation in water supply infrastructure in future. Equitable and efficient water supply management for Freetown is becoming an increasingly complex and complicated task for water managers and governments. Superimposing water scarcities and pollution events in harbour cities on issues dealing with urbanisation, limited investment on the construction and maintenance of water infrastructures, inadequate managerial capacities and weak leadership, poor governance or water governance, inappropriate institutional frameworks and inadequate legal and regulatory mechanisms, high national debts, poor water quality, ineffective resources allocation, inefficient methods of water conservation and treatment techniques, make water supply management in the harbour city of Freetown challenging in the future.

The challenges of water supply management in meeting multiple sustainability criteria like social, economic and environmental for harbour city of Freetown are crucial. Water supply to the city is unable to meet the requirement of the rising population in Freetown. In order for the water supply to meet the required demand, water conservation methods and better treatment plan need to be deployed. Water users are supposed to invest in rainwater harvesting techniques. Making this mandatory and compulsory with good governance can increase water storage in the harbour city of Freetown. The government and utility managers can devise a legal and regulatory mechanism to protect the water environment and biodiversity.

## References

Batten J (2016) Sustainable cities water index. Which cities are the best placed to harness water for future success? https://www.arcadis.com/media/4/6/2/%7B462EFA0A-4278-49DF-9943-C067182CA682%7DArcadisSustainableCitiesWaterIndex_003.pdf

He B, Wang Y, Takase K, Mouri G, Razafindrabe BH (2009) Estimating land use impacts on regional scale urban water balance and groundwater recharge. Water Resour Manage 23(9):1863–1873

Howard G, Calow R, Macdonald A, Bartram J (2016) Climate change and water and sanitation: likely impacts and emerging trends for action. Annu Rev Environ Resour 41:253–276

Koop SH, van Leeuwen CJ (2017) The challenges of water, waste and climate change in cities. Environ Dev Sustain 19(2):385–418

The Government of Republic of Sierra Leone and Ministry of Energy (2010) Ministry of Energy and Water Resources

UNICEF/GWP (2015) WASH Climate Resilient Development 27

# Chapter 28
# Comparison of Methods for Evapotranspiration Computation in the Tana Basin, Ethiopia

Hailu Birara, S. K. Mishra, and R. P. Pandey

## 28.1 Introduction

In many parts of the world, where rainfall amount necessary to meet the crop water requirement is limited, irrigation is a significant component of agricultural planning. In irrigation agriculture, it is very important to determine when and how much water applies to meet the water demand of the crops. The water demand is determined through proper estimation of evapotranspiration procedures. Evapotranspiration is described as the sum of evaporation and plant transpiration from the land and ocean surface to the atmosphere.

Evaporation accounts for the movement of water from the land surface (such as soil, canopy interception, and water bodies) to the air. It is an essential parameter in the hydrological cycle. Many researchers have projected that water resources will be influenced due to ET0 as the general effect of global climate change. High ET0 due to temperature increase will affect the watershed hydrological system and water resources of the globe (Shahid 2010). Thus, reliable and accurate estimation of ET0 is essential for the long-term water resources and irrigation management (Pour et al. 2016).

ET crop estimation for a given crop can be estimated as grass reference crop evapotranspiration (ET0) and crop coefficient, and several methods are available for the estimation of ET0 which depends on the availability of climate data. The method ranges from the most complex equation, which requires detailed climate data (Penman–Monteith [PM], Allen et al. 1989) to simple equations, which require

H. Birara (✉) · S. K. Mishra
Department of Water Resources Development and Management, Indian Institute of Technology Roorkee, Roorkee 247667, Uttarakhand, India
e-mail: hailubirara@gmail.com

R. P. Pandey
National Institute of Hydrology, Roorkee 247667, Uttarakhand, India

A. Pandey et al. (eds.), *Hydrological Extremes*, Water Science and Technology Library 97, https://doi.org/10.1007/978-3-030-59148-9_28

less meteorological data (Blaney–Criddle, Hargreaves). These methods can be also grouped into combination methods (PM, FAO-24 Penman, 1982 Kimberly-Penman, 1972 Kimberly-Penman, FAO-24 Corrected Penman), radiation-based methods (Turc, Jensen–Haise, Priestley–Taylor, and FAO-24 Radiation), temperature-based methods (Thornthwaite, Blaney–Criddle, and Hargreaves), mass transfer-based methods (WMO 1966), and pan coefficient-based methods.

The major factors in the estimation of ET0 are reliability and accuracy (Burnash 1995). A number of methods have been developed for given purposes and specific climate conditions. They may provide poor estimates of ET for other climatic conditions due to their different assumptions and input data requirements. However, various studies conducted in different parts of the world have used some specified models (such as PM, Priestley–Taylor, Turc, Thornthwaite, and Hargreaves) as the standard approach to estimate ET0 (Rácz et al. 2013; Wang et al. 2012). Nowadays, the PM method proposed by Allen et al. (1998) has become more reliable and provides accurate estimates of ET0 over a wide range of climatic regions (De Bruin et al. 2010; Rácz et al. 2013). The PM method requires many meteorological parameters such as temperature, humidity, wind speed, sunshine, and solar radiation, among others. However, measurement of all these parameters is not often available, particularly in developing countries like Ethiopia (Hubbard 1994; Pielke et al. 2007).

This study presents a performance comparison of seven widely used ET0 estimation methods, namely, Hargreaves, Thornthwaite, Blaney–Criddle, Priestley–Taylor, Turc, Abtew, and, the new method recently proposed by Temesgen and Melesse (2013), Enku, against the PM method in the study area as it is one of the most reliable methods to compute ET0 (Hubbard 1994; Irmak et al. 2002; Pielke et al. 2007; Rahimikhoob 2009). Table 28.1 shows the different methods with the required meteorological variables.

**Table 28.1** Data requirement for the methods used

| Methods | Temperature | Relative humidity | Solar radiation | Wind speed | Atmospheric pressure |
|---|---|---|---|---|---|
| Penman–Monteith | ✓ | ✓ | ✓ | ✓ | ✓ |
| Priestley–Taylor | ✓ | ○ | ✓ | ○ | ○ |
| Turc Method | ✓ | ✓ | ✓ | ○ | ○ |
| Hargreaves Method | ✓ | ○ | ○ | ○ | ○ |
| Makkink Method | ✓ | ○ | ✓ | ○ | ○ |
| Blaney–Criddle Method | ✓ | ✓ | ✓ | ✓ | ✓ |
| Abtew Method | ✓ | ○ | ✓ | ○ | ○ |
| EnkuTemperature Method | ✓ | ○ | ○ | ○ | ○ |

## 28.2   Study Area

The Amhara National Regional State (ANRS) is located in Ethiopia's north-western and north-central parts (latitude 8°–13° 45′ N and 36° and 40° 30′ E). According to central Statistical Agency of Ethiopia (CSA 2008), the region has a total area of around 170,000 km² and categorized into 12 administrative zones and 105 woredas with different characteristics of the physical landscape, i.e., valleys, rugged mountains, and gorges with elevation ranging from 700 m a.s.l to 4,600 m a.s.l in the eastern and the northwest part, respectively. The lake Tana Basin is the largest sub-basin in the Amhara region, covering an area of 15,096 km², including the lake area (Fig. 28.1). The average annual precipitation and evapotranspiration of the basin are approximately 1,280 mm and 1,036 mm, respectively (Allam et al. 2016).

The annual climate classified into two major seasons, viz. the rainy and the dry season. The rainy season is also divided into a minor and major rainy season, which lasts from March to May (Belg), and June to September (kiremt), respectively, and the dry season lasts from October to February (Bega). As a result of its diverse nature of the region with altitudes ranging from 1,327 to 4,009 m.a.sl, the basin contributes a national importance because of its high potentials for irrigation development, high-value crops, hydroelectric power development, livestock production, and ecotourism (CSA 2008).

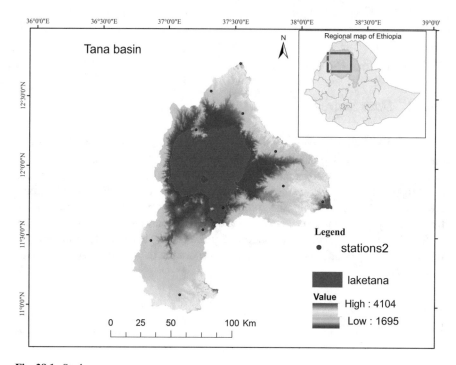

**Fig. 28.1** Study area map

Lake Tana, among one of the Blue Nile River's main source, is Ethiopia's largest lake and the third-largest in the Nile Basin. It is about 84 km long and 66 km wide in the north-western highlands of the country. The lake is one of a natural freshwater, at an elevation of 1,800 m, covering an area of 3,000–3,600 km$^2$. Gumera, Ribb, Gilgel Abay, and Megech Rivers are among the main feeding of the lake Tana. These four rivers contribute to the annual water budget of the lake to more than 65% inflow (Setegn et al. 2008). The only surface outflow is the Blue Nile River, measured at the Bahirdar gauge station with an annual flow volume of 4BCM is sourced from lake Tana.

In the study area, land use is classified on the basis of Abay River master plan study conducted by BCEOM's (1998). Approximately 51.37% of the watershed area is covered by agriculture, 21.94% by agriculture, 0.15% by agrisilviculture, 0.03% by sylvopastoral, and 0.11% by urban use.

The main tributaries of the lake Tana are Gilge Abay, Gumera, Ribb, and Megech Rivers. The present study shows that these four rivers contribute to more than 45% inflow of the annual lake water budget. The only surface outflow is the Blue Nile (Abba) River with an annual flow volume of 4 billion cubic meters measured at the Bahir Dar gauge station.

Land use of the study area is classified based on Abay River master plan study conducted by BCEOM (1998), about 51.37% of the watershed area is covered by agriculture, 21.94% by agro-pastoral, 0.15% by silviculture, 0.03% by sylvopastoral, and 0.11% of urban.

## 28.3  Methodology

### 28.3.1  Data Source

For this particular study, daily weather data from 1980 to 2015 were obtained from the National Meteorological Service of Ethiopia. Meteorological variables including rainfall, maximum and minimum air temperature, relative humidity, wind speed, and bright sunshine hour has been collected. The climatic data used in this study consisted of daily maximum and minimum temperature (Tmax and Tmin), relative humidity (RH), wind speed (WS), and sunshine hours (N). Six stations average (referred in Table 28.2) climatic data for 35 years is used to calculate evapotranspiration. However, 10 (2004–2013) years pan evaporation data is used to estimate pan coefficient.

**Table 28.2** Detail of selected stations

| S. no. | Name of station | Lat. (N) | Long. (E) | Alt. (m a.s.l) | Duration |
|--------|-----------------|----------|-----------|----------------|----------|
| 1 | Bahir Dar | 11° 71′ | 37° 50′ | 1800 | 1980–2015 |
| 2 | Gondar | 12° 63′ | 37° 45′ | 2133 | 1980–2015 |
| 3 | Woreta | 11° 55′ | 37° 42′ | 1828 | 1980–2015 |
| 4 | DebreTabor | 11° 86′ | 38° 02′ | 2506 | 1980–2015 |
| 5 | Dangla | 11° 25′ | 36° 74′ | 2122 | 1980–2015 |
| 6 | Injibara | 11° 70′ | 38° 43′ | 2372 | 1980–2015 |

## 28.3.2 Estimation of Evapotranspiration and (ET0) and Pan Coefficient ($K_P$)

Evapotranspiration can be directly measured using lysimeters, but it can also be measured using empirical equations or simply be estimated by multiplying observed standard pan evaporation data by the pan coefficient (Grismer et al. 2002). The pan evapotranspiration (ET0) is obtained by the following formula:

$$ET_o = E_p * K_p \tag{28.1}$$

where ET0 = Reference evapotranspiration (mm); $E_p$ = Observed Pan evaporation data for class A pan (mm); $K_p$ = Pan coefficient.

Equations have been developed by Cuenca (1989), Orange (1998), Allen and Pruitt (1991), and Snyder (1992) to estimate $K_P$ for a class A pan with green vegetation on the surrounding condition. The calculated $K_p$ value obtained from the above equation was compared and correlated with the observed value (ET0/Ep as ET0 calculated by the FAO-56 PM equation) since the FAO-56 PM is confirmed as a standard and suitable method to estimate ET0 for different climates (Allen et al. 1998, 2005; Irmak et al. 2003; Temesgen et al. 2005; Zhao et al. 2005; Garcia et al. 2007; Gundekar et al. 2008). Hence, one of the empirical formulas which are more correlated with the observed is considered as accurate method to estimate $K_P$ in the study area.

The following four approaches have been considered to determine $K_p$ as well as ET0 from class A pan evaporation:

Orange (1998)

$$K_p = 0.5126 - 0.00321u + 0.002889RH + 0.031886\ln(F) \tag{28.2}$$

where U = Wind speed (km/day), RH = relative humidity (%), and F = upwind fetch distance around the pan.

Snyder (1992)

$$K_P = 0.482 - (0.0003768u_2) + (0.0245 * 0.0045 * RH) \tag{28.3}$$

where RH is mean daily relative humidity in %
   Cuenca (1989)

$$K_P = (0.475 - 92.4 * 10^{-4} * u_2) + (5.16 * 10^{-3} * RH)$$
$$+ (1.18 * 10^{-3} * F) - (1.16 * 10^{-3} RH^2) - (1.01 * 10^{-6} * F^2)$$
$$- (8 * 10^{-8} * RH^2 * U^2) - (0.1 * 10^{-7} * RH^2 * F) \tag{28.4}$$

Allen and Pruitt (1991)

$$K_P = 0.108 - 3.31 * 10^{-4} * U_2 + 4.22 * 10^{-2} * ln(F) + 10^{-1}$$

$$*ln(RH) - 6.31 * 10^{-4} * (F)^2 * ln(RH) \tag{28.5}$$

The class A pan evaporimeter is sited on a short green grass cover and the value of F is 70 m.


### 28.3.3  Description of ET0 Estimation Equations

**Penman–Monteith Equation** The Penman–Monteith method is a combination method developed by Penman (1948). It combines the energy balance with mass transfer method and proposes an equation to estimate ET0 on daily basis using climatic variables viz., temperature, sunshine hours, relative humidity, and wind speed. It is expressed as below

$$ETo = \frac{0.408\Delta(Rn - G) + Y\frac{900}{T+278}U_2(p_s - p_a)}{\Delta + \gamma(1 + 0.34u_2)} \tag{28.6}$$

where ETo is reference evapotranspiration (mm day$^{-1}$); $\Delta$ is the slope vapor curve (Kpa °C$^{-1}$); RN is the net radiation of the crop surface (MJm$^{-2}$ day$^{-1}$); $G$ is the soil heat flux density (MJm$^{-2}$ day$^{-1}$); $T$ is the air temperature (°C); U2 is the wind speed at 2 m height (m s$^{-1}$); $\rho_s$ is the saturation vapor (Kpa); $\rho_a$ is the actual vapor pressure (Kpa); and $\gamma$ is the psychometric constant (Kpa °C$^{-1}$).

**Priestley–Taylor Method** Priestley–Taylor method (Priestley and Taylor 1972) is a radiation-based method to estimate reference evapotranspiration (ET0). They establish that the potential evaporation is 1.26 times lesser than the actual evaporation and thus they replace the aerodynamic terms with constant (1.26). Therefore, the method needs only long-wave radiation and temperature for the assessment of ET0. The equation for calculating ET0 is given below

$$ETo = 1.26 * \frac{\Delta}{\Delta + \gamma}(Rn - G) * \frac{1}{\lambda} \tag{28.7}$$

where $\Delta$ is the slope vapor curve (Kpa $°C^{-1}$); $\gamma$ is the psychometric constant (Kpa $°C^{-1}$); RN is the net radiation of the crop surface (MJm$^{-2}$ day$^{-1}$); $G$ is the soil heat flux density (MJm$^{-2}$ day$^{-1}$); and $\lambda$ is the latent heat of vapor (MJ kg$^{-1}$).

**Turc Method** Turc method (Turc 1961) provides an easy equation for calculating ET0 by using only a few climatic variables (relative humidity, solar radiation, and mean temperature). The Turc method gives reliable estimates of ET0 under humid conditions (Jensen et al. 1990). The equation is given as follows:

$$ETo = 0.0133 \frac{Tm}{Tm} (Rs + 50) \tag{28.8}$$

when RH > 50%

$$ETo = 0.0133 \frac{Tm}{Tm + 15} (Rs + 50) \left(1 + \frac{50 - RH}{70}\right) \tag{28.9}$$

when RH < 50%

where Tm is mean temperature (°C); Rs is the solar radiation of the crop surface (MJm$^{-2}$ day$^{-1}$); and RH is the relative humidity (%).

**Hargreaves Method** Hargreaves is the temperature-based method proposed by Hargreaves and Samani in 1982. The equation is given as

$$ETo = 0.0023(T_{mean} + 17.8)(T_{max} - T_{min})^{0.5} Ra \tag{28.10}$$

where $T_{max}$, $T_{min}$, and $T_{mean}$ denote maximum, minimum, and mean temperatures (°C), respectively; and Ra is the extra-terrestrial radiation of the crop surface (MJm$^{-2}$ day$^{-1}$).

**Makkink Method**
$$ETo = 0.61 \left(\frac{\Delta}{\Delta + \gamma}\right) * \frac{Rs}{58.5} - 0.12 \tag{28.11}$$

**Blaney–Criddle Method** The Blaney–Criddle is the simple temperature-based method for the assessment of ET0. It is widely used method applied before Penman–Monteith method. This equation only considers changes in temperature for specific conditions for estimating ET0. The Blaney–Criddle equation is given below

$$ETo = p(0.46Tm + 8) \tag{28.12}$$

where $p$ = percentage of average daily annual daytime hours due to the latitude of the region.

**Abtew Method** Abtew (1996) used a simple model to estimate ET from solar radiation. This method requires only solar radiation, and is less subject to local variations

$$ETo = K \frac{Rs}{\lambda} \tag{28.13}$$

where $K$ is a coefficient (0.53).

**Enku Temperature Method** The method is developed by Temesgen and Melesse (2013) by replacing Abtewk and by a single constant $k*$. They used power form of maximum temperature ($T_{mx}$) to estimate ET0 and the method is hereafter denoted as $ET_{Tm}$

$$ET_{Tm} = \frac{(T_{mx})^n}{k*} \tag{28.14}$$

where $n = 2.5$ and they used maximum temperature dependent $k*$ of $48\ T_{mx} - 330$ for dry and wet conditions or season.

### 28.3.4 Model Evaluation Methods

To evaluate the performance of the ET0 estimation model and $K_P$ estimation equations, different statistical measures were used. Most of these methods are proposed to capture the degree of good agreement between observed and calculated (modeled) values. Some performance measures are described below (Jachner et al. 2007).

**Root Mean Square Error (RMSE)**

$$RMSE = \sqrt{\frac{\sum\limits_{i=1}^{n} (X_{observed} - X_{model})^2}{n}} \tag{28.15}$$

where
$X_{observe} = $ observed ET0 value
$X_{model} = $ calculated ET0 value

**Mean Square Error (MSE)**

$$MSE = \frac{\sum\limits_{i=1}^{n} (E_{observe} - E_{model})^2}{n} \tag{28.16}$$

**Coefficient of Determination ($R^2$)**

$$R^2 = 1 - \frac{\sum\limits_{i=1}^{n} (E_{\text{observe}} - E_{\text{model}})^2}{\sum\limits_{i=1}^{n} (E_{\text{observe}} - \overline{E_{\text{model}}})^2} \tag{28.17}$$

It is defined as the degree of collinearity between observed and calculated value data. The value of $R^2$ lies between 0 and 1

***Efficiency Factor (EF)***

$$\text{EF} = \frac{\sum\limits_{i=1}^{n} (O_i - \bar{O})^2 - \sum\limits_{i=1}^{n} (O_i - E_i)^2}{\sum\limits_{i=1}^{n} (O_i - \bar{O})^2} \tag{28.18}$$

where EF = efficiency factor; $O_i$ = observed value; $E_i$ = estimated value; $\bar{O}$ = mean observed value.

## 28.4  Results and Discussion

### 28.4.1  Meteorological Parameters

A summary of average daily, monthly, and seasonal climatic data from 1980 to 2015 is described as follows: Maximum and minimum temperature occurred in April ($T_{\text{max}}$ 33 °C) and July ($T_{\text{min}}$ 9 °C) on the study area, respectively. The summer (kiremt) season is wet compared with winter (Bega) and spring (Belg), and the relative humidity is higher during July (81%). However, the bright sunshine hour of the summer season is less (from 0 to 4.6). The highest pan evaporation also occurred in April (8.1 mm/day), which seems to be related to the higher temperature, lower humidity, and longest bright sunshine hours. The reverse is true for July, which gives the lowest pan evaporation (2.4 mm/day). To develop the pan coefficient for different seasons of the study area, pan evaporation data should first relate to the PM equation. Since the pan coefficient is very dependent on local conditions, $K_p$ should be determined by comparing the observed pan data with the PM ET0. Figure 28.2 shows a good relationship ($R^2 = 0.84$) between ET0 (FAO-56 PM) and standard class A pan evaporation (average from 2004 to 2013).

Such a relationship of comparing the observed pan data with PM indicates that with an appropriate pan coefficient value, the pan evaporation can be quite useful in estimating evaporation in the study area. The results also agreed with previous studies that showed a good relationship between pan evaporation and PM-based evapotranspiration values (Jensen et al. 1961; Pruitt 1966; Doorenbos and Pruit 1975).

## 28.4.2  Estimation of Pan Coefficient (K_P)

Pan evaporation data are important only if a suitable pan coefficient with local conditions is used to relate the pan data with different ET0 estimation methods. The mean monthly $K_p$ values between the observed (FAO-56 PM/Ep) and the estimated value using different methods (Eqs. 28.1–28.4) are shown in Fig. 28.3. The monthly Kp values varied between 0.73–0.84, 0.75–0.84, 0.74–0.81, 0.7–0.83, and 0.71–0.84 for the observed, Snyder, Orange, Cuenca, and Allen and Pruitt methods, respectively.

The observed (FAO-56 PM/Ep) $K_p$ value indicates the lowest value during January–February and June–August, which might be due to less sunshine hours that may decrease air temperature impacts.

The statistical test of the different equations to compute $K_P$ is presented in Table 28.3. The statistical criteria of $R^2$ indicate that, except Allen and Pruitt, all tested equations' coefficient of determination ($R^2$) was above 0.8 (Table 28.3). It is also clear from Table 28.3 that the Snyder method showed a good correlation with the

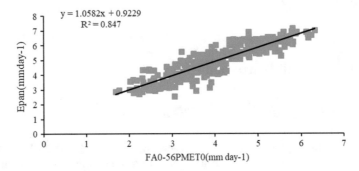

**Fig. 28.2**  Calculated ET0(FAO-56PM) versus measured Epan (average of 2004–2013)

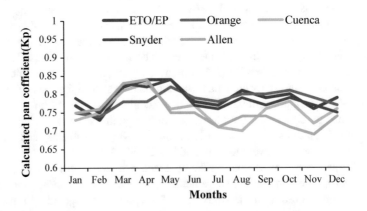

**Fig. 28.3**  Observed (ET0/EP) and calculated monthly pan coefficient($K_p$) of the study area

**Table 28.3**  Ranking of $K_p$ equations on the basis of statistical test with the observed value

| Rank | Equations | $R^2$ | RMSE | MAD |
|------|-----------|-------|------|-----|
| 1 | Snyder | 0.93 | 0.46 | 0.97 |
| 2 | Orange | 0.89 | 0.51 | 1.07 |
| 3 | Cuenca | 0.86 | 0.68 | 1.18 |
| 4 | Allen and Pruit | 0.74 | 0.84 | 1.12 |

observed value followed by Orange. The Snyder value of the coefficient of determination ($R^2$) showed the highest (0.93) and lowest RMSE (0.46). Thus, this analysis
suggests that Snyder can be used to compute reasonably accurate $K_p$ values as far as
monthly estimation is concerned, whereas the Allen and Pruitt method showed the
weakest ability to estimate $K_p$ in the study area.

Similar results were reported by Pradhan et al. (2013) on their studies of evaluation
of $K_p$ methods to compute FAO-56 crop evapotranspiration in the semi-arid region,
which indicates that Snyder provides more accurate estimations of regional-based
$K_p$ values.

To estimate the accurate and consistent pan evaporation on the study area, ET0
between the observed (FAO-56 PM/$E_p$) with the pan value was compared. The
comparison of monthly pan ET value between the observed and using different $K_p$
estimation equations for 10-year data is plotted in Fig. 28.4. The closest relationship
was observed between ET using the Snyder method and the observed, and Orange
showed nearly the same performance (Fig. 28.4a–d). Accordingly, the annual ET
value using Snyder, Orange, Cuenca, and Allen and Pruitt differed by 3.12%, 7.6%,

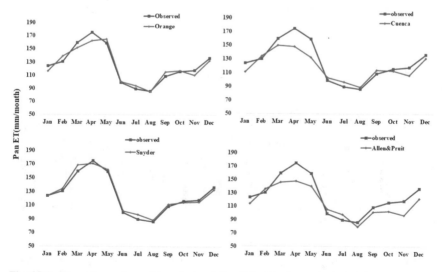

**Fig. 28.4**  Comparison of monthly calculated Pan ET with observed (FAO-56PM/Epan) using
different $K_p$ estimation methods

−9.2%, and −17.7% from the observed ET0, respectively (Fig. 28.4). Hence, Snyder provided by far a good correlation and is therefore recommended as a suitable method to compute $K_p$ in the study area.

### 28.4.3 Cross Comparison of ET0 Estimation Methods

The comparison of the results of ET0 estimated from various methods against PM is presented in Table 28.4 and Figs. 28.5 and 28.6. ET0 was estimated by all the above-described methods (Eqs. 28.6–28.14) and used for comparison with the PM results. The calculated ET0 with various methods showed that the average yearly estimated ET0 for the study area ranges from 1240 to 1860 mm/year (Fig. 28.6). However, the total calculated annual ET0 value obtained by PM was 1501.6 mm on the study

**Table 28.4** Daily Pearson's cross-correlation analysis between ET0 estimation methods

|      | PM   | Turc | Enku | Mkn  | Abt  | B&C  | Pt   | Hg   |
|------|------|------|------|------|------|------|------|------|
| PM   | –    | 0.98 | 0.95 | 0.91 | 0.87 | 0.88 | 0.81 | 0.64 |
| Turc | 0.98 | –    | 0.96 | 0.93 | 0.87 | 0.81 | 0.83 | 0.69 |
| Enku | 0.95 | 0.96 | –    | 0.96 | 0.92 | 0.74 | 0.88 | 0.68 |
| Mkn  | 0.91 | 0.93 | 0.96 | –    | 0.84 | 0.79 | 0.84 | 0.71 |
| Abt  | 0.87 | 0.87 | 0.92 | 0.84 | –    | 0.92 | 0.83 | 0.74 |
| B&C  | 0.88 | 0.81 | 0.74 | 0.79 | 0.92 | –    | 0.78 | 0.76 |
| Pt   | 0.81 | 0.83 | 0.88 | 0.84 | 0.83 | 0.78 | –    | 0.74 |
| Hg   | 0.64 | 0.69 | 0.68 | 0.71 | 0.74 | 0.76 | 0.74 | –    |

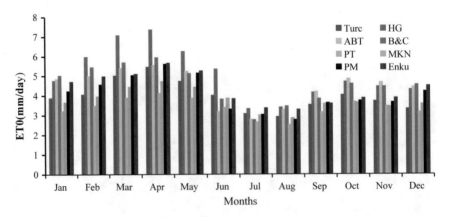

**Fig. 28.5** Monthly average reference evapotranspiration(ET0) of the study area. Abbreviations: *PM* = Penman–Monteith method; *HG* = Hargreaves; *Abt* = Abtew simple method; *BC* = Blaney–Criddle method; *PT* = Priestley–Taylor method; *Mkn* = Makkink method; *Tur* = Turc Method

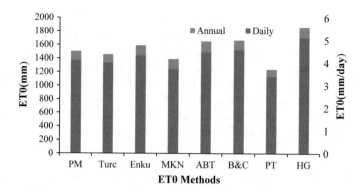

**Fig. 28.6** Mean monthly and annual reference evapotranspiration

area. The result also showed that the peak annual value of ET0 was found to be higher during March–May as the period is associated with the highest temperature, low relative humidity, and highest wind speeds. The value of ET0 on these months accounts for 41.2% of the total annual ET0 and maximum mean monthly ET0 that has been observed during March (169 mm), while the minimum values occurred on August (51 mm), which is almost a decrease of one-third from the maximum value.

Using the PM method as observed (standard), the performance evaluation and correlation of the tested methods that compute ET0 are presented in Tables 28.4 and 28.5 using statistical parameters. The correlation was calculated for each pair of models (Table 28.4).

The closest correlation was observed between PM and Turc followed by Enku and Makkink. Hargreaves showed the weakest correlation with PM and other tested methods, whereas Blaney–Criddle showed the weakest correlation (<0.8) with Enku, Priestley–Taylor, and Hargreaves methods.

Another statistical analysis was also performed. The mean deviation (%) coefficient of variation, efficiency factor, and root mean square error of the calculated mean daily ET0 of the various methods are presented in Table 28.5. All the methods

**Table 28.5** Statistical evaluation of daily average ET0 estimation methods against PM

| Rank | Methods | Mean deviation | * | $R^2$ | * | RMSE | * | EF | * | MSE | * |
|---|---|---|---|---|---|---|---|---|---|---|---|
| 1 | Turc | −4.2 | 1 | 0.96 | 1 | 0.87 | 1 | 0.88 | 1 | 0.53 | 2 |
| 2 | Enku | +5.7 | 2 | 0.89 | 2 | 0.90 | 3 | 0.86 | 3 | 0.51 | 1 |
| 3 | Makkink | −7.2 | 3 | 0.87 | 3 | 0.89 | 2 | 0.88 | 2 | 0.61 | 3 |
| 4 | Abtew | +9.9 | 4 | 0.86 | 4 | 1.20 | 4 | 0.83 | 4 | 0.64 | 4 |
| 5 | Blaney–Criddle | +11 | 5 | 0.84 | 5 | 1.51 | 5 | 0.72 | 5 | 0.74 | 5 |
| 6 | Priestly–Taylor | −17.6 | 6 | 0.84 | 6 | 1.62 | 6 | 0.67 | 6 | 1.16 | 7 |
| 7 | Hargreaves | +22.5 | 7 | 0.67 | 7 | 2.91 | 7 | 0.53 | 7 | 1.11 | 6 |

*Daily estimate rank number for each statistical index

that showed underestimated/overestimated of ET0 from the observed (PM) value present as negative and positive signs. Hence, the methods were ranked against the PM value using all the above-mentioned statistical indices (Table 28.5).

Based on the mean absolute deviation of the daily ET0 value against the observed value, the Turc method underestimates the mean daily ET0 by 4.2% with the coefficient of determination (0.96) and showed considerably lowest RMSE (0.87) among the rest of the methods. It underestimates ET0 particularly from January to May, as presented in Fig. 28.5. Though Turc slightly underestimates the daily ET0 value, it shows the closest value of mean daily and yearly ET0 value with the observed value (PM) followed by Enku and Makkink, whereas the Hargreaves and Priestley–Taylor methods provide relatively the highest positive and negative mean deviation value from the PM with 1.58 and 1.62 highest values of RMSE, respectively (Table 28.5). Therefore, the Turc method showed the best performance to estimate ET0 in the study area. Thus, estimating ET0 using the Turc method is considered more accurate and reliable than the other tested empirical methods in the study area. Studies by Trajkovic and Kolakovic (2009) state that the Turc performs well to compute reference evapotranspiration at humid regions. Similarly, Lu et al. (2005) state that radiation-based methods that were developed for warm and humid climate conditions (Priestley–Taylor and Turc methods) perform well for the southeastern United States.

Kashyap and Panda (2001) and Tukimat et al. (2012) also declare the same conclusion, that is, the Turc method provides the closest results to the PM model for sub-humid and humid climate conditions when weather data are insufficient to apply the PM equation. The Enku method also performs well because it yielded relatively least overestimated values by 5.7% with reasonable $R^2$ (0.894) and RMSE (0.90 mm/day) errors (Table 28.5 and Fig 28.7). It also slightly overestimated evapotranspiration almost seventh months of the year.

The Priestley–Taylor method gives the highest underestimated values by $-17.6\%$. Conversely, the Hargreaves method overestimated by as much as 22.5%, giving the worst estimates among all the tested methods followed by Priestley–Taylor method. Similar performance of the Hargreaves method under humid and sub-humid conditions has been reported by different researchers (Droogers and Allen 2002; Temesgen et al. 2005; Alexandris et al. 2008) (Table 28.5).

## 28.5  Conclusions

With various methods and applications described in the study, the following conclusion was made. Estimation of ET0 using the class A pan evaporimeter is the simplest and most direct way, but if an appropriate pan coefficient is not used, the accuracy of ET0 estimation will not be satisfactory. Therefore, the monthly $K_p$ values were estimated for the study area. To estimate accurate and reliable $K_p$, the different approaches were compared, and statistical indices resulted in the ranking from the

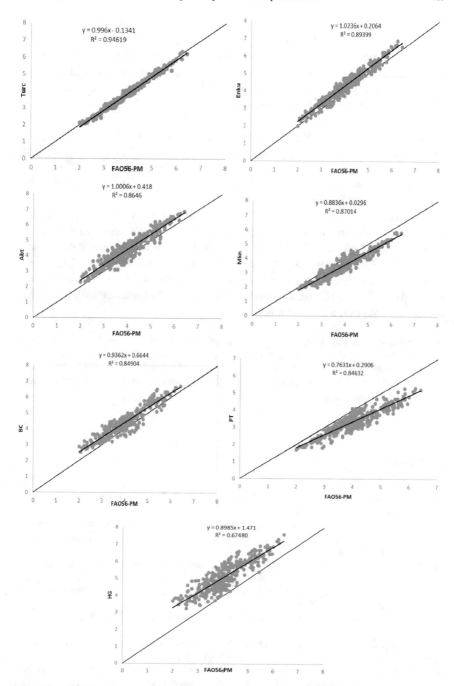

**Fig. 28.7** Comparison of daily FAO-56PM versus the other tested methods

most to the least accurate. The comparison analysis between the different $K_p$ estimation equations resulted in the following order according to prediction accuracy: Snyder > Orange > Cuenca > Allen and Pruitt. Therefore, the computed $K_p$ of the Snyder method closely agreed with the observed $K_p$ values (FAO-56 PM/$E_p$). As a result, the Snyder method is recommended as a good estimator of $K_p$ followed by Orange. Allen and Pruitt showed the weakest ability to predict $K_p$.

The complexity and inaccuracies in ET0 estimation often appear as major constraints in developing effective water management strategies for maintaining crop water requirement. Therefore, in this present study, seven ET0 estimation methods have been compared with FAO-56 PM values to show the reliability of different ET0 estimation methods. The analysis revealed that the ET0 estimates using different methods significantly vary.

From the comparison between the above methods and PM ET0 values, a relative underestimated/overestimated of the calculated ET0 value has been found. The Turc method is found to be suitable to calculate ET0 as it provided the closest values with FAO-56 PM ET0 followed by Enku and Makkink. Due to large overestimated and poor statistical indices, the Hargreaves method is found to be the least to estimate ET0 in the study area. Therefore, the Hargreaves method should be used with caution if only less data availability conditions.

**Acknowledgements** First author is thankful to the Ministry of Education, the Federal Democratic Republic of Ethiopia, for financial support during the period of study. The scholarship provided by the University of Gonder collaborate with the Ministry of Education of the FDRE (Ethiopia) is greatly acknowledged. The authors are grateful to the Ethiopian National Meteorological Agency for providing climatic data.

# References

Abtew W (1996) Evapotranspiration measurement and modeling for three wet-land systems in South Florida. Water Resour Bull 32:465–473

Alexandris S, Stricevic R, Petcock S (2008) Comparative analysis of reference evapotranspiration from the surface of rainfed grass in central Serbia, calculated by six empirical methods against the Penman-Monteith formula. Eur Water 21(22):17–28

Allam M, Jain A, McLaughlin D, Eltahir E (2016) Estimation of evaporation over the upper Blue Nile basin by combining observations from satellites and river flow gauges. Water Resour Res

Allen RG, Jensen ME, Wright JL, Burman R (1989) Operational estimates of reference evapotranspiration. Agron J 81(4):650–662

Allen RG, Pruitt WO (1991) FAO-24 reference evapotranspiration factors. J Irri Drain Eng 117:758–773

Allen RG, Pereira LS, Raes D, Smith M (1998) Crop evapotranspiration: guide-lines for computing crop water requirements. FAO Irri Drain Paper, 56, FAO, Rome, Italy

Allen RG, Clemmens AJ, Burt CM, Solomon KO, Halloran T (2005) Prediction accuracy for project wide evapotranspiration using crop coefficients and reference evapotranspiration. J Irri Drain Eng 131(1):24–36

BCEOM (1998) Abay River Basin integrated development master plan project to ministry of water resources. The Federal Democratic Republic of Ethiopia

Burnash RJC (1995) The NWS River forecast system: catchment modeling. In: Singh VP (ed) Computer models of watershed hydrology. Water Resources Publications, Highlands Ranch CO, pp 311–366

Central Statistical Agency (2008) Statistical abstract of Ethiopia. Central Statistical Authority, Addis Ababa, Ethiopia

Cuenca RH (1989) Irrigation system design: an engineering approach. Prentice Hall, Englewood Cliffs, 552. Evapotranspiration. Water Resour Bull 10(3):486–498

De Bruin HAR, Trigo IF, Jitan MA, Temesgen EN, Van Der Tol C, Gieske ASM (2010) Reference crop evapotranspiration derived from geo-stationary satellite imagery: a case study for the Fogera flood plain, NW-Ethiopia and the Jordan Valley, Jordan. Hydrol Earth Syst Sci 14(11):2219–2228. https://doi.org/10.5194/hess-14-2219-2010

Doorenbos J, Pruit WO (1975). Guidelines for prediction of crop water requirements. FAO Irrigation Drainage Paper, 24, FAO, Rome, Italy

Droogers P, Allen RG (2002) Estimating reference evapotranspiration under inaccurate data conditions. Irrigat Drain Syst 16(1):33–45

Garcia M, RaesD Jacobsen SE, Michel T (2007) Agro climatic constraints for rainfed agriculture the Bolivian Altiplano. J Arid Environ 71:109–121

Grismer ME, Orang M, Snyder R, Matyac R (2002) Pan evaporation to reference evapotranspiration conversion methods. J Irrigat Drain Eng 128(3):180–184

Gundekar HG, Khodke UM, Sarkar S, Rai RK (2008) Evaluation of pan coefficient for reference crop evapotranspiration for semiarid region. Irrigat Sci 26:169–175

Hubbard KG (1994) Spatial variability of daily weather variables in the high plains of the USA. Agric For Meteorol 68(1–2):29–41

Hargreaves GH, Samani ZA (1982) Estimating potential evapotranspiration. J Irrigat Drain Div 108(3):225–230

Irmak S, Haman DZ, Jones JW (2002) Evaluation of class A pan coefficients for estimating reference evapotranspiration in humid location. J Irrigat Drain Eng 128(3):153–159

Irmak S, Allen RG, Whitty EB (2003) Daily Grass and alfalfa-reference evapotranspiration estimates and alfalfa to grass evapotranspiration ratios in Florida. J Irrig Drain Eng 129:360–370

Jachner S, Boogaart G, Petzoldt T (2007) Statistical methods for the qualitative assessment of dynamic models with time delay (R Package qualV). J Stat Softw 22(8):1–30

Jensen MC, Middleton JE, Pruitt WO (1961) Scheduling irrigation from pan evaporation. Circular 386, Washington Agricultural Experiment Station

Jensen MC, MiddlET0n JE, Pruitt WO (1990) Scheduling irrigation from pan evaporation. Circular 386, Washington Agricultural Experiment Station

Kashyap PS, Panda RK (2001) Evaluation of evapotranspiration estimation methods and development of crop-coefficients for potato crop in a subhumid region. Agric Water Manag 50(1):9–25

Lu JB, Sun G, McNulty SG, Amatya DM (2005) A comparison of six potential evapotranspiration methods for regional use in the Southeastern United States. Am Water Resour Assoc 41(3):621–633

Orange M (1998) Potential accuracy of the popular nonlinear regression equations for estimating crop coefficient values in the original and ALLEN AND PRUIT tables. Unpublished Report, California Department of Water Resources

Penman HL (1948) Natural evaporation from open water, bare soil and grass. Proc R Soc Lond Ser A Math Phys Sci 193(1032):120–145

Pour SH, Shahid S, Chung ES (2016) A hybrid model for statistical downscaling of daily rainfall. Procedia Eng 154:1424–1430. https://doi.org/10.1016/j.proeng.2016.07.514

Priestley CHB, Taylor RJ (1972) On the assessment of surface heat flux and evaporation using large–scale parameters. Mon Weather Rev 100(2):81–92

Pielke R, Prins G, Rayner S, Sarewitz D (2007) Climate change 2007: lifting the taboo on adaptation. Nature 445(7128):597–598

Pradhan S, Sehgal VK, Das DK, Bandyopadhyay KK, Singh R (2013) Evaluation of pan coefficient methods for estimating FAO-56 reference crop evapotranspiration in a semi-arid environment

Pruitt W O (1966). Empirical method of estimating evapotranspiration using primary evaporation pans. In: Proceedings of the Conference on evapotranspiration and its role in water resources management, American Society of Agricultural Engineers, St. Joseph

Rácz C, Nagy J, Dobos AC (2013) Comparison of several methods for calculation of reference evapotranspiration. Acta SilvaticaLignariaHungarica 9(1):9–24

Rahimikhoob A (2009) Estimating daily pan evaporation using artificial neural network in a semi-arid environment. Theor Appl Climatol 98(1–2):101–105

Setegn SG, Srinivasan R, Dargahi B (2008) Hydrological modelling in the Lake Tana Basin, Ethiopia using SWAT model. Open Hydrol J 2(1)

Shahid S (2010) Rainfall variability and the trends of wet and dry periods in Bangladesh. Int J Climatol 30(15):2299–2313. https://doi.org/10.1002/joc.2053

Snyder RL (1992) Equation for evaporation pan to evapotranspiration conversions. J Irri Drain Eng 118(6):977–980

Temesgen B, Eching S, Davidoff B, Frame K (2005) Comparison of some reference evapotranspiration equations for California. J Irrig Drain Eng 131:73–84

Temesgen E, Melesse AM (2013) A simple temperature method for the estimation of evapotranspiration. Hydrol Process 28(6):2945–2960

Trajkovic S, Kolakovic S (2009) Evaluation of reference evapotranspiration equations under humid conditions. Water Resour Manage 23(14):3057–3067

Turc L (1961) Estimation of irrigation water requirements, potential evapotransipartion: a simple climatic formula evolved up to date. Ann Agronomy 12:13–49

Tukimat NNA, Harun S, Shahid S (2012) Comparison of different methods in estimating potential evapotranspiration at Muda irrigation scheme of Malaysia. J Agric Rural Dev Tropics Subtropics 113(1):77–85

Wang W, Shao Q, Peng S, Xing W, Yang T, Luo Y, Yong B, Xu J (2012) Reference evapotranspiration change and the causes across the Yellow River Basin during 1957–2008 and their spatial and seasonal differences. Water Resour Res 48(5)

WMO (1966) Climatic change. WMO Technical Note, No. 79, Geneva: Secretariat of the World Meteorological Organization (WMO), 79 pp

Zhao C, Nan Z, Cheng G (2005) Evaluating methods of estimating and modelling spatial distribution of evapotranspiration in the Middle River Basin, China. Am J Environ Sci 1:278–285

# Chapter 29
# Performance Evaluation of Four Models for Estimating the Capillary Rise in Wheat Crop Root Zone Considering Shallow Water Table

**Arunava Poddar**⊕, **Navsal Kumar**⊕, **and Vijay Shankar**

## 29.1 Introduction

Root water uptake from shallow water table (*SWT*) constitutes an important component of water balance, which is essential to understand the contribution of *SWT* in managing irrigation systems and optimizing irrigation needs (Poddar et al. 2018b; Kumar et al. 2020b). With the increase in population year by year, the demand of food, drinking water, and water requirement for many other purposes is increasing, of which approximately 80% is required for irrigation purpose (Babajimopoulos et al. 2007; Poddar et al. 2017; Goel et al. 2019; Kumar et al. 2019a). Since water resources are limited and the amount of water used for irrigation is maximum in comparison to other purposes, there is a need to develop and understand an alternative source of water supply for irrigation to have an effective water supply system.

Many lysimeter-based studies have stated that crops extract water when required from available *SWT* (Ayars and Hutmacher 1994). In water-scarce regions, appropriate use of *SWT* can be of great importance in crop water uptake (Xu et al. 2015; Poddar et al. 2018c). The capillary rise from *SWT* contributes to restore the soil moisture in the crop root zone lost due to root uptake. (Yang et al. 2000; Loheide and Steven 2008; Wu et al. 2015).

Gardner (1958) studied the theoretical solution of the flow equation for the capillary rise from *SWT* passing through the unsaturated soil profile. Schoeller (1961)

A. Poddar (✉) · N. Kumar · V. Shankar
Civil Engineering Department, National Institute of Technology Hamirpur, Hamirpur 177005,
Himachal Pradesh, India
e-mail: arunava.nithrs@gmail.com

N. Kumar
e-mail: navsal.happy@gmail.com

V. Shankar
e-mail: vsdogra12@gmail.com

© The Editor(s) (if applicable) and The Author(s), under exclusive license to Springer Nature Switzerland AG 2021
A. Pandey et al. (eds.), *Hydrological Extremes*, Water Science and Technology Library 97, https://doi.org/10.1007/978-3-030-59148-9_29

recommended an empirical Averianov formula for the estimation of the capillary rise from *SWT*. McDonald and Harbaugh (1988) used a finite-difference groundwater model MODFLOW with an assumption that evaporation possesses linear variation with *SWT* depth. Prathapar et al. (1992) defined capillary rise from *SWT* as the amount of water uptake from an *SWT* due to evaporation plant transpiration and soil evaporation. Furthermore, the performance of a transient state analytical model (*TSAM*), a numerical model (*NM*), and a quasi-steady-state model (*QSSAM*) was compared with the experimental values and the *NM* performed satisfactorily. But *NM* becomes inapplicable in complex conditions as it is inconvenient to have temporal and spatial data and application of analytical models is easy when input data is sparse and uncertain. Jorenush and Sepaskhah (2003) improved the *TSAM* for estimation of the capillary rise and the soil profile salinity under non-irrigated and irrigated conditions and stated that whenever application of *NM* to large irrigation areas is not possible, modified *TSAM* can be applied to estimate the rate of capillary rise.

Zammouri (2001) mentioned that field observation of the capillary rise is difficult, nearly impractical, and economically unfeasible and performed a comparative study between Gardner's analytical solution, MODFLOW approach, and Averianov formula to estimate the capillary rise. Averianov formula presented better results and it required easily measurable data to estimate capillary rise. Yang et al. (2011) improved the Averianov formula by incorporating soil moisture, depth to the *SWT*, and leaf area index (*LAI*), to estimate capillary rise formulating the modified Averianov formula (*MAF*). Liu et al. (2014) established a solution for estimation of the maximum height of the capillary rise utilizing soil parameters and reported that theoretically obtained solutions presented more realistic results. Wang et al. (2016) assessed capillary rise with many influencing variables in crop growing seasons for multi-soil profiles and observed compelling results but suggested further investigation in different agro-climatic conditions.

A thorough investigation of the literature mentioned above indicates that most of the studies were conducted in plain terrain. This puts a question mark on the applicability of these models to estimate the capillary rise from *SWT* for crops grown in hilly terrain. Hence, in the present study, we evaluated the performance of four existing capillary rise models. The precision of different capillary rise models is examined by comparing estimated cumulative capillary rise with the experimentally obtained capillary rise in wheat crop root zone in mid-hills of Himachal Pradesh, (India).

### 29.1.1 Capillary Rise Models

Four models, which are compared to estimate the capillary rise in this study are as follows:

(a) Jorenush and Sepaskhah (2003)

Jorenush and Sepaskhah (2003) proposed estimation of the capillary rise (CR) by modifying the TSAM (hereafter denoted as M1) considering non-uniform water uptake pattern and variable root depth including water balance equation, given in Eq. 29.1 as

$$CR = \frac{\Phi_0 - \Phi_{rz}}{L - DRZ} - K_{\Phi,rz} \tag{29.1}$$

where $DRZ$ = root zone depth (m), $\Phi_{rz}$ = root zone flux matric potential (m$^2$ s$^{-1}$), $\Phi_0$ = surface flux matric potential (m$^2$ s$^{-1}$), $L = SWT$ depth (m), and $K_{\Phi,rz}$ = saturated hydraulic conductivity (m s$^{-1}$) at $\Phi_{rz}$.

(b) Yang et al. (2011)

Yang et al. (2011) developed MAF (hereafter denoted as M2) to estimate capillary rise considering LAI, soil moisture, and SWT depth. The empirical formula developed in Eqs. 29.2–29.3 as

$$CR = LAI \times \left(\frac{\theta}{\theta_{FC} - \theta_{wp}}\right)^{-0.2} \times \left(1 - \frac{h}{2.5}\right)^{1.89} \quad \text{if } h < h_0 \tag{29.2}$$

$$CR = 0 \quad \text{if } h < h_0 \tag{29.3}$$

where $LAI$ = leaf area index (m$^2$ m$^{-2}$), $\theta_{FC}$ = field capacity of soil (m$^3$ m$^{-3}$), $\theta_{WP}$ = permanent wilting point of soil (m$^3$ m$^{-3}$), $\theta$ = soil moisture (m$^3$ m$^{-3}$), and $h$ = average SWT depth (m).

(c) Liu et al. (2014)

Liu et al. (2014) proposed a method (hereafter denoted as M3) to estimate the capillary rise using Eq. 29.4 having easily obtainable laboratory parameter as

$$CR = \frac{\sigma n}{4\eta k_s} \frac{r \cos \alpha}{f(z)} + \left(1 - \frac{\rho_w g n}{8\eta k_s} \frac{r^2}{f(z)}\right) z \tag{29.4}$$

where $\alpha$ = advancing contact angle, $k_s$ = saturated hydraulic conductivity (m s$^{-1}$), $\eta$ = viscosity of water, $z$ = negative water head (m), $n$ = porosity of water (m$^3$ m$^{-3}$), $r$ = equivalent radius of tube (m), $\sigma$ = surface tension of water (kg s$^{-2}$), $\rho_w$ = water density (kg m$^{-3}$), $g$ = gravitational acceleration (m s$^{-2}$), and $f$ = mathematical model.

(d) Wang et al. (2016)

Wang et al. (2016) developed an equation (hereafter denoted as M4) (Eq. 29.5) to estimate capillary rise based on multiple influencing factors, i.e., soil moisture, crop growth stage effects, climatic variables, and *SWT* depth as

$$CR = K_c \times ET_c \times \left(1 - \frac{H}{H_{max}}\right)^n \times \frac{\theta_{FC} - \theta}{\theta_{FC} - \theta_{wp}} \qquad (29.5)$$

where $H_{max}$ = potential maximum depth (m) beyond which no capillary rise occurs, $ETc$ = crop evapotranspiration (mm day$^{-1}$), $H$ = actual *SWT* depth (m), $K_c$ = crop coefficient, $n$ = soil characteristics parameter, and $\theta$ = actual averaged soil moisture content in the root zone (cm$^3$ cm$^{-3}$).

### 29.1.2  Field Crop Experiments

The field crop experiments were executed at an agricultural experimental station of the National Institute of Technology (NIT) Hamirpur for two consecutive years 2016 and 2017. The study area is located at 76° 52′ E longitude and 31° 68′ N latitude, and the average elevation is 890 m. The study area comes under the humid sub-tropical agro-climate of north-western mid-hills.

Reference evapotranspiration ($ET_0$) values were determined using daily climatic values obtained from an automatic weather station located at the agricultural experimental station, NIT Hamirpur. FAO-56 Penman–Monteith equation was used to compute $ET_0$ (Allen et al. 1998; Sharma et al. 2016; Poddar et al. 2018a). Table 29.1 shows the mean annual rainfall ($P$), minimum and maximum temperature ($T_{min}$, $T_{max}$), relative humidity ($RH$), wind speed ($u$) (2 m height), and solar radiation ($R_s$) recorded throughout the study period.

### 29.1.3  Lysimeter Setup

Lysimeters were employed for the field observation of the capillary rise. The detailed Lysimeter setup is shown in Fig. 29.1, i.e., Lysimeter 1 and Lysimeter 2 were equipped with *SWT* and gravity drainage, respectively. The bottom of the Lysimeters was equipped with a 30 cm thick filter. A glass piezometer to observe the water table depth was provided on the outside of the Lysimeter which was internally connected to a vertical filter tube. The drainage was controlled at the bottom of the Lysimeters 1 to allow the water table formation and was attached to the mess-cylinder and

**Table 29.1** Details of mean annual values of climatic variables

| Years | $P$ (mm) | $T_{max}$ (°C) | $T_{min}$ (°C) | $RH$ (%) | $R_s$ (MJ m$^{-2}$ d$^{-1}$) | $U$ (m s$^{-1}$) |
|-------|----------|----------------|----------------|----------|-------------------------------|------------------|
| 2016  | 390      | 25.1           | 14.8           | 75       | 17.4                          | 2.0              |
| 2017  | 440      | 27.2           | 15.8           | 69       | 17.2                          | 2.2              |

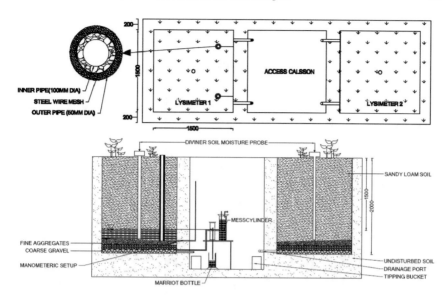

**Fig. 29.1** Experimental Setup to observe the depth of the water table and capillary rise

Mariotte's bottle to sustain a constant *SWT* at 150 cm depth and observe the capillary rise occurring from the *SWT*. The bottom of the Lysimeter 2 had gravity drainage, which represented percolation to the deep-water table. The soil in the Lysimeters and the surrounding agricultural field was sandy loam in texture with % of silt, sand, and clay as 23.83, 54.98, and 21.19, respectively. A soil moisture capacitance probe was used to measure the daily soil moisture at 10 cm interval till 150 cm depth.

To determine the capillary rise using the model given by Liu et al. (2014), three parameters required are the contact angle ($\alpha$), porosity ($n$), and saturated hydraulic conductivity ($K_s$). Porosity for the soil samples was determined through a routine laboratory test. $K_s$ was determined using the KSAT apparatus (M/S METER Environment, Pullman, WA, USA). The value of the contact angle was obtained from Kumar and Malik (1990). The pressure plate apparatus was used to determine the permanent wilting point ($\theta_{WP}$) and field capacity ($\theta_{FC}$). The values of unsaturated soil hydraulic parameters $K_s$, saturated moisture content ($\theta_s$), residual moisture content ($\theta_r$), $\theta_{FC}$, and $\theta_{wp}$ are 2.96 cm hr$^{-1}$, 0.36 cm$^3$ cm$^{-3}$, 0.056 cm$^3$ cm$^{-3}$, 0.23, and 0.08, respectively (Kumar et al. 2020a, b).

### 29.1.4 Crop Parameters

Wheat (*T. aestivum*) was grown in two replications during the study period, details of which are given in Table 29.2.

**Table 29.2** Details of crop during two crop seasons with irrigation events

| Crop | Variety Sown | Date of sowing | Date of harvesting | Duration | Growth stages (days) | | | | Irrigation provided (Day) | Spacing (cm) |
|---|---|---|---|---|---|---|---|---|---|---|
| | | | | | I | II | III | IV | | |
| Wheat (*T. aestivum*) | Super (6776/PB) | January 10, 2016 | May 9, 2016 | 119 Days | 22 | 34 | 39 | 24 | 25,54, 67,79, and 98 | 20 × 5 |
| | Super (6776/PB) | January 31, 2017 | May 20, 2017 | 110 Days | 19 | 30 | 47 | 14 | 18,49, 64,77, and 95 | 20 × 5 |

**Fig. 29.2** Plant height, root depth, and leaf area index during the crop period of 2016

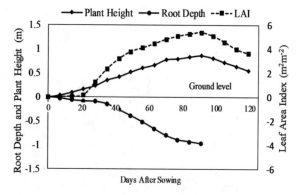

**Fig. 29.3** Plant height, root depth, and leaf area index during the crop period of 2017

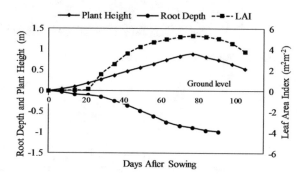

Root depth, plant height, and *LAI* are the three crop parameters observed at regular intervals during both years. Observation of root depth was conducted using a trench profile method (Bohm 1979). A Digital planimeter was used to determine *LAI*. Plant height was obtained by having an average of ten selected plants at random. Figures 29.2 and 29.3 show the variation of crop parameters of wheat for both the years and Fig. 29.4 represents wheat grown in experimental site. $K_c$ was determined with the modification procedure suggested in FAO-56 (Allen et al. 1998).

### 29.1.5    Performance Evaluation Indicators

The performance of the models was evaluated using root mean square (*RMSE*), coefficient of determination ($R^2$), and Nash–Sutcliffe efficiency (*NSE*) given in Eqs. 29.6–29.8, respectively. The deviation of estimated values from observed values is understood by *RMSE* values. $R^2$ denotes the degree of linear dependency, whereas *NSE* is used to verify the credibility of models and maximum value for both evaluation indices is 1. The performance evaluation of the model is considered better when *RMSE* values are smaller and $R^2$ and *NSE* are higher. The equations are

**Fig. 29.4** Wheat crop grown at the experimental site (root depth, plant height, and *LAI*)

$$RMSE = \sqrt{\frac{1}{n} \sum_{cr=1}^{n} (Obs_{cr} - Est_{cr})^2} \qquad (29.6)$$

$$R^2 = \frac{\sum_{cr=1}^{n} (Obs_{cr} - \overline{Obs_{cr}})(Est_{cr} - \overline{Est_{cr}})}{\sqrt{\sum_{cr=1}^{n} (Obs_{cr} - \overline{Obs_{cr}})^2} \sqrt{\sum_{cr=1}^{n} (Est_{cr} - \overline{Est_{cr}})^2}} \qquad (29.7)$$

$$NSE = 1 - \frac{\sum_{cr=1}^{n} (Obs_{cr} - Est_{cr})^2}{\sum_{cr=1}^{n} (Obs_{cr} - \overline{Obs_{cr}})^2} \qquad (29.8)$$

where $Obs_{cr}$, $Est_{cr}$ = observed and estimated values. $\overline{Obs_{cr}}$, $\overline{Est_{cr}}$ = observed and estimated average values $n$ = no. of observations.

## 29.2   Results and Discussion

Field crop experiments were performed on wheat during relevant crop seasons of 2016 and 2017. Both the crop replications responded well in the presence of *SWT*. Capillary rise for two years was observed and simulated, but due to brevity, observations of 2017 are discussed in this chapter.

## 29.2.1 Measurements Using Lysimeter

In the presence of *SWT*, less capillary rise was observed for the initial 2–3 weeks (Table 29.3). The reduction in the capillary rise was also observed when irrigation was applied. This was the result of the decrease in variation of matric potential ($\psi$) between the root zone and *SWT*. $\psi$ in the root zone increased due to irrigation, while $\psi$ for subsoil is increased because of capillary rise from the *SWT*.

From 4 to 9 week, it was found that the capillary rise for wheat was approximately 3.5 mm week$^{-1}$. Pan evaporation during these weeks increased from 10.6 to 18.3 mm week$^{-1}$. Two irrigations during this period were required to fulfill the soil water deficit. Due to the application of frequent irrigations, the rate of capillary rise increased between 10 and 16 week to a magnitude of 7.3 mm week$^{-1}$. During the crop period, five irrigations were applied. Weekly capillary rise values from the *SWT* in the lysimeter were measured, and it was found that every day a measurable amount of capillary rise occurs apart from the day of irrigation. It was observed that during the end season, capillary rise decreased due to the decreased root water uptake.

**Table 29.3** Details of evaporation, irrigation, and capillary rise for wheat

| Week after sowing | Evaporation (Pan) (mm) | Irrigation (mm) | Capillary rise (observed) (mm week$^{-1}$) | Capillary rise (estimated) (mm week$^{-1}$) | | | |
|---|---|---|---|---|---|---|---|
| | | | | M1 | M2 | M3 | M4 |
| 1 | 10.9 | | 4.7 | 2.3 | 5.9 | 1.3 | 1 |
| 2 | 10.2 | | 1.9 | 0.5 | 2.6 | 1 | 1.7 |
| 3 | 10.7 | | 2.9 | 0.9 | 3.6 | 1.8 | 2.6 |
| 4 | 14.5 | 22 | 3.8 | 1 | 4.8 | 0.9 | 6.5 |
| 5 | 16.8 | | 14.1 | 5.1 | 16.7 | 8.2 | 12.4 |
| 6 | 21.1 | | 6.2 | 2.8 | 7.8 | 2 | 2.4 |
| 7 | 24.6 | 34 | 1.6 | 0.3 | 1.8 | 0.5 | 0.5 |
| 8 | 18.9 | | 15.2 | 5.1 | 15.5 | 7.4 | 13.5 |
| 9 | 14.1 | | 12.6 | 4 | 13.9 | 9.3 | 7.2 |
| 10 | 17.2 | 56 | 3.7 | 1.2 | 2.3 | 2 | 0.4 |
| 11 | 29.9 | 52 | 2.8 | 0.6 | 3.5 | 1.6 | 0.9 |
| 12 | 34 | | 6.3 | 1.1 | 5.2 | 3.5 | 2.4 |
| 13 | 35.7 | 48 | 10.1 | 2.5 | 9.6 | 4.8 | 7.2 |
| 14 | 40.8 | | 8.1 | 3.2 | 7.5 | 6.8 | 6.9 |
| 15 | 44.9 | | 7.8 | 3.8 | 6.8 | 7.2 | 5.7 |
| 16 | 47.2 | | 6.7 | 5.9 | 6 | 9.7 | 7.4 |

**Fig. 29.5** Measured and
Estimated cumulative
capillary rise for wheat

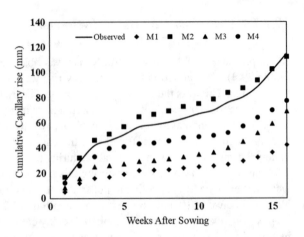

### 29.2.2 Comparison of Models for Capillary Rise Simulation

Measured and estimated capillary rise for wheat is shown graphically in Fig. 29.5.

M1, M3, and M4 underestimated the capillary rise during the complete crop period because the possessed lesser values, which resulted in a lesser amount of capillary rise for M1 and M3. Although the trend of observed capillary rise is followed by both the models but the examined values are consistently found to be lesser than the observed values. The fluctuation in the M4 estimated values is probably due to the variation of $K_c$ values in the model. M2 estimated results presented better agreement with the observed values, which is also evident from the indicative trend of the observed values for wheat (Fig. 29.5) in the same pattern.

Error statistics including $RMSE$, $R^2$, and $NSE$ values for the models are given in Table 29.4. $RMSE$ values are 0.04(2016) and 0.06(2017) for the Modified Averianov formula (Yang et al. 2011) model (M2) which are smaller than other models. Similarly, $R^2$ and $NSE$ are 0.71(2016); 0.67(2017) and 0.74(2016); 0.72(2017), respectively, for M2 which are comparatively higher than other models. Results indicate that M2 performed better to estimate capillary rise for wheat in the study area (hilly terrain).

**Table 29.4** Error statistics of estimated values in comparison with observed values

| Models | Evaluation indicators | | | | | |
|--------|-----------------------|------|---------|------|------|------|
|        | RMSE                  |      | $R^2$   |      | NSE  |      |
|        | 2016                  | 2017 | 2016    | 2017 | 2016 | 2017 |
| M1     | 0.17                  | 0.18 | 0.58    | 0.55 | 0.43 | 0.47 |
| M2     | 0.04                  | 0.06 | 0.71    | 0.67 | 0.74 | 0.72 |
| M3     | 0.11                  | 0.12 | 0.64    | 0.69 | 0.44 | 0.51 |
| M4     | 0.08                  | 0.09 | 0.66    | 0.62 | 0.63 | 0.64 |

## 29.3   Conclusion

The exchange between the groundwater and soil water is significantly affected by the irrigation provided in the presence of *SWT*. Modified Averianov formula which considered *LAI* as a parameter for capillary rise estimation performed better in the present study conducted on wheat. The study establishes that the Modified Averianov formula model is highly reliable for the prediction of capillary rise in sandy loam soil in hilly terrain and hence proposes its suitability in predicting capillary rise through for wheat crop root zone. The findings of the study indicate that the presence of a crop parameter improves the prediction capability for capillary rise estimation. Thus, by only obtaining the *LAI* experimentally (which is easily observable/measurable) for each crop in an agro-climate, the capillary rise can be predicted using the Modified Averianov formula. Further work is suggested to establish the significance of the Modified Averianov formula model for a variety of crops in predicting the values of the capillary rise to determine the impact of *SWT* in irrigation scheduling.

**Acknowledgements**   The financial support was received through Ministry of Earth Sciences sponsored project titled "Sustaining Himalayan water resources in a changing climate (2016–2020)" and Department of Biotechnology sponsored project titled "Social-economic-environmental tradeoffs in managing Land-river interface (2019–2021)".

## References

Allen RG, Pereira LS, Raes D, Smith M (1998) Crop evapotranspiration-Guidelines for computing crop water requirements-FAO irrigation and drainage paper 56. FAO, Rome 300(9):D05109

Ayars JE, Hutmacher RB (1994) Crop coefficients for irrigating cotton in the presence of groundwater. Irrig Sci 15(1):45–52

Babajimopoulos C, Panoras A, Georgoussis H, Arampatzis G, Hatzigiannakis E, Papamichail D (2007) Contribution to irrigation from shallow water table under field conditions. Agric Water Manag 92(3):205–210

Bohm W (1979) Container methods. In: Methods of studying root systems. Ecological studies (analysis and synthesis), vol 33. Springer, Berlin, Heidelberg

Gardner WR (1958) Some steady-state solutions of the unsaturated moisture flow equation with application to evaporation from a water table. Soil Sci 85(4):228–232

Goel L, Shankar V, Sharma RK (2019) Investigations on effectiveness of wheat and rice straw mulches on moisture retention in potato crop (Solanum tuberosum L.). Int J Recycl Org Waste Agric, 1–12

Jorenush MH, Sepaskhah AR (2003) Modelling capillary rise and soil salinity for shallow saline water table under irrigated and non-irrigated conditions. Agric Water Manag 61(2):125–141

Kumar N, Poddar A, Shankar V (2019a, August) Optimizing irrigation through environmental canopy sensing–A proposed automated approach. In AIP conference proceedings, vol 2134, no 1. AIP Publishing LLC, p 060003

Kumar N, Poddar A, Shankar V (2020a) Nonlinear regression for identifying the optimal soil hydraulic model parameters. In: Numerical optimization in engineering and sciences. Springer, Singapore, pp 25–34

Kumar N, Poddar A, Dobhal A, Shankar V (2019b) Performance assessment of PSO and GA in estimating soil hydraulic properties using near-surface soil moisture observations. Compusoft 8(8):3294–3301

Kumar N, Poddar A, Shankar V, Ojha CSP, Adeloye AJ (2020b) Crop water stress index for scheduling irrigation of Indian mustard (*Brassica juncea*) based on water use efficiency considerations. J Agron Crop Sci 206(1):148–159. https://doi.org/10.1111/jac.12371

Kumar S, Malik RS (1990) Verification of quick capillary rise approach for determining pore geometrical characteristics in soils of varying texture. Soil Sci 150(6):883–888

Liu Q, Yasufuku N, Miao J, Ren J (2014) An approach for quick estimation of maximum height of capillary rise. Soils Found 54(6):1241–1245

Loheide II, Steven P (2008) A method for estimating subdaily evapotranspiration of shallow groundwater using diurnal water table fluctuations. Ecohydrology 1(1):59–66

McDonald MG, Harbaugh AW (1988) A modular three-dimensional finite-difference ground-water flow model, vol 6. Reston, VA: USGS,, p A1

Poddar A, Gupta P, Kumar N, Shankar V, Ojha CSP (2018a) Evaluation of reference evapotranspiration methods and sensitivity analysis of climatic parameters for sub-humid sub-tropical locations in western Himalayas (India). ISH J Hydraul Eng, 1–11. https://doi.org/10.1080/09715010.2018.1551731

Poddar A, Kumar N, Shankar V (2018b) Evaluation of two irrigation scheduling methodologies for potato (Solanum tuberosum L.) in north-western mid-hills of India. ISH J Hydraul Eng, 1–10. https://doi.org/10.1080/09715010.2018.1518733

Poddar A, Kumar N, Shankar V (2018c, February) Effect of capillary rise on irrigation requirements for wheat. In: Proceedings of international conference on sustainable technologies for intelligent water management (STIWM-2018), IIT Roorkee, India

Poddar A, Sharma A, Shankar V (2017, December) Irrigation Scheduling for Potato (solanum tuberosum l.) Based on daily crop coefficient approach in a sub-humid sub-tropical region. In Proceedings of hydro-2017 international, L. D. College of Engineering Ahmedabad, India

Prathapar SA, Robbins CW, Meyer WS, Jayawardane NS (1992) Models for estimating capillary rise in a heavy clay soil with a saline shallow water table. Irrig Sci 13(1):1–7

Schoeller H (1961) Les eaux souterraines. Masson et Cie Editeurs, Paris

Sharma V, Poddar A, Shankar V (2016, September). Performance evaluation of data-driven and statistical rainfall runoff models for a mountainous catchment. In: Proceedings of national conference: civil engineering conference–innovation for sustainability (CEC–2016), vol 1, pp 247–261

Wang X, Huo Z, Feng S, Guo P, Guan H (2016) Estimating groundwater evapotranspiration from irrigated cropland incorporating root zone soil texture and moisture dynamics. J Hydrol 543:501–509

Wu Y, Liu T, Paredes P, Duan L, Pereira LS (2015) Water use by a groundwater dependent maize in a semi-arid region of inner mongolia: evapotranspiration partitioning and capillary rise. Agric Water Manage 152:222–232

Xu X, Sun C, Qu Z, Huang Q, Ramos TB, Huang G (2015) Groundwater recharge and capillary rise in irrigated areas of the upper yellow river basin assessed by an agro-hydrological model. Irrig Drain 64(5):587–599

Yang F, Zhang G, Yin X, Liu Z, Huang Z (2011) Study on capillary rise from shallow groundwater and critical water table depth of a saline-sodic soil in western Songnen plain of China. Environ Earth Sci 64(8):2119–2126

Yang J, Li B, Shiping L (2000) A large weighing lysimeter for evapotranspiration and soil-water–groundwater exchange studies. Hydrol Process 14(10):1887–1897

Zammouri M (2001) Case study of water table evaporation at Ichkeul Marshes (Tunisia). J Irrig Drain Engrg 127(5):265–271

# Chapter 30
# SCADA Based Rainfall Simulation and Precision Lysimeters with Open Top Climate Chambers for Assessing Climate Change Impacts on Resource Losses in Semi-arid Regions

**Konda S. Reddy, Maddi Vanaja, Vegapareddy Maruthi, and T. Saikrishna**

## 30.1  Introduction

Climate change and variability is the major problem across the world today. The average temperatures have increased 0.89 °C over the last century and as per AR5 report of IPCC (2012, 2013), the predictions indicated that there would be an average increase of temperatures by 2–5 °C across the globe. There was an increase up to 3 °C since 1900 in the observed annual mean temperatures (Cruz et al. 2007). Observed annual mean precipitation trends indicate extreme variability in different regions, including areas of increased and decreased precipitation (Cruz et al. 2007). Globally and in Asia, there is an observed increase in the intensity and frequency of extreme weather events (IPCC 2007, 2012). In coastal Asia, the sea level has been rising at a rate of 1–3 millimetres per year. This Climate change results in continued changes in temperature, precipitation, extreme events and sea level, both globally and in Asia (IPCC 2012).

Temperature and annual precipitation with both seasonal and regional variations within Asia are foreseen to increase within the continent (Christensen et al. 2007). The largest increases are projected for North and East Asia with a decreasing trend of mean precipitation in Central Asia. There can be an increase in the extreme weather events in South and East Asia (Cruz et al. 2007). On the basis of increase in the sea surface temperature by 2–4 °C, the intensity of tropical cyclones is projected

K. S. Reddy (✉) · M. Vanaja
Division of Resource Management, ICAR- Central Research Institute for Dryland Agriculture, Santhoshnagar, Hyderabad 500059, Telangana, India
e-mail: ks.reddy@icar.gov.in

V. Maruthi · T. Saikrishna
Division of Crop Sciences, ICAR- Central Research Institute for Dryland Agriculture, Santhoshnagar, Hyderabad 500059, Telangana, India

© The Editor(s) (if applicable) and The Author(s), under exclusive license to Springer Nature Switzerland AG 2021
A. Pandey et al. (eds.), *Hydrological Extremes*, Water Science and Technology Library 97, https://doi.org/10.1007/978-3-030-59148-9_30

to increase 10–20%. By the end of twenty-first century, there is an expected rise of global sea level by 3.8 mm.

India has shown an increasing trend, as the annual mean surface temperature has increased by 0.65 °C over the past century as indicated by IMD, Pune. The temperature in India is likely to be more in rabi than in kharif season and the kharif rainfall is likely to increase by as much as 10 per cent. With such climate changes, there will be an impact on agriculture considerably through its direct and indirect effect on crops, livestock, pest and diseases and soils, thereby threatening the food security, an important problem for most of the developing countries. The impact of climate change on agriculture at various crop management levels and water management, fertilizer, increasing water availability and water use efficiency needs to be assessed. Also, there is a need to study the effect of climate change impacts as variations in precipitation pattern, increased air and soil temperatures on the soil processes viz., available soil water content, runoff and erosion. Rainfall simulators have been used for experimentation in rangelands since 1930 ranging from small to bigger size systems for studying the resource loss of water, soil and nutrients (Stone et al. 1992). These systems are extensively used in various hydrology laboratories to assess the erosion and runoff and to generate for the data for calibration and validation. Rainfall for a particular known duration and intensity can be generated using rainfall simulator in controlled manner, which allows to quantify runoff and soil loss (Martínez-Mena et al. 2001). The advantage of using simulated rainfall is the collection of data under uniformly natural conditions. Rainfall Simulator can be used for accurate reproduction of natural rainfall drop sizes with nearly continuous, uniform application over an area, the ability to apply rainfall of varying durations and intensities of interest (Hignett et al. 1995).

Lysimeters are extensively used in the world to basically study the soil water dynamics and solute transport in the soil (Kim et al. 2010). The concept of lysimeters varies from simple to high precision of data acquisition of various soil water balance parameters of soil moisture, crop evapotranspiration, effective rainfall, percolation from the soil. The present system has provision of open type climate chambers with control of $CO_2$ gas release and temperature increase. The present system is designed to study the resource losses at extreme weather events of the rainfall intensity varying from 25 to 150 mm/h in different soils and up to 10% slope with moving type simulator over the soil bins and for assessing the changes in soil water balance with respect to increased temp and $CO_2$ levels through lysimeters in open-top climate chambers. The data generated through experiments are acquired through sensors into the SCADA excel sheet and SCADA has the provisions to operate the system from the computer terminal itself and the control of the system parameters.

## 30.2 Methodology

### 30.2.1 Rainfall Simulator

State of the art instrumentation for measuring soil moisture, temperature, rainfall intensity, runoff, silt load and hydraulic power lift has been integrated to SCADA software. The process controls are imposed by the Programmable Logic Controller (PLC) through programming. The rainfall simulator is designed for the rainfall intensity of ≤150 mm/h (extreme events). It can be moved from one soil bin to another through SCADA. Three soil bins of size 6 × 3 × 1.2 m are provided with all instrumentation and percolation tubes and hydraulic lift. The system is configured for maximum slope of 10%, and it can be operated from SCADA and rainfall intensity control through PLC (Fig. 30.1).

A simulator was designed with six nozzles of varying discharges at different pressures. Three different nozzle sizes of 1, 3 and 4 mm were used for simulating the rainfall intensities varying from 30 to 150 mm/h. It has an oscillating mechanism for swinging the nozzle heads covering the entire width of the soil bin of 3 m. It is achieved by using DC motor and limit switches. This oscillation is controlled by relay and contact type of sensors. The simulator can move on the iron rail track

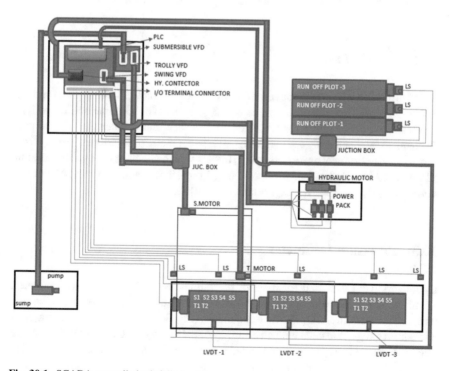

**Fig. 30.1** SCADA controlled rainfall simulator

from one soil bin to other so that the experiments can be run continuously with varying slope. The nozzles have 45 degree spray angle. Six nozzles will have a linear spray covering the entire length of the soil bin. As per international standards and literature, the height of the simulator was kept at 2 m so that the particle distribution is uniform throughout the soil bin. SCADA operates the DC motor with limit switches for controlling oscillation of the rainfall simulation bar.

The facility of rainfall simulator with soil bins was designed with provision of rain shelter having GI tubular structure in semi-circular shape with 6 m width and height of 4 m. The shelter has the provisions of opening for allowing sufficient radiation for crop growth and with opening on both sides of the shelter along the length of 22 m. The shelter frame is covered with 100 micron PE UV stabilized transparent sheet grouted at the bottom of structure for its stability against the wind and natural rain.

### 30.2.2    Calibration of Rainfall Simulator

Nozzles and the uniformity of the rainfall are tested in the facility. Christiansen's uniformity coefficient was used to characterize the homogeneousness of simulated rainfall in the calibration. A submersible pump controlled through PLC and drive for varying discharge and hydraulic pressures was used in the water sump.

### 30.2.3    Soil Bin

Soil bins are fabricated with rugged structure of MS channel and MS angle as support below the bins. The size of the bin is $6 \times 3 \times 1$ m having 3 tonne soil by weight. The bins have inner lining for corrosion-free surface under water. An oil pumping mechanism has been designed to lift the soil bins for maintaining slopes varying from 1 to 10%. The lift mechanism has two cylinders which are operated by the hydraulic oil pumping installed at the centre of the soil bins. These soil bins can be operated simultaneously at a time for experiment (Fig. 30.2).

## 30.3    Precision Lysimeter

Lysimeters were designed for most of the rainfed crops (specifications: diameter; 1.2 m and length; 2 m) with state of the art instrumentation for measuring soil moisture, soil temperature, $CO_2$ and air temperature (Fig. 30.3). Weighing type lysimeters with strain gauge load cells with an accuracy of $\pm 1\%$. Physical controls of increasing temperatures and $CO_2$ in the open-top chambers was done through SCADA control to SCADA. It can generate the reports of the data and can prompt screen. Total data

**Fig. 30.2** Rainfall simulator system with soil bins and maize as a test crop

was acquired through sensor network through data loggers, and the user can diagnose any fault in the measurement system for correction through alarms.

### 30.3.1   Lysimeter Tank

Precision lysimeter tanks were fabricated using steel cylinders of 1.2 m dia and 2 m deep and 6.4 mm thick. The soil is filled up to 1.8 m deep allowing 20 cm for irrigation and crop manoeuvring. These cylinders have the provision of collecting leachate at the bottom of the cylinder through porous pipe with sand filter around the pipe. The drain tubes are connected to collection tanks installed in the service well. Load cell is installed between the concrete base and soil tank so that the changes

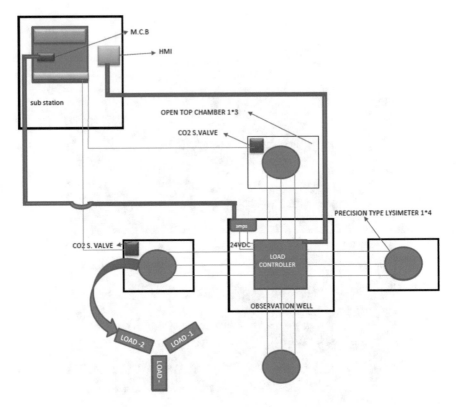

**Fig. 30.3** Precision lysimeter system

in the weight of the soil can be observed. Real-time TDR soil moisture along with temperature sensors in soil tank at different depths varying from 15 to 90 cm. All these soil moisture sensors work on standard signal of 4–20 mA signal with an accuracy of ±1%. These sensors are connected through data logger to PLC for online data acquisition in MS excel format in SCADA.

### 30.3.2  Service Well

The central control sump well of 3 m dia and 2.5 m depth with step ladder was constructed for mounting data loggers, collection of water samples of leachate or percolated water. Data logger having 64 channels were used for storing the data and its transfer to SCADA. It has sufficient memory and data retrieval ports and compatibility to data portability.

**Fig. 30.4**  Open-top climate chambers with maize as test crop

### 30.3.3   Open-Top Climate Chambers for Controlling CO$_2$ and Temperature

Open-top climate chambers were constructed of size 4 × 4 × 4 m with GI pipe covered with Polycarbonated sheet of 4 mm thick with 85% transparency with 100% opening at the top of the chambers (Fig. 30.4). Three such chambers were constructed taking the lysimeter at the centre of the open-top chamber. One chamber is provided with supply and control of CO$_2$, and the IR heaters were used for control and increase in temperature in the chambers. As per IPCC guideline, 550 ppm was taken as a standard protocol to maintain the CO$_2$ level throughout the crop growth season in the chamber and lysimeter. The provision of on and off for the heaters and solenoid valves for controlling the supply of CO$_2$ was ensured for the precision. CO$_2$ analyser was used to measure the CO$_2$ concentration in the chamber and 550 ppm was maintained with spatial variation of ±50 ppm in the chamber. Out of three chambers, one chamber is provided of for elevated CO$_2$ and elevated temperature, second chamber is provided with elevated CO$_2$ alone and third chamber is provided with elevated temperature alone. However, one lysimeter without open-top chamber is taken as control at ambient conditions of climate. One has to enter the temperature rise between 1 and 5 °C in the SCADA, and it has the dynamic change in the data sets of the temperature with respect to ambient temperature. The data is acquired into SCADA through PLC.

## 30.4   Results and Discussion

Maize crop (DHM117) was grown as a test crop in the soil bins under the rain shelter with artificial rainfall simulator. The slope of the soil bins was kept at 0, 2, 4% for soil bins 1, 2 and 3, respectively. The rainfall simulator was used for generating extreme rainfall event of 100 mm per hour over the soil bins simultaneously at a critical stage of silking and tasseling of maize crop. Due to the extreme rainfall maximum damage was obtained in soil bin 3 with reduction in the cob dimensions and the seed yield, followed by soil bin 2 and soil bin 1. The maize seed yield under the normal rainfall in the control condition was 1.33 kg/m$^2$. The maximum reduction in the yield (95%) was observed in the soil bin with 4% slope (Table 30.1). It was due to maximum washout of the pollination in the maize cobs by reducing in number of cobs and its dimensions (Figs. 30.5 and 30.6). It is evident from the results that the maize yields are reduced due to change in the soil moisture in the soil bin with runoff and soil loss. The nutrient availability was also reduced in the case of 4% slope.

The experiment conducted in the open-top chambers indicated that the maize seed yield was maximum in the elevated $CO_2$ chamber (1.77 kg/m$^2$) (Fig. 30.7) as compared to control (1.33 kg/m$^2$). The $CO_2$ maintained in the chamber was 550 $\pm$ 50 ppm through PLC from the SCADA. The temperature was raised by 1 °C with respect to ambient temperature and varies dynamically. This temperature was maintained throughout the growth period of maize. There was reduction in the yields in $eCO_2$ + eTemp (1.42 kg/m$^2$) and eTemp (1.33 kg/m$^2$) (Table 30.2) chambers, respectively. Similarly, there was reduction in the dry matter yields of maize except in $eCO_2$ chamber.

## 30.5 Conclusions

SCADA-based rainfall simulation facility and precision type lysimeters with open-top climate chambers have been designed and developed for studying the climate change impacts on the crop, soil, water and nutrient losses with state of art process automation and data acquisition through PLC. The high accuracy $\pm$1% soil moisture and turbidity sensor, CO2 sensors, load cells, temperature have been used to measure the various parameters in the system continuously. A simple experiment conducted with maize as test crop showed that there was maximum reduction in the crop yield and its attributes over the control. The reduction in the yield increases with respect to increase in the slope of the soil bin. In the open-top climate chambers the elevated CO2 at 550 ppm showed the maximum yields as compared to $eCO_2$ + Temp and Temp alone.

**Table 30.1** Impact of extreme rainfall on maize yield and its attributes

| % slope | Rainfall intensity (mm/hr) | Crop growth stage | % filling | % reduction in filling | Average Cob size | | % reduction in dimensions | | Yield (kg/m$^2$) | % reduction in yield |
|---|---|---|---|---|---|---|---|---|---|---|
| | | | | | Length | Diameter | Length | Diameter | | |
| Control | Rainfed | – | 95 | | 12 | 4.5 | – | – | 1.33 | – |
| 0 | 100 | Silking and tassling | 83 | 12.63 | 10 | 3.7 | 16.67 | 17.78 | 0.139 | 89.56 |
| 2 | 100 | Silking and tassling | 71 | 25.26 | 8 | 3.1 | 33.33 | 16.22 | 0.111 | 91.65 |
| 4 | 100 | Silking and tassling | 68 | 28.42 | 7 | 2.5 | 41.67 | 19.35 | 0.069 | 94.78 |

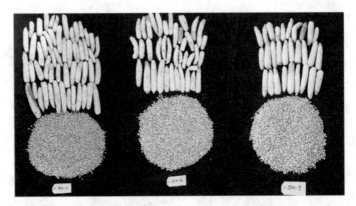

**Fig. 30.5** Maize seed yield and cobs in different soil bins of rainfall simulator

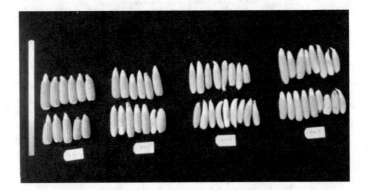

**Fig. 30.6** Maize cob size under extreme rainfall

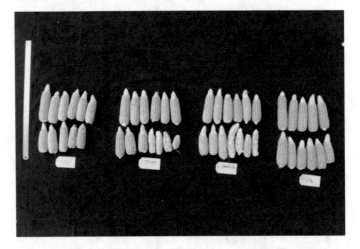

**Fig. 30.7** Maize cobs under eCO$_2$ and temperature in open-top climate chambers

**Table 30.2** Effect of elevated $CO_2$ and temperature on maize seed yield and dry matter yield

| Treatments | Yield (kg/m$^2$) | % change in yield | Dry matter yield (kg/m$^2$) | % change in dry matter yield |
|---|---|---|---|---|
| Control | 1.33 | – | 2.21 | – |
| $CO_2$ + temp | 1.42 | 6.67 | 2.65 | 20 |
| $CO_2$ | 1.77 | 33.33 | 3.54 | 60 |
| Temp | 1.33 | 0 | 2.65 | 20 |

**Acknowledgements** The authors are thankful to the NICRA project for financing the study and the infrastructure developed for climate change studies in soil and water conservation at ICAR-CRIDA. We are also thankful to the Director ICAR-CRIDA and PI NICRA for their encouragement and cooperation.

# References

Christensen JH, Hewitson B, Busuioc A, Chen A, Gao X, Held I, Jones R, Kolli RK, Kwon W-T, Laprise W-T, Magaña Rueda V, Mearns L, Menéndez L, Räisänen J, Rinke A, Sarr A, Whetton P (2007) Regional climate projections. In: Solomon S, Qin D, Manning M, Chen Z, Marquis M, Averyt KB, Tignor M, Miller HL (eds) Climate change 2007: the physical science basis. Contribution of working group I to the fourth assessment report of the intergovernmental panel on climate change. Cambridge University Press, Cambridge, United Kingdom and New York, NY, USA

Cruz RV, Harasawa H, Lal M, Wu M, Anokhin Y, Punsalmaa B, Honda Y, Jafari M, Li C, Huu Ninh N (2007) Asia. In: Parry N, Canziani OF, Palutikof OF, van der Linden PJ, Hanson CE (eds) Climate change 2007: impacts, adaptation and vulnerability. Contribution of working group II to the fourth assessment report of the intergovernmental panel on climate change. Cambridge University Press, CE

Hignett CT, Gusli S, Cass A, Besz W (1995) An automated laboratory rainfall simulation system with controlled rainfall intensity, raindrop energy and soil drainage. Soil Technol 8:31–42

IPCC (2007) Climate change 2007: impacts, adaptation and vulnerability. In: Parry ML, Canziani OF, Palutikof JP, van der Linden PJ, Hanson CE (eds) Fourth assessment report of the intergovernmental panel on climate change. Cambridge University Press. Cruz, Cambridge, UK

IPCC (2012) Managing the risks of extreme events and disasters to advance climate change adaptation. In: Field CB, Barros V, Stocker TF, Qin D, Dokken DJ, Ebi KL, Mastrandrea MD, Mach KJ, Plattner G-K, Allen SK, Tignor M, Midgley PM (eds) A special report of working groups i and ii of the intergovernmental panel on climate change. Cambridge University Press, Cambridge, UK and New York

IPCC (2013) The physical science basis. Contribution of working group I to the fifth assessment report of the intergovernmental panel on climate change. Cambridge University Press, Cambridge, United Kingdom and New York, NY, USA, 1535 pp. Climate Change 2013: The Physical Science Basis. Cambridge University Press, U.K

Kim Y, Jabro JD, Evans RG (2010) Wireless lysimeters for real time online soil water monitoring. Irrig Sci. https://doi.org/10.1007/s00271-010-0249-x

Martínez-Mena M, Abadía R, Castillo V, Albaladejo J (2001) Diseño experimental mediante lluvia simulada para el estudio de los cambios en la erosión del suelo durante la tormenta. Rev C G. 15(1–2):31–43

Stone JJ, Lane LJ, Shirley ED (1992) Infiltration and runoff simulation on a plane. Trans ASAE 35(1):161–170

Printed in the United States
by Baker & Taylor Publisher Services